Aquaculture Production Systems

Aquaculture Production Systems

Editor

James H. Tidwell
Kentucky State University
Division of Aquaculture
Frankfort, Kentucky, USA

A John Wiley & Sons, Ltd., Publication

This edition first published 2012 © 2012 by John Wiley & Sons, Inc.

Wiley-Blackwell is an imprint of John Wiley & Sons, formed by the merger of Wiley's global Scientific, Technical, and Medical business with Blackwell Publishing.

Editorial offices: 2121 State Avenue, Ames, Iowa 50014-8300, USA
The Atrium, Southern Gate, Chichester, West Sussex, PO19 8SQ, UK
9600 Garsington Road, Oxford, OX4 2DQ, UK

For details of our global editorial offices, for customer services, and for information about how to apply for permission to reuse the copyright material in this book please see our website at www.wiley.com/wiley-blackwell.

Library of Congress Cataloging-in-Publication Data

Aquaculture production systems / editor, James Tidwell.
       p. cm.
    Includes bibliographical references and index.
    ISBN 978-0-8138-0126-1 (hardcover : alk. paper) 1. Aquaculture.   I. Tidwell, James.
    SH135.A76 2012
    639.8–dc23
                                                                    2011048340

A catalog record for this book is available from the British Library.

Wiley also publishes its books in a variety of electronic formats. Some content that appears in print may not be available in electronic books.

Set in 10/12.5pt Sabon by Aptara® Inc., New Delhi, India

1   2012

Contents

Contributors

Geoff Allan
Department of Primary Industries
NSW Department of Trade and Investment
Regional Infrastructure and Services
Port Stephens Fisheries Institute
New South Wales, Australia

Yoram Avnimelech
Department of Civil and Environmental Engineering Technician
Israel Institute of Technology
Haifa, Israel

Craig L. Browdy
Novus International Inc.
Charleston, South Carolina, USA

D. E. Brune
Department of Agricultural Systems Management
Columbia, Missouri, USA

Jesse Chappell
Fisheries and Allied Aquacultures
Auburn, Alabama, USA

James M. Ebeling
Aquaculture System Technologies
New Orleans, Louisiana, USA

Gary Fornshell
University of Idaho Extension
Twin Falls, Idaho, USA

John Hargreaves
Aquaculture Assessments LLC
Baton Rouge, Louisiana, USA

Jeff Hinshaw
North Carolina State University
Department of Zoology
Raleigh, North Carolina, USA

Richard Langan
University of New Hampshire
Coastal and Ocean Technology Programs
Durham, New Hampshire, USA

John W. Leffler
Waddell Mariculture Center
Marine Resources Research Institute
South Carolina Department of Natural Resources
Charleston, South Carolina, USA

Michael P. Masser
Texas Co-op Extension
College Station, Texas, USA

Michael Massingill
Kent BioEnergy Corporation
San Diego, California, USA

Steven D. Mims
Kentucky State University
Division of Aquaculture
Frankfort, Kentucky, USA

Richard J. Onders
Kentucky State University
Division of Aquaculture
Frankfort, Kentucky, USA

James E. Rakocy
University of the Virgin Islands
Agricultural Experiment Station
St. Croix, US Virgin Islands

Andrew J. Ray
The University of Southern Mississippi
Gulf Coast Research Laboratory
Ocean Springs, Mississippi, USA

Robert Rheault
Shellfish Environmental Services, Ltd.
Wakefield, Rhode Island, USA

Robert R. Stickney
Texas Sea Grant
Texas A&M University
College Station, Texas, USA

James H. Tidwell
Kentucky State University
Division of Aquaculture
Frankfort, Kentucky, USA

Michael B. Timmons
Cornell University
Biological and Environmental Engineering Department
Ithaca, New York, USA

Granvil D. Treece
Texas Sea Grant
Texas A&M University
College Station, Texas, USA

Craig Tucker
National Warmwater Aquaculture Center
Stoneville, Mississippi, USA

Preface

Aquaculture. A simple word but a complex story. It's also a story of contradictions. In some ways aquaculture is very old, having been around in some regions for 4,000 to 5,000 years. However, as a major industry, and source food for mankind, it's been around only about fifty to sixty years. While aquaculture is an industry of several hundred species, the vast majority of production is dominated by less than ten. Also, unlike other livestock crops, we not only raise herbivores and omnivores but also carnivores and even filter feeders. It is a complex story indeed.

The idea for this book began in the 1990s. At Kentucky State University (KSU) aquaculture was initially entirely a research area. We received approval to teach our first course in 1991 and I developed Principles of Aquaculture as an experimental course. Gradually my colleagues and I at KSU developed additional courses to fill out a curriculum. In the Principles of Aquaculture course, I gave an overview of concepts, and then worked through a short but comprehensive overview of some major aquaculture species. However, the systems used to raise the fish were given a very cursory overview of one or two lectures. The more I thought about it the more it seemed to me that the aquaculturist's real job is to manage the environment, and that is the job of the production system. Wouldn't it be productive to develop another course that approached aquaculture not from the direction of the culture species, but from the direction of the culture system itself? The fact is that all species from shellfish to blue fin tuna have certain things they all need. Primary among them is a suitable water temperature, sufficient dissolved oxygen, and a way to remove or detoxify their waste products. The theme of this book is to explain how all of the different production systems we use provide these services, in many diverse ways.

To provide the best coverage of the subject, and a comprehensive explanation of each system, my job was to try to convince one of the most knowledgeable experts on each system to provide a chapter covering that system. To do this I tapped into a network of colleagues and friends, many of whom I had gotten to know during my years or while working with the World Aquaculture Society (WAS) in a number of different roles. If you go through the list of contributors, you will find that there are no less than six former WAS presidents contributing to the book.

The book is intended as a resource for students and researchers. Even within aquaculture there are individuals who know a tremendous amount about one system, but have had limited exposure to other systems. It is also intended as a resource for those outside of aquaculture who wish to understand the industry better. In two of my chapters I have tried to explain in simple terms the basic concepts of the different systems. I have also used extreme examples to help those from other professions appreciate just how hard our job can be with some aquatic species. Examples of non-aquaculture professionals that I hope can benefit from this book include entrepreneurs, investment bankers, feed and equipment salesmen, engineers, and environmentalists.

Environmental groups often use the broad term "aquaculture" when referring to issues related to one particular species or production system. They often paint with a very "broad brush." With a greater knowledge of the many different systems encompassed by this term, they might better understand aquaculture and all it represents. They might also better understand that the system they take issue with is only a very small portion of the larger aquaculture industry while their comments and criticisms negatively impact ALL parts of the industry. They might also become better able to appreciate the continuing efforts to improve the system's efficiencies and sustainability credentials. They can then come to understand that some of these systems are actually able to *improve* the environment by filtering out excess nutrients from whatever source.

A final theme of the book is a look ahead. What new types or combinations of systems might we see down the road? How will climate change affect aquaculture and its ability to provide increasing amounts of high quality protein to human populations, especially in regions of the world that need it the most?

I hope this text can serve as a resource for students and practitioners for many years to come and that it inspires them to develop new systems in the future. The Blue Revolution is really just beginning.

Jim Tidwell

Acknowledgments

A first of many thanks goes to Ms. Leigh Anne Bright. Her organizational skills and keen eye as a reviewer/editor of all of the chapter manuscripts kept the project moving forward. Also, her patience in times of crisis kept me from losing mine. Thanks to Ms. Karla Johnson for typing, retyping, and re-retyping my chapters through their many stages of evolution, while keeping our other duties on track as well. My appreciation to Mr. Charles Weibel for his good-natured assistance with figures. He improved many and actually recreated several to ensure the best quality for publication. Thanks to Mr. Shawn Coyle for keeping more than his share of our research responsibilities on track while this project demanded a significant percentage of my attention. My appreciation to the faculty, staff, and students of the Division of Aquaculture at Kentucky State University for their support while this project came together. Many, many thanks to the contributors of the chapters in the book. They have endured hundreds of e-mails and requests with patience and quick responses. I appreciate their support, endurance, and perseverance. I also look back and thank my mentors and fellow students at Mississippi State University in years past who helped me develop a real devotion to this discipline that has not diminished. I thank my friends and colleagues in the World Aquaculture Society who have helped me appreciate how diverse and dynamic this industry is and will continue to be. Finally, I thank my family. This includes my big brother, Bill, who has shown a real interest in the project; my wife, Vicki; and my children, Will, Chandler, and Patrick, who have shared me, and have often helped me, with many aquaculture endeavors over the years.

Aquaculture Production Systems

Chapter 1

The Role of Aquaculture

James H. Tidwell and Geoff Allan

Fish represent both a vital contribution to the human food supply and an extremely important component of world trade. The trend in both of these areas is toward increasing importance. This chapter discusses the current status of seafood supply, world trade in fisheries products, and the relative contributions of aquaculture and capture fisheries. It addresses the question "Can we continue to meet the increasing global demand for seafood?"

1.1 Seafood demand

Fish is a vital component of the human food supply and man's most important source of high-quality animal protein. (As used here, the general term "fish" includes fish, mollusks, and crustaceans consumed by humans). It is estimated that worldwide about 1 billion people rely on fish as their primary source of animal protein (FAO 2001) and it provides more than 3 billion people with at least 15% of their average per capita animal protein intake (FAO 2009). It is a particularly important protein source in regions where high-quality protein from terrestrial livestock is relatively scarce. For example, in 2005, fish supplied less than 10% of animal protein consumed in North America and Europe (7.6%) but 19% of animal protein in Africa and 21% in Asia (FAO 2009).

Consumption of food fish is increasing, having risen from 40 million tonnes in 1970 to 86 million tonnes in 1998 (FAO 2001), and then to 115 million tonnes

Aquaculture Production Systems, First Edition. Edited by James Tidwell.
© 2012 John Wiley & Sons, Inc. Published 2012 by John Wiley & Sons, Inc.

in 2008 (FAO 2010). Large increases in international meat prices in 2004 and 2005 continued to push consumers toward alternative protein sources, such as fish. Global per capita fish consumption has increased over the past four decades, rising from 9.0 kg/person in 1961 to an estimated 17.1 kg/person in 2008 (FAO 2010). Based on projected increases in consumption rates alone (assuming no increase in the human population) it is estimated that the demand for seafood will increase by more than 10 million tonnes per year by 2020 (Diana 2009). However, fish consumption is not distributed evenly. In 2008 Low Income Food Deficit Countries (LIFDCs) had a per capita fish consumption rate of 13.8 kg/person/year, which is about half that of industrialized countries (28.7 kg/person/year; FAO 2010). In Africa in 2007, per capita fish consumption was 8.5 kg, Latin America 9.2 kg, and Asian countries other than China, 14.6 kg. On the higher end, per capita consumption in 2007 averaged 22.2 kg in Europe, 25.2 kg in Oceania, 24.0 kg in North America, and 26.7 kg in China (FAO 2010).

How much seafood is consumed varies not only by region but also by the type of seafood. In northern Europe and North America demersal (bottom living) fish are preferred, while in Asia and the Mediterranean cephalopods, such as squid, are preferred. Crustaceans (like crabs and shrimp, which are relatively expensive) are mostly consumed in affluent economies. Of the 16.5 kg of fish products available for consumption per person worldwide in 2007, 12.8 kg (75%) were finfish, 1.6 kg were crustaceans and 2.5 kg were molluscs (FAO 2010). These figures represent an over three-fold increase in consumption of crustaceans and molluscs over the past forty years.

While increases in per capita consumption account for a small portion of the increase in total demand, it is the growing human population that is the main driving force for this steadily increasing demand for food fish. In fact, although the total amount of fish available for human consumption has increased, the supply per capita has remained at about the same levels as those in 2004 because the human population is growing at about the same rate as seafood supplies. The global population reached 6 billion in 1999 with predictions that it may exceed 9 billion by 2050 (Duarte *et al.* 2009). That figure is approaching the maximum human population that some research calculates the earth can sustain (Cohen 1995). Contributing to that conclusion are analyses that indicate that shortages in both food and water will constrain the growth of terrestrial agriculture in the future (Duarte *et al.* 2009). Disturbingly, most of the population growth is predicted to occur in less developed countries such as Asia, Africa, and South America.

1.2   Seafood supply

In 2008 the total world supply of fish was about 142 million tonnes (FAO 2010). Capture fisheries (inland and marine) produced about 90 million tonnes with about 80 million tonnes being from marine capture and a record 10 million tonnes being captured from freshwater (FAO 2010). Of this, about 27 million tonnes (roughly 19% of the total) was destined for nonfood uses, primarily as fish meal in animal feeds (20.8 million tonnes). The other 81% of total fishery production (115 million tonnes in 2008) was used for human food (FAO 2010).

Today, fish is the only important food source where a large portion is still gathered from the wild rather than produced from farming. While some marine and freshwater capture fisheries may have individual populations that could support additional exploitation, it appears unlikely that large increases from either of these sources will be forthcoming on a sustainable basis. For marine capture fisheries, FAO reports that in 2008 only 3% of the stock groups were under exploited and 12% were moderately exploited and could perhaps produce greater yields (FAO 2010). However, 53% were fully exploited, 28% overexploited, 3% depleted, and 1% were recovering from depletion (FAO 2010). This means that 85% of marine fisheries are biologically incapable of sustainably supporting increased yields (FAO 2010).

The FAO reports that the percentage of overexploited, depleted, and recovering stocks is consistently increasing. In fact, global marine capture fisheries production has been, at best, stagnant for over twenty-five years. The 80 million tonnes produced by global marine capture fisheries in 2008 is less than the 85 million tonnes produced in 1992 (FAO 2010). The maximum wild capture fisheries potential for the world's oceans has likely been reached. In fact, by some estimates, current ocean harvests may already be greater than levels considered sustainable (Coll *et al.* 2008) and it does not appear likely that we can turn to increased capture yields from freshwater. The FAO states that "globally, inland fishery resources appear to be continuing to decline as a result of habitat degradation and overfishing" and that this trend "is unlikely to be reversed" (FAO 2007).

As marine capture fisheries have become depleted and fish harder to catch, many fishermen and governments have responded with increased investment in equipment and technology. These changes have actually put increased pressure on wild-fish stocks. More efficient fishing technology also decreases the reproductive capacities of fisheries, thus exacerbating the effects of overharvesting. Based on the assessment of overexploitation of many fish stocks, and overcapacity and overcapitalization of many fishing fleets, by the mid 1970s it was widely concluded that many capture fisheries were not commercially viable without significant government subsidies (Mace 1997). The solution appeared to be to reduce the size of the fishing fleets. However, with advances in technology and increased mechanization, the ability of each remaining boat to catch fish (its "fishing power") increased. So while the number of fishers in industrialized countries has steadily declined, dropping 24% between 1990 and 2009 (FAO 2009), the pressure on the fish stocks largely has not decreased.

However, not all the news for capture fisheries is bad. Consistent increases in catches of certain species have been observed in the Northwest Atlantic and Northeast Pacific. These two regions are considered among the most regulated and managed in the world and this probably indicates that with proper management these fisheries can effectively continue producing significant levels of harvest without depleting the populations. However, in summary, there is widespread agreement that the supply from the wild, be it of freshwater or marine origin, is *not* likely to increase substantially in the future.

1.3   Seafood trade

Fish not only makes important contributions to food security but also has tremendous economic importance, being one of the most highly traded food and feed commodities globally. Total world exports of fish and fishery products reached a record value of US$85.9 billion in 2006 and are predicted to reach US$92 billion for 2007. This represents a 57% increase in exports since 1996 (FAO 2009). In 2008, 44.9 million people were directly engaged in primary production of fish either through fishing or aquaculture (FAO 2010). This represents a 167% increase since 1980 (16.7 million people; FAO 2010).

Table 1.1 lists the top-ten exporters and importers of fish and fish products in 1998 and 2008. In 2008 China was the world's largest exporter, shipping fish products valued at US$10.1 billion. This represents an almost four-fold increase in export values in ten years. However, the most *rapid* growth of the

Table 1.1 Top-ten exporters and importers of fish and fishery products in 1998 and 2008 in terms of value (USD) and annual rate of growth (APR; FAO 2010).

	1998	2008	APR (%)
	Millions (USD)		
EXPORTERS			
China	2,656	10,114	14.3
Norway	3,661	6,937	6.6
Thailand	4,031	6,532	4.9
Denmark	2,898	4,601	4.7
Vietnam	821	4,550	18.7
United States	2,400	4,463	6.4
Chile	1,598	3,831	8.4
Canada	2,266	3,706	5.0
Spain	1,529	3,465	8.5
Netherlands	1,364	3,394	9.5
TOP TEN SUBTOTAL	23,225	51,695	8.3
REST OF THE WORLD TOTAL	28,226	50,289	5.9
WORLD TOTAL	51,451	101,983	7.1
IMPORTERS			
Japan	12,827	14,947	1.5
United States	8,576	14,135	5.1
Spain	3,546	7,101	7.2
France	3,505	5,836	5.2
Italy	2,809	5,453	6.92
China	991	5,143	17.9
Germany	2,624	4,502	5.5
United Kingdom	2,384	4,220	5.9
Denmark	1,704	3,111	6.2
Republic of Korea	569	2,928	17.8
TOP TEN SUBTOTAL	39,534	67,377	5.5
REST OF THE WORLD TOTAL	15,517	39,750	9.9
WORLD TOTAL	55,051	107,128	6.9

period actually occurred in Vietnam, whose exports increased 450% over the same ten-year period. Between 2006 and 2008 Vietnam moved from eighth to fifth on the list of top exporters. On the import side, Japan has remained the world's largest importer of fish products for twenty-five years, importing approximately US$15 billion per year. However, Japan's rate of increase has slowed in recent years, increasing only US$500 million from 1998 to 2008. The second largest importer has historically been the United States, whose imports increased US$5.5 billion during the same period, and who will likely overtake Japan as the world's top importer (ARUSSI 2009). Paradoxically, despite being the world's largest *exporter,* China also had the most rapid increase in *imports* during this period, with a 420% increase in value between 1998 and 2008 (FAO 2010). It is predicted by some that China will actually become a net *importer* of fish and fish products in coming years as per capita incomes there continue to rise. South Korea also showed substantial increases, with a greater than 400% increase in imports over the ten-year period.

Fish products are extremely important to the economies of many countries, and the past four decades have seen major changes in the geographical patterns of the fish trade, much of it benefiting the developing world. In 1976, developing countries accounted for approximately 37% of fisheries exports. By 2008, developing countries were responsible for about 50% of exports (FAO 2010). These changes are further supported by the fact that developing countries had a trade surplus of US$4.6 billion in 1984 which grew to US$24.6 billion in 2006, a 434% increase in just over twenty years (FAO 2009). This is a much faster increase than we see in other agricultural commodities such as rice, tea, or coffee. The poorest countries (Low Income Food Deficit Countries, or LIFDCs) have also shown considerable growth in exports accounting for 20% of fishery exports in 2006 with a trade surplus of US$10.7 billion (FAO 2009).

Another major trend that is occurring is in *what* is being traded. In the past, developing countries exported raw materials that were then processed into value-added product forms in developed countries. Increasingly, the processed or value-added products are being generated *within* the developing country for export, capitalizing on low labor and operating costs. This is often done with processing infrastructure developed with outside investments from developed countries. The quantity of fish exported by developing countries for human consumption increased from 46% in 1998 to 55% in 2008 (FAO 2010).

However, an important share of the exports of developing countries is still in lower value nonfood products. A large portion of this is in the form of fish meal, destined for use as a feed ingredient or fertilizer. In 2008, of the fish products exported by developing countries, fish meal represented 36% by quantity but only 5% by value.

1.4   Status of aquaculture

As we have shown, the demand for food fish increases each year. As we have also shown, the supply from wild harvest is not expected to increase substantially

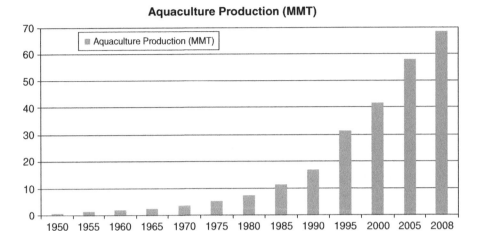

Figure 1.1 Annual world aquaculture production (in million tonnes) since 1950.

in the future. The only other source for the human population to produce food fish is aquaculture and global aquaculture growth has been extraordinary (fig. 1.1). Aquaculture production was only 1 million tonnes in the 1950s (FAO 2007). In the 1970s aquaculture contributed less than 4% of total seafood production. However, by 1997 aquaculture contributed about 27% of the food fish supply, by 2004 it contributed 32%, and by 2008 it contributed more than 47% (fig. 1.2). By 2015, aquaculture will pass capture fisheries as the leading source of food fish for the human population and the proportion contributed by aquaculture will continue to increase each year thereafter (Lowther 2007).

Aquaculture is growing more rapidly than any other animal food-producing sector, with an annual growth rate of 6.6% since 1970 (FAO 2010). This is contrasted with a growth of only 1.2% for capture fisheries and 2.8% for terrestrial farmed meat production over the same period (fig. 1.3). It is estimated that

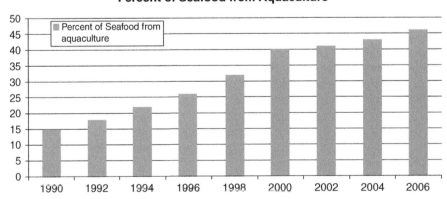

Figure 1.2 Aquaculture production as a percentage of total seafood supply.

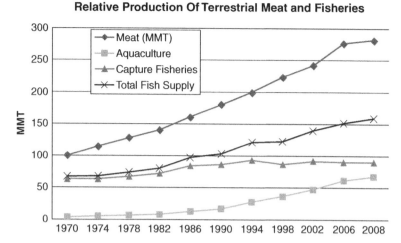

Figure 1.3 Relative production of terrestrial meat production, total seafood supply, capture fisheries, and aquaculture (in million tonnes).

the land devoted to row crop and grazing will have to increase by 50 to 70% by 2050 to meet food requirements for the projected increases in the human population (Molden 2007). However, the amount of land devoted to terrestrial crop production actually decreased from 0.5 ha/person to 0.25 ha/per person during the period 1960 to 2000 (Molden 2007). Extrapolation of population growth estimates and estimates of the availability of cultivable lands create "a likely scenario in which Earth's capacity to support the human population may be reached within the next decades, at population levels below currently proposed estimates" (Duarte *et al.* 2009). This raises the real question—can the human population feed itself in the coming decades?

These conditions only bolster the case that a prudent development of aquaculture is essential. In 2008 total aquaculture production (including plants) was reported to be 68.3 million tonnes with a value of US$106 billion, of which 53 million tonnes was for food fish with a value of US$98.4 billion (FAO 2010). It is anticipated that to keep pace with demand, aquaculture production of food fish will need to increase to 85 million tonnes (more than 75% growth) in the next twenty years (Subasinghe 2007).

So where is aquaculture production occurring? Currently, Asia dominates production. In 2009, Asia accounted for 89% of world aquaculture production by quantity and 79% by value (FAO 2010). China alone produces more than 62% of the world's aquaculture volume and 51% by value (FAO 2010). Of the top-ten countries in aquaculture production in 2006, only two (Chile and Norway) are not in the Asian region and they account for less than 3% of world production (table 1.2). However, as illustrated by table 1.3, there are very rapid increases in production occurring in some countries outside of Asia.

Aquaculture is extremely varied in terms of what species are raised. Based on tonnage, if we include aquatic plants, the individual species with the highest

Table 1.2 Top-ten aquaculture producers of food fish supply in 2008 in quantity and growth.

	2000	2008	APR (%)
	Thousand Tonnes		
China	21,522	32,736	5.4
India	1,943	3,479	7.1
Vietnam	499	2,462	16.4
Indonesia	789	1,690	7.0
Thailand	738	1,374	8.1
Bangladesh	657	1,006	5.5
Norway	491	844	7.0
Chile	392	843	10.1
Philippines	394	741	8.2
Japan	763	732	0.5

Note: Data exclude aquatic plants.

aquaculture production in 2005 was the Japanese kelp (*Laminaria japonica*) at 4.9 million tonnes followed by the Pacific cupped oyster (*Gassostrea gigas*) at 4.5 million tonnes (Lowther 2007), silver carp (*Hypopthal michthys molitrix*) at 4.0 million tonnes, grass carp (*Ctenopharyngodon idellus*) at 3.9 million tonnes, and common carp (*Cyprinus carpio*) at 3.4 million tonnes (FAO 2007). Bighead carp (*H. nobilis*) and crucian carp (*Carassius carassius*) also exceeded 2 million tonnes (Lowther 2007).

If we look at value-based species groups as defined by FAO, the highest reported values were for carps (US$18.2 billion), followed by shrimp and prawns (US$10.6 billion; Lowther 2007) and salmonids (US$7.6 billion). While crustaceans (such as shrimp) rank fourth in terms of quantity produced, they rank second in terms of total value, reflecting their relatively high selling prices. In fact, aquaculture production of shrimp increased 165% from 1997 to 2004, driving

Table 1.3 Top-ten aquaculture producers ranked in terms of their annual percentage rates (APR) of growth over a two-year period.

	2004	2006	APR (%)
	Tonnes		
Uganda	5,539	32,392	141.83
Guatemala	4,908	16,293	82.20
Mozambique	446	1,174	62.24
Malawi	733	1,500	43.05
Togo	1,525	3,020	40.72
Nigeria	43,950	84,578	38.72
Cambodia	20,675	34,200	28.61
Pakistan	76,653	121,825	26.07
Singapore	5,406	8,573	25.93
Mexico	104,354	158,642	23.30

supply up but prices down. The highest reported value for a single species was US$5.9 billion for the Pacific white shrimp (*Litopenaeus vannamei*) followed by the Atlantic salmon (*Salmo salar;* Lowther 2007).

Compared to terrestrial agriculture, aquaculture is extremely diverse with over 449 species of plants and animals being raised (Duarte *et al.* 2009). Production trends indicate that the diversity of species being produced in aquaculture is still on the increase. Duarte *et al.* (2009) estimated that the number of species being cultured increases 3% per year. Some of these new species groups have shown very large increases in production. Examples include sea urchins and echinoderms (4,833% increase), abalones, winkles, conchs (884%), and frogs and other amphibians (400%) in only a two-year period (2002 to 2004). However, a few species dominate production with the top-five species accounting for 62% of total aquaculture production and the top-ten species accounting for 87% (FAO 2007).

Aquaculture also varies by environment, utilizing marine, freshwater, and brackish water environments. When considered in terms of total weight, in 2005 mariculture accounted for approximately 51% of production while freshwater accounted for 43% (FAO 2007). However, these values include a substantial tonnage of aquatic plants, which are primarily produced in marine systems. When we look specifically at food animal production, freshwater becomes more important, accounting for 60% of production by quantity (fig. 1.4) and 48% by value (fig. 1.5), compared to 32% and 31%, respectively for mariculture and 8% and 13%, respectively for brackish water (FAO 2009).

Worldwide in 2008, freshwater fishes were the dominant group (table 1.4) in terms of productions (28.8 million tonnes) and most of this is composed of different species of carps (FAO 2010). In fact, carps accounted for approximately

Production of Fish and Shellfish

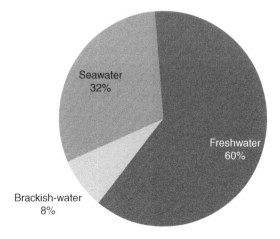

Figure 1.4 Percentage by weight of edible fish and shellfish products produced in freshwater, seawater, or brackish water.

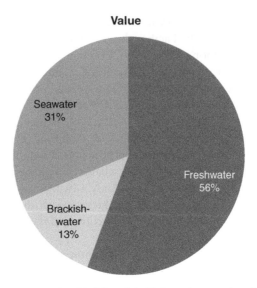

Value

Figure 1.5 Percentage by value of edible fish and shellfish products produced in freshwater, seawater, or brackish water.

71% of all freshwater fish production (FAO 2010). A snapshot of global aquaculture in 2008 shows that over one half of its production (55%) was freshwater finfish with a value of US$40.5 billion. The next largest group was molluscs at 25% of total production worth US$13 billion. Crustaceans accounted for 9.5% of total aquaculture production by weight, but 23% by value. Like crustaceans, the relative value of marine fish is quite high representing only 3% of global aquaculture production but 7% of value (FAO 2009).

1.5   Production systems

Although data on production systems are not yet widely tracked, it would be safe to say that the majority of fish and crustaceans produced for food by aquaculture are currently raised in ponds. In China in 2008, 70.4% of freshwater aquaculture

Table 1.4 World aquaculture production: major species groups[1] by percent of total quantity and total value in 2006 (FAO 2010).

Species Groups	Quantity (%)	Value (%)
Freshwater fishes	54.7	41.2
Mollusks	24.9	13.3
Crustaceans	9.5	23.1
Diadromous fishes	6.3	13.3
Marine fishes	3.4	6.7
Aquatic animals—NEI[2]	1.2	2.4

[1] Does not include plants.
[2] NEI = not otherwise included.

relied on ponds, with 11.7% conducted in reservoirs, 7.7% in natural lakes, 5.6% in rice paddies, 2.7% in canals, and 2.6% in other systems (FAO 2010).

1.6   The future and the challenge

As we have seen, the demand for fish increases each year. Per capita consumption shows slight increases but the more important factor is continuing population growth. It is estimated that to even maintain the current level of per capita consumption, the fish supply will have to almost double in the next twenty years. That translates into almost 40 million tonnes of additional supply per year. As discussed, it is unlikely that significant increases in wild harvests will occur. So where will this additional fish come from? There is only one answer. It has to come from aquaculture.

As we can see, aquaculture is no longer just a promising line of research or a promising theory. As Melba Reantso of FAO described it, "aquaculture is now known as the emerging new agriculture, the catalyst of the 'blue revolution,' the answer to the world's future fish supply, the fastest food producing sector, the future of fisheries." Still, the task ahead is daunting. Aquaculture is expected to supply global seafood security, nutritional well-being, poverty reduction, and economic development by meeting all of these demands, but also accomplishing this with a minimum impact on the environment and maximum benefit to society. The remainder of this book will be devoted to helping the reader follow the development of aquaculture over time, truly understand and appreciate how the diverse systems used to raise these aquatic animals operate, and grasp the evolution of new systems and the changes that are sure to be wrought by climate change.

1.7   References

Annual Report on the United States Seafood Industry (ARUSSI; 2009) 16th Edition. Urner Barry, Tom's River, New Jersey.

Cohen, J.H. (1995) *How Many People Can The Earth Support?* W.W. Norton & Co., New York.

Coll, M., Libralato, S., Tudela, S., Palomera, I. & Pranovi, F. (2008) Ecosystem in the ocean. *PlosOne* 3(12):e3881.

Diana, J.S. (2009) Aquaculture production and biodiversity conservation. *BioScience* 59(1):27–38.

Duarte, C.M. *et al.* (2009) Will the oceans help feed humanity? *BioScience* 59(11): 967–76.

FAO (2001) *State of the World Fisheries and Aquaculture.* FAO, Rome.

FAO (2007) *State of the World Fisheries and Aquaculture.* FAO, Rome.

FAO (2009) *State of the World Fisheries and Aquaculture.* FAO, Rome.

FAO (2010) *State of the World Fisheries and Aquaculture.* FAO, Rome.

Lowther, A. (2007) Highlights from the FAO database on aquaculture statistics. *FAO Aquaculture Newsletter* 38:20–1.

Mace, R.M. (1997) Developing and sustaining world fisheries resources: The state of the science and management. In *Second World Fisheries Congress* (Ed. by D.A. Hancock, D.C. Smith, A. Grant & J.P. Beumer), pp. 98–102. CSIRO Publishing, Collingwood, Victoria.

Molden, D. (Ed.) (2007) *Water for Food; Water for Life: A Comprehensive Assessment of Water Management in Agriculture*. Earthscan, London.

Subasinghe, R.P. (2007) Aquaculture: Status and prospects. *FAO Aquaculture Newsletter* 38:4–6.

Chapter 2

History of Aquaculture

Robert R. Stickney and Granvil D. Treece

For purposes of this historical account the concentration is on finfish, shrimp, and mollusks that are reared as food for human consumption. The reader should be aware that aquaculture is much more expansive. In the finfish category alone there is a considerable amount of culture of ornamental fishes, particularly freshwater species. There is also a good deal of interest in marine ornamentals but the number of species that have been successfully cultured is considerably smaller than is the case for their freshwater counterparts. In some parts of the world there is also aquaculture of baitfish, such as minnows, for marketing to the recreational fishing community.

There are a large number of mollusks, crustaceans, and other invertebrate groups that include species currently being produced in aquaculture. Oysters, clams, mussels, and abalone are examples of mollusks being cultured, while various species of marine shrimp and one species of freshwater shrimp, *Macrobrachium rosenbergii*, are the most important groups of crustaceans being cultured for the human food market. Crabs are also cultured to a limited extent, and there is some culture of lobster. There is also some culture of sea urchins and sea cucumbers as well as a few other invertebrates.

Captive spawning and rearing of fingerlings for stocking recreational fishing waters has a relatively long history. The foundations upon which the science of aquaculture are based can be traced in large part to the pioneering fish culturists who developed the various processes associated with spawning and rearing a variety of both marine and freshwater fishes. Fish continue to be produced

---

Aquaculture Production Systems, First Edition. Edited by James Tidwell.

for recreational purposes but there are also hatcheries producing fish in many countries to enhance wild capture fisheries. Japan has a history over the past several years of also producing shrimp for enhancement stocking.

Ocean ranching is another form of aquaculture. It most commonly involves salmon that are released from hatcheries as smolts to find their way to the ocean where they grow to maturity and then return to their natal waters to spawn. Some of the returning fish are used as broodstock, with the majority being taken by commercial fishermen. In Japan, a modified form of ocean ranching has been developed with non-salmonid marine fishes. Young fish are trained in the hatchery to respond to a sound source at feeding time. Once the behavior is ingrained in the fish, they are released into the wild near a feeding station that emits the same sound that had been used for conditioning. Feeding continues using the sound attractant until the fish reach market size, after which they are captured for processing.

A variety of invertebrates are cultured as food for other cultured species. Examples are brine shrimp, rotifers, and copepods. Various species of phytoplanktonic algae are cultured as food for those invertebrates, thus requiring the aquaculturists to maintain at least three different culture systems: phytoplankton, zooplankton, and the fish or invertebrate species that will ultimately be marketed.

Phytoplanktonic species are not the only members of the plant kingdom that are being cultured. Macroscopic algae are also of interest to aquaculturists. For example, nori is used to make sushi wrappers and is also used in other products that are consumed by people. A few higher plants, such as water chestnuts, are also cultured for human consumption.

In most cases invertebrates and aquatic plants have a long history of consumption by humans but a relatively short history of being cultured. Most of that culture activity began in the twentieth century. One notable exception is oysters, which were apparently cultured during the days of the Roman Empire nearly 2,000 years ago (Beveridge & Little 2002).

Hydroponics is another form of aquaculture in which terrestrial plants are grown in nutrient-enriched water. Various vegetables can be grown hydroponically and the approach has been used in conjunction with fish or invertebrate aquaculture in polyculture.

2.1   Beginnings of aquaculture

The dawn of aquaculture is shrouded in mystery, although most who have delved into that history agree that the art of aquaculture began in China. Various authors have indicated that the beginnings of aquaculture can be traced as far back as 3,500 to 4,000 years Before Present Era (BPE). Among them were Hickling (1962, 1968), Rabanal (1988), Bardach *et al.* (1972), and Ling (1977). Jessé & Casey undated posited that aquaculture may go as far back as 5,000 years BPE. The latter indicated that pond culture of grey mullet (*Mugil cephalus*) and carp (presumably common carp, *Cyprinus carpio*) began with the first of China's

Emperors in the period from 2852 to 2737 BPE, but provided no details on the source of that information. Ling (1977) indicated that carp culture developed from simply catching and holding fish in baskets to holding them in traps and finally, feeding them to grow them to larger sizes before harvesting.

While the precise period when aquaculture developed in China may be in question, there is general agreement that the first published work on the subject was a small volume by Fan Li that appeared about 475 BPE (for example, see Borgese 1980). Ling (1977) indicated that Fan Lee described how to spawn fish including how to select ripe brooders. Pillay & Kutty (2005) indicated that Fan Li was a politician turned fish culturist and that he made his fortune growing fish. Other early Chinese publications on collecting carp fry from rivers and rearing them in ponds were also produced according to Pillay & Kutty (2005).

Ling (1977) reported that when the Lee Dynasty was founded in 618 BPE, a thousand years or more of common carp culture was threatened. That was because the word Lee is pronounced the same as lee, which was the word used for carp. It was considered sacrilege to catch and eat lee because it was an insult to the Emperor. To get around the problem, culture methods for grass (*Ctenopharyngodon idella*), silver (*Hypophthalmichthys molitirix*), bighead (*Aristichthys nobilis*), and mud carp (*Cirrhinus molitorella*) were developed, which led to polyculture.

If one follows the chronology of aquaculture development produced by Jessé & Casey undated, the Egyptians may have developed some form of aquaculture at about the same time as the Chinese. Those authors attribute Egyptian hieroglyphics that appear to show pond culture being developed between 2357 and 1786 BPE. Another source places the Egyptians practicing intensive fish culture during the period 2052 to 1786 BPE, corresponding with the Middle Kingdom period (Anonymous, undated). In addition to reports of hieroglyphics depicting fish culture, a bas relief from the tomb of Thebaine, an Egyptian nobleman, depicts some type of tilapia being fished from an artificial pond (Chimits 1957). That bas relief was redrawn and also mentioned by Beveridge & Little (2002) as well as Pillay & Kutty (2005). Another bas relief from around 2500 BPE was reproduced in a book by Maar *et al.* (1966) that shows a pond with a variety of fish, at least one of which appears to be tilapia.

2.2   Expansion prior to the mid-1800s

Common carp culture was introduced from China to various parts of Southeast Asia by Chinese immigrants (Ling 1977). Carp appear to have been raised in ponds in Japan at least 1900 years ago (Drews 1951). However, fish culture was not important in Japan before the seventeenth century.

The Romans reportedly introduced carp from Asia Minor into Greece and Italy (Maar *et al.* 1966). The Romans also appear to have introduced common carp culture to England during the first century BPE (Beveridge & Little 2002). Carp culture was said by the same authors to have been established in central Europe by the seventh century. Pillay & Kutty (2005) indicated that carp culture was initiated when the fish were introduced to monastery ponds which occurred in the Middle Ages in western and central Europe (Beveridge & Little 2002). Over the centuries carp culture expanded throughout much of Europe. Today, the culture of various species of carp accounts for about 80% of the aquaculture production from marine and freshwaters in central and Eastern Europe (Szücs *et al.* 2007). Worldwide, carp and other cyprinids clearly dominate world production with respect to weight of cultured finfish produced (FAO 2006), with most of that production taking place in China.

Carp culture appears to have been conducted on at least part of the Indian subcontinent since the eleventh century (Pillay & Kutty 2005). The tradition in India has always been to culture the local carp species catla (*Catla catla*), rohu (*Labeo rohita*), and mrigal (*Cirrhinus mrigala*).

Brackishwater aquaculture in Southeast Asia reportedly began during the fifteenth century, probably in Indonesia (Pillay & Kutty 2005). Milkfish were produced at that time in coastal ponds called tambaks. Milkfish culture expanded to other parts of Southeast Asia including the Philippines and Malaysia, which continue to be major producing nations. Mullet also were probably produced in Southeast Asia 1400 years BPE (Ling 1977). Pillay & Kutty (2005) placed the origins of walking catfish (*Clarias* spp.) culture in Cambodia.

Native Hawaiians had been growing fish in ponds for as long as 500 years before the Hawaiian Islands were discovered by Captain James Cook in 1778. The Hawaiians would allow seawater to enter their ponds with the rising tide, then block off the pond entrances thereby trapping whatever creatures were present. The animals would then be allowed to grow to harvest size (Costa-Pierce 2002).

Trout were among the early groups of fishes to be cultured in Europe. Davis (1956) attributed the first successful artificial insemination of trout eggs to a monk in France who lived in the fourteenth century. Over the centuries, trout culture spread across much Europe and, ultimately, around the world. The European brown trout (*Salmo trutta*) was the subject of the early activity, though rainbow trout (*Oncorhynchus mykiss*) introduced from North America ultimately became more popular with fish culturists. The British introduced trout to some of their colonies in Asia and Africa mainly for recreational fishing (Pillay & Kutty 2005).

2.3   The explosion of hatcheries

While fish culture expanded over the centuries that followed the middle ages, that expansion was largely restricted to a small number of species and often depended largely on the capture of wild fry or fingerlings that were then confined for growout. Milkfish is an excellent example and that industry still relies largely on wild captured fry.

In the latter half of the nineteenth century, enthusiasm for recreational fishing coupled with increasing numbers of reports that many freshwater and marine stocks were being negatively impacted by overfishing, led to the development

Figure 2.1 State of Oregon railroad car used to transport fish.

of private and public hatcheries in Europe and North America. In the United States the initiative began when several states in the northeast established fish commissions that were charged with managing their fisheries. Some of the commissions turned to stocking as a means of enhancing their recreational fisheries. In most cases they initially turned to private fish producers to supply the fish, most commonly brook trout (*Salvelinus fontinalis*).

In 1872, the United States Commission on Fish and Fisheries was established by Congress and Spencer F. Baird, who conceived the idea and was named the first Commissioner. Baird was concerned about declining fish populations and directed a lot of the activity of the Commission toward developing hatcheries for both freshwater and marine species. As a result, over a period of several decades billions of eggs were hatched and their fry were distributed by the United States Commission on Fish and Fisheries, which later became the United States Bureau of Fisheries, Bureau of Commercial Fisheries, and, ultimately, the National Marine Fisheries Service (Stickney 1996a–e, 1997a–c). Millions of railroad miles passed under the wheels of cars hauling fish, first by the United States Commission and soon followed by state fish commissions or fisheries agencies (fig. 2.1). Hatcheries on the east coast produced Atlantic salmon (*Salmo salar*) for shipment to the west coast and hatcheries on the west coast (initially in northern California and later in Oregon and Washington) produced Pacific coho (*Oncorhynchus kisutch*) and Chinook salmon (*O. tshawytscha*) for shipment to the east. While no survival of salmon appears to have occurred in either case, hundreds of millions of eggs and fry were involved in the effort. This may be due to the natural tendency of smolts would be to turn the wrong direction if they survived long enough to reach the sea. It wasn't only salmon that were involved in the transfers. There were also largemouth bass (*Micropterus salmoides*), walleye (*Stizostedion vitreum vitreum*), cod (*Gadus morhua*), rainbow trout, and striped bass (*Morone saxatilis*), to name but a few. Striped bass from the east became established in California and continue to thrive there. Baird's successors continued to expand the numbers of hatcheries and species produced. Baird himself promoted the introduction of common carp into United States waters perhaps thinking that many of the European immigrants who made up a large percentage of the population at the time would have a preference for a species from their native countries with which they were familiar. The Commission stocked carp around the nation and territories but discontinued the process after only a few years because the waters were soon teeming with reproducing populations. Ultimately, carp were seen by most Americans as trash fish and their introduction is not generally considered to be one of Baird's crowning achievements.

Of the billions or perhaps trillions of eggs and larvae of marine species and some freshwater species that were stocked during the 1800s, there is little or no evidence of survival. The fish culturists of the late nineteenth and much of the twentieth century were unsuccessful in finding ways to feed the very tiny fry that are so common among marine species and the animals were typically released before their digestive systems were developed. The same was true for freshwater fishes with small eggs. Salmon and trout were exceptions in that they had large eggs and fry that survived on a yolk sac for an extended period following hatching. When they did begin exogenous feeding, the fry were more advanced and it was possible to successfully provide them with food of one kind or another—chicken egg yolk being one popular item. Details of this fascinating period of fish culture development in the United States can be found in a book by Stickney (1996a).

2.4   Art becomes science

It wasn't until the mid-twentieth century that the art of aquaculture developed into what exists today as a complex multidisciplinary science that now features over 200 species under culture (including invertebrates as well as finfish). However, significant progress in the development of the techniques required to spawn and rear a variety of species had been developed by the end of the nineteenth century as detailed by Bowers (1900). In fact, if one compares a hatchery from the late nineteenth century with one of today and overlooks modern materials such as fiberglass, PVC plumbing, electric pumps, and all sorts of electronic gadgets and just looks at the basics, it is not difficult to imagine that someone from the time of Spencer F. Baird would recognize a lot of what he would see. An earthen pond is still a pond. The linear wooden raceways of old are now fabricated of concrete, fiberglass, or aluminum and in many instances circular raceways have replaced linear ones, but the concept has not changed. I can see a Livingston Stone (who developed salmon and trout hatcheries in California in the 1870s) as finding himself right at home as soon as he learned how to use a dissolved oxygen meter and become adjusted to a few other gadgets that didn't exist in his time.

It is not possible to put a precise date as to when the art became science as it was an evolutionary process. Each hatchery manager probably developed some new techniques and those were shared among peers at meetings such as those of the American Fish Culturists Association, which was established in 1870 and later became the American Fisheries Society.

In the United States, university courses associated with fish culture were rare before 1950, and any that existed would have undoubtedly focused on techniques associated with the culture of species for stocking, which would mean almost exclusively salmonids and non-salmonid freshwater species. The University of Washington in Seattle, Washington, established a position in fish culture in 1920 (Stickney 1989). Lauren Donaldson, who took a position at the university in 1932, constructed a salmon return pond on the campus and established a salmon run. Salmon were literally returning to the classroom.

Ichthyologic research was being conducted at many universities in the early twentieth century but the focus was usually not on aiding in the development of fish culture. It is probably safe to say that most of the applied fish culture research that was being conducted took place in government hatcheries where the methodology was probably of the trial and error variety. Since the approach involved producing fry and fingerlings for stocking public waters, there was not a great deal of interest by the government hatcheries in developing the technology associated with growout to foodfish size. Those relatively few private foodfish producers were largely on their own.

Today aquaculture courses (which typically include fish culture, but are usually not restricted to the study of finfishes) can be found at colleges and universities across the world. Certificate, undergraduate, and graduate programs abound. A quick search for "aquaculture institutions" on the Internet will provide page after page of hits.

While Hickling (1962, 1968) authored short books on fish farming, there was no comprehensive treatment of global aquaculture available until the classic book by Bardach *et al.* (1972) appeared. That work was based on the information gathered by the two senior authors who traveled the world looking at the status of aquaculture. The book provides good documentation of the state of the art that existed early in the third quarter of the twentieth century. Much of the aquaculture at the time was artisanal in developing countries, and artisanal aquaculture continues to provide much of the fish available to local communities in developing nations today (fig. 2.2).

Much of the published information on fish culture prior to the 1970s consisted of gray literature and in such government publications as *The Progressive Fish-Culturist,* which was published by the United States government for many years before being taken over by the American Fisheries Society. That publication is now being published as *The North American Journal of Aquaculture.*

The growth of information on commercial fish culture over the past few decades has been exponential. A large number of journals dedicated to

Figure 2.2 Artisanal tilapia pond in the Philippines (Photo by Robert R. Stickney).

aquaculture in the broad sense, fish culture, or one of the many aquaculture subdisciplines—nutrition, disease, engineering, genetics and water quality to name a few—have been established and the list continues to grow. Similarly, books on aquaculture and the various specialized branches of the field have proliferated.

The number of societies that are solely focused on aquaculture or include aquaculture within the broader context of fisheries and/or aquatic sciences has also expanded. The formation of the World Mariculture Society in 1969 (the name was changed to the World Aquaculture Society in 1986) was seen by some observers as an audacious move. How dare a small group of Americans (and one individual from Great Britain) gathered in the state of Mississippi have the gall to create a global organization? While initiated by a group of about forty, the society has thrived and came to live up to its name. It has been joined by a number of other regional or national societies (e.g., the European Aquaculture Society) and chapters, meetings of which often attract over 1,000 attendees. Virtually all of these associations or groups also publish a magazine or newsletter, many produce published proceedings of their meetings, and some also have developed book series. Such organizations typically have members or at least attendees at meetings from academia and government agencies as well as from the commercial production and supply sectors.

Specialty organizations that focus on a particular area of interest have also been formed. Examples are organizations devoted to the culture of channel catfish, tilapia, trout, salmon, engineering, and economics.

2.5   Commercial finfish species development

Commercial finfish culture was limited in the United States until rapid development began in the 1960s. Initially, species selection was often based upon local interest and availability (e.g., channel catfish, trout) but in many cases was also driven by declines in wild fisheries (e.g., Atlantic salmon, striped bass, cod, tuna) or finding alternatives to more traditionally consumed species (e.g., tilapia in the Americas). The following are some examples to illustrate how development took place with respect to a few species or species groups.

2.5.1   Channel catfish

The prediction that channel catfish (*Ictalurus punctatus*), which had been grown for stocking for many years, could be reared to market size and sold for a profit was first put forth in papers by Swingle (1957, 1958). Swingle speculated that a farmer would need to get US$1.10/kg at the farm gate to make a profit. Now, a half-century later, the farm gate value of the fish has not even doubled. The early work with channel catfish, as was true of many other native fish species being cultured in the United States, began at state and federal government hatcheries. Channel catfish were found to be difficult to spawn but success in pond spawning was finally achieved in the early twentieth century (Shira 1917). The species is native to streams where it spawns under logs, in depressions in riverbanks, or in some other convenient hiding place. Once the cavity spawning requirement was recognized and appropriate nests were placed in ponds, spawning became routine and is practiced today much like it was a century ago (Hargreaves & Tucker 2004).

It was in the state of Alabama where Swingle (1957, 1958) conducted his research that the industry was born. The history of catfish farming in that state was chronicled in a fascinating book by Perez (2006). After World War II, some farmers in the southern United States turned to fish culture. They were first interested in buffalo fish (*Ictiobus* sp.) and later turned to polyculture of buffalo fish with catfish. By the early 1960s, buffalo fish culture was on the way out and was being replaced with channel catfish monoculture (Hargreaves & Tucker 2004). Because of a lack of abundant ground water Alabama catfish farmers had to rely for the most part on precipitation runoff to fill their ponds. A seemingly much better situation with respect to water availability was seen in central Arkansas where rice had been the staple crop. Rice paddies could be fairly easily turned into catfish ponds on the flatlands of the region located to the north of the Ozark Mountains. The US Department of Interior set up one of two laboratories dedicated to fish culture research in 1958 when the Fish Farming Experimental Station was established in Stuttgart, Arkansas. The other laboratory was the Southeastern Fish Cultural Laboratory located in Marion, Alabama. The Stuttgart laboratory was moved to the Department of Agriculture in 1996 and became the Harry K. Dupree Stuttgart National Aquaculture

Research Center, named after a scientist who had been the director at both laboratories, first in Marion, and later in Stuttgart.

The first author of this chapter had the opportunity to spend the summer at the Stuttgart laboratory in 1969. It was at that time that the water table was dropping rapidly due to the massive withdrawals required to meet the needs of both the rice and catfish industries. Catfish farming was rapidly being developed in Mississippi, the state that quickly took over domination of the industry. However, Arkansas, Alabama, and Louisiana still produce considerable volumes of catfish with lesser amounts grown in many other American states.

Once popular only in the southern region of the United States, the demand for channel catfish is now nationwide due in part to an excellent marketing campaign that was launched during the rapid growth phase of the industry. Other factors that contributed to the growth of the industry included the formation of processing and feed cooperatives by the farmers that allowed them to share in the profits from those sectors of the industry while the farm gate value of the fish remained low.

One major problem affecting sector growth had been the seasonality of catfish availability. The fish require about eighteen months to grow from hatching to the typical harvest weight of about 0.5 kg. Initially, harvests occurred in the fall and were associated with complete draining of the ponds. In response to the market demand for year-round availability of fresh catfish, the procedure was changed to one in which partial harvesting is practiced at intervals throughout the year. Fingerlings are added to replace the fish that are harvested in a process known as understocking. The ponds are not drained during the harvesting process. Seines of certain mesh sizes selectively harvest the size of fish that the buyer prefers. Those of the preferred size—which can vary depending upon the market to which the fish are to serve—are loaded on trucks for live-hauling to the processing plant and those not needed are released back into the open pond. Ponds are often operated in this manner for several years without being drained.

The 1970s was a period of consolidation as the number of farms was reduced while the area under culture increased greatly as many farmers failed and the successful ones expanded their holdings (Hargreaves & Tucker 2004). Government and university research led to the development of high-quality formulated feeds containing little or no fish meal or other animal protein; it also addressed various disease problems, developed aeration devices that effectively maintain dissolved oxygen in heavily stocked ponds, and addressed a series of other issues that faced the industry. Off-flavor was (and continues to be) a major issue particularly during the warm months. A process has been developed whereby a few fish from a pond scheduled for harvest are taken to the processing plant for taste testing a week or two in advance of harvesting, the day before and when the truckload of fish reaches the processing plant. If the fish have off-flavor (often described as an earthy-muddy flavor), they are rejected. Blooms of certain algae are the source of the chemicals that cause the off-flavor, and given time, the blooms will dissipate and the fish will metabolize the off-flavor producing chemicals.

In 2005 the industry produced 275,757 metric tonnes of channel catfish, down from the high of over 300,000 metric tonnes in 2003. Production has continued to fall with the 2007 production estimated at 224,500 to 229,000 metric tonnes (Anonymous 2008). According to the National Agricultural Statistics Service (NASS 2008), the amount of channel catfish processed through November 2008 was up several percent over 2007. Imported basa (*Pangasius bocourti*) from Vietnam compete with channel catfish in the marketplace and for some time were available at a lower price. A regulation was put in place that prohibits marketing of basa as catfish in the United States. In addition, and more broadly, the United States has adopted country of origin labeling (COOL) on imported fishery products. Those regulatory changes may have contributed to increases in the price of basa relative to that of channel catfish.

Brazil is also now producing channel catfish, some of which are exported from that country to the United States. China is also growing channel catfish and sent over 260,000 kg to the United States in 2007 (NASS 2008). Imports of frozen basa and channel catfish to the United States during the first nine months of 2007 amounted to 28,760 tonnes (Anonymous 2008). Other nations that export catfish of various species to the United States include Cambodia, Indonesia, Malaysia, Spain, and Thailand (NASS 2008). The plight of catfish farmers from foreign competition has been exacerbated by current high feed costs. Still, channel catfish represent the largest component of commercial fish farming in the United States. Rainbow trout run a distant second with 27,504 metric tonnes of production in 2005 (United States Department of Commerce 2007).

2.5.2   Tilapia

Maar *et al.* (1966) indicated that the first attempts to culture tilapia occurred in Kenya in 1924, followed by the Congo in 1937. Pond trials were initiated in Zambia in 1942 followed by Rhodesia in 1950. While native to Africa and a portion of the Middle East (Philippart & Ruwet 1982), tilapia were introduced to parts of Southeast Asia in the 1930s and are thought by today's inhabitants of such nations as the Philippines to be native. The majority of the tilapia produced for human consumption are in the genus *Oreochromis,* with Nile (*O. niloticus*) and blue (*O. aureus*) and red hybrid tilapia (produced from various crosses) being the most popular. Mozambique tilapia (*O. mossambicus*) are still produced in some countries but are no longer the most highly preferred species by culturists. Mozambique tilapia were introduced to the United States in the 1960s (Stickney 1996) followed by various other species, including blue and Nile tilapia. One or more species were later spread into various countries in Latin America and the Caribbean region often by scientists involved in foreign aid projects in the tropics aimed at increasing the human food supply. Rural farmers could easily be taught to rear tilapia and could do so without the need for any sophisticated methods or technology. In addition to being easy to produce, tilapia are usually readily accepted by consumers. Most of the distribution of tilapia around the world was made before issues associated with exotic introductions developed.

Stunting and overpopulation are significant negatives associated with tilapia pond culture. Hand sexing is possible once the fish reach the appropriate size but is time consuming and subject to human error. One means of dealing with the problem was found the 1960s when monosex populations of tilapia were produced in Israel through hybridization (Sarig 1989). Following that break-through, Israeli fish culturists began using tilapia in polyculture with carp and mullet. The creation of all-male tilapia populations by feeding small amounts of male sex hormone to fry tilapia was developed in the 1970s (Guerrero 1975).

Over a period of several years a genetically improved farmed tilapia (GIFT) strain was developed through selective breeding. The effort, largely conducted in the Philippines, has produced fish of exceptional quality and rapid growth that have been widely distributed across the world including back to Africa, the natal home of tilapia. While there is some production of tilapia in the United States—7,803 metric tonnes in 2005 (United States Department of Commerce 2007)—that production cannot begin to meet the demand that has developed in the nation in recent years and continues to grow. Most of the United States experiences winter temperatures that are lethal to tilapia, which limits locales where the fish can be produced year-round to the southernmost regions and Hawaii, with the exception of places that have access to warm water (geothermal or from an industrial source such as power plant cooling water). Considerable numbers of tilapia are being produced in Idaho using geothermal water in a state where the air temperature is well below freezing during extended periods in the winter (fig. 2.3). There is also seasonal production of tilapia in ponds in certain areas of the United States as well as indoor production in recirculating systems. Much of the tilapia imported to North America comes from the Caribbean and

Figure 2.3 Tilapia tank culture facility near Boise, Idaho, uses geothermal water to maintain the proper temperature in a high-temperate climate. (Photo by Robert R. Stickney)

Latin America, and recently, China. Details of tilapia culture in the Americas can be found in a two-volume set of books edited by Costa-Pierce & Rakocy (1997, 2000). Tilapia are produced on a large scale in many Southeast Asian nations. A good deal of effort has gone into tilapia culture in Africa, largely on an artisanal scale.

2.5.3 Flatfish

Publications began to appear in Europe in the late 1800s and early 1900s with respect to plaice (*Pleuronectes platessa*), sole (*Solea solea*), turbot (*Scophthalmus maximus*), and perhaps other flatfish species (Dannevig 1897, 1898; Malard 1899; Fabre-Domergue & Bietrix 1905). Those studies demonstrated that flatfishes could be spawned and reared under hatchery conditions. Additional rearing studies associated with plaice were conducted in later years (Dawes 1930; Rollefsen 1939). The short paper by the latter author also involved plaice hybrids (*P. platessa* × *P. pseudoflesus*) and appears to have been the first to show that brine shrimp could easily be reared as food for fish larvae.

In the late 1940s there was a report on marine fish rearing from the west coast of the United States that included flatfish (McHugh & Walker 1948), while Japan became involved in the culture of flatfish by the mid-1950s (Kurata 1956).

Activity in conjunction with European flatfish culture increased and major advances were made in the 1960s and 1970s. Nash (1968) described his attempts to rear plaice and sole in the thermal effluents of power plants and reported on the best rearing temperature range for each species. A series of publications on plaice rearing in Britain appeared in the 1960s (e.g., Shelbourne 1963a–c; Riley & Thacker 1963; Shelbourne *et al.* 1963, 1964). Cowey *et al.* (1970a, b, 1971) conducted early work on the nutritional requirements of plaice, while Kirk & Howell (1972) described the growth and food conversion rates of plaice fed prepared and natural feeds. An annotated bibliography on attempts to rear marine fish larvae in the laboratory, which included flatfish, was written by May (1971) and a partially annotated bibliography on flatish was produced by White & Stickney (1973a).

With the foundations of flatfish culture having been laid, research activity grew rapidly and additional species were the subject of investigation. Many of them are shown in table 2.1. Citations on research are too numerous to include here but many references to each of the species can be found in Stickney (1997d), which covers the period from the late 1970s to 1997, and in Stickney (2005). For an extensive bibliography focused on Atlantic and Pacific halibut, see Stickney and Seawright (1993), which captures some of the research that led to development of the commercial halibut culture. A special issue of the journal *Aquaculture* is a collection of papers on the culture of flatfish that includes species under study as possible candidates for culture (Stickney *et al.* 1999).

The first author's involvement with flounder culture began in the early 1970s at the Skidaway Institute of Oceanography in Savannah, Georgia USA with

Table 2.1 Species of flatfish that have been the focus of aquaculture research, enhancement stocking, and/or commercial culture, along with country or region of greatest interest or research activity.

Scientific Name	Common Name	Region/Country
Ammotretis rostratus	Longsnout flonder	Indo-Pacific
Heteromycteris japonicus	Bamboo sole	Japan
Hippoglossus hippoglossus	Atlantic halibut	Norway, North America
Hippoglossus stenolepis	Pacific halibut	United States
Kareius bicoloratus	Stone flounder	Japan
Limanda punctatissima	Longsnout flounder	Japan
Limanda yokohamae	Marbled sole	Japan
Liopsetta putnami	Smooth flounder	United States
Microstomus kitt	Lemon sole	Europe
Paralichthys spp.	Chilean flounder	Latin America
Paralichthys adspersus	Fine flounder	Latin America
Paralichthys californicus	California halibut	United States
Paralichthyts dentatus	Summer flounder	United States
Paralichthys lethostigma	Southern flounder	United States
Paralichtys microps	Smalleye flounder	Latin America
Paralichtys olivaceous	Olive flounder[a]	Japan
Platichthys flesus	European flounder	Europe
Pleuronectes ferrugineus	Yellowtail flounder	United States
Pleuronectes platessa	Plaice	Europe
Psuedopleuronectes americana	Winter flounder	United States
Rhombosolea tapirina	Greenback flounder	Australia New Zealand
Scophthalmus maximus[b]	Turbot	Europe
Scophthalmus maeoticus	Black Sea turbot	Europe
Solea solea	Sole	Europe
Verasper variegates	Spotted flounder[c]	Asia

[a] Also known as the Japanese flounder and bastard flounder.
[b] *Psetta maxima* is synonymous.
[c] Also known as the barfin flounder.

support from a Sea Grant project to culture southern and summer flounders. Postlarvae were captured from a tidal river since the technique for spawning the species of interest had not been developed. The methodology for spawning southern flounder was published by Arnold *et al.* (1977).

One result of the experiments was development of ambicoloration in tank-cultured flounders (Stickney & White 1975). A study was conducted on the effects of salinity on southern flounder growth (Stickney & White 1973). There was also a report on the occurrence of lymphocystis in cultured flounder (Stickney & White 1974). A manual on flatfish rearing was produced by White and Stickney (1973), which included information from the literature, but also development of the flounder culture system at Skidaway, which included secondary filters and UV sterilization to reduce the chances of disease organisms entering the system with the incoming water. The manual also discussed collection of postlarvae and juveniles, first feeding of postlarvae with brine shrimp and later conversion to chopped frozen shrimp, then freeze-dried shrimp, and

Table 2.2 Atlantic halibut production in Europe annually from 1998–2005 in metric tonnes (Federation of European Aquaculture Producers 2006).

Year	1998	1999	2000	2001	2002	2003	2004	2005
Production	20	503	135	389	350	84	855	905[a]

[a] Estimated.

finally prepared feeds. Summer flounder were grown from less than a gram to approximately 130 g in 24 weeks during 1972 and to about 120 g in 20 weeks during 1973.

What basically led to the discontinuation of the research on flounders at Skidaway was the fact that they were worth only about US$0.55/kg. The possibility of rearing them profitably appeared to be remote. Interest arose once again in the 1990s as flounder stocks along the east coast of the United States and in the northern Gulf of Mexico became depleted and the value of flounders in the marketplace had greatly increased. Various research groups began producing information after about a twenty-year hiatus. Summer flounders were first stocked in commercial quantities in offshore cages by the University of New Hampshire's Atlantic Marine Aquaculture Center in 1999 (Atlantic Marine Aquaculture Center 2007a).

Research on Atlantic halibut culture was initiated in the mid-1980s, initially in Norway and Iceland, and later in the Maritime Provinces of Canada and in Maine and New Hampshire in the United States. Commercial culture in Europe began in the late 1990s, with Norway and Iceland being the producing nations. Production figures for 1998 to 2005 are presented in table 2.2. A commercial hatchery in Nova Scotia, Canada was the source of fingerlings stocked into an offshore cage off New Hampshire by the Atlantic Marine Aquaculture Center (2007b) and grown for two years from an average of 7.6 cm to weights of 3.2 to 4.1 kg.

The amount of research that has been conducted on Pacific halibut pales in comparison with that on the Atlantic species (Stickney 1997d, 2005). To date, there is no commercial Pacific halibut culture and no indication the situation will change in the near future.

2.5.4 Atlantic salmon

Salmon hatcheries were present in both North America and Scotland by the end of the nineteenth century (Laird 1996). The first in North America was built in 1866 in Ontario, Canada. The Craig Brook salmon hatchery in Maine USA was established a few years later in 1871 (Kirk 1987). Many hatcheries were constructed after the establishment of the United States Commission on Fish and Fisheries in 1872. Until 1965, when the first salmon cage farm was established in Norway (Laird 1996), production was aimed at replacement or enhancement, and in relation to the United States Commission also in attempts to establish populations in areas where the fish were not native.

Figure 2.4 Atlantic salmon net pen culture facility in a Norwegian fjord (Photo by Robert R. Stickney).

Salmonid foodfish production in Norway actually began with the culture of rainbow trout introduced from North America, but in the mid-1960s Norwegian fish farmers turned to Atlantic salmon, which was deemed to be more profitable (Tilseth *et al.* 1991). Various types of sea enclosures, beach enclosures, and cages were in production by the late 1970s. In Norwegian fjords, salmon culture facilities today use net pens (fig. 2.4). The 287 Norwegian salmonid farmers in 1973 produced 171 metric tonnes of salmon and 101 metric tonnes of rainbow trout. By 1989 there were 791 farmers responsible for a total of over 118,000 tonnes of salmon and trout (Tilseth *et al.* 1991), with salmon accounting for the majority. Norwegian production rose from about 4,000 metric tonnes in 1980 to 200,000 metric tonnes in 1994 (Laird 1996). Norway was producing about 70% of the world's farmed salmon in 1989 (Tilseth *et al.* 1991).

In North America, Atlantic salmon net pen facilities had been established on both coasts by 1990. Commercial culture occurs in the Maritime Provinces of eastern Canada and in British Columbia in the west. In the United States, salmon are produced in Maine and Washington. The production is primarily Atlantic salmon in Washington and exclusively Atlantic salmon in Maine. There are some Coho and Chinook salmon being produced in Washington and British Columbia.

Atlantic salmon were also being produced in Scotland, Australia (Tasmania), Chile, and the Faroe Islands by 1990. Chile has been on a trajectory to reach the production levels of Norway within the foreseeable future, however recent disease problems and environmental degradation concerns have arisen, which may forestall that from taking place.

Figure 2.5 Salmon net pens in Japan near a resort hotel.

Atlantic salmon hatcheries and smolt rearing facilities are land-based and typically use flow-through tank technology for smolt production. Smolts are typically stocked into marine facilities at around 40 g. Those facilities are for the most part net-pens located in protected coastal areas (fig. 2.5). The technology for open ocean production has been developed but is currently being utilized more for the production of marine fishes rather than anadromous Atlantic salmon. That situation may change in the future as the carrying capacity of protected coastal waters is reached, and as already has been the case in some areas, surpassed.

Ocean ranching of Pacific salmon is practiced in the state of Alaska USA. Pacific salmon farming and ranching is also practiced in the northern part of Japan. The ocean ranching activity in Japan is focused on chum salmon (*O. keta*). Salmon culture and ranching also occurs in Russia (Dushkina 1994).

2.5.5   Additional species and regions

At the risk of leaving out a number of important species and centers of activity, the mention of several additional commercially cultured finfishes and production centers is appropriate. Space does not permit further elaboration; however, interesting stories could easily be developed about how culture methods and technologies were developed around each of the species mentioned.

European sea bass (*Dicentrarchus labrax*) and gilthead sea bream (*Sparus aurata*) are produced in most countries bordering the Mediterranean Sea (Rad

2007). Red sea bream (*Pagrus major*) are cultured in Japan, which is also known for the culture of yellowtail (*Seriola quinqueradiata*).

Tuna (*Thunnus* sp.) are cultured in the Mediterranean using the approach of collecting several kilogram juveniles in purse seines and hauling them to net pen facilities for growout to tens of kilograms over a period of several months. That approach is also used in Australia and on the Pacific coast of Mexico. This production method is covered in more detail in Chapter 7.

Walking catfish of various species in the genus *Clarias* are being produced in Southeast Asia, India, parts of Europe, and various nations in Africa, as well as in Hawaii USA. Most walking catfish are produced in ponds. Those are often constructed with vertical levees and may have low fences around them to keep the fish from escaping.

Carp of various species continue to be the primary finfishes produced in the world (FAO 2006), with China leading the way by a considerable margin. While once favorites throughout the country, carp consumption is now largely confined to rural areas in China since the emergence of a middle class in the cities that is populated by people who have a preference for higher value fishes.

The Indian carp species known as catla, mrigal, and rohu are cultured on the Indian subcontinent. Together, carp and other cyprind species accounted for over 18 million metric tonnes of aquaculture production in 2004 (FAO 2006).

Rainbow trout were mentioned above with respect to their production in Norway having largely been displaced by Atlantic salmon. Rainbow trout are also cultured in many other nations, including seemingly unlikely places such as Nepal, New Zealand, and South and Central America. Rainbow trout were shipped to New Zealand from the United States, along with Chinook salmon, over 100 years ago. New Zealand has prohibited the import or sale of rainbow trout either live or processed to avoid the introduction of diseases to the trout that exist in the island nation today as a result of the century old introductions. In the United States, 95% of the trout that reach the market are produced in the Hagerman Valley of Idaho.

Striped bass (*Morone saxatilis*) and striped bass hybrid (*M. saxatilis* × *M. chrysops*) culture is confined to the United States, while red drum (*Sciaenops ocellatus*), native to the Gulf of Mexico and southeastern United States, are being cultured in parts of Asia. Cobia (*Rachycentron canadum*) is also being cultured in parts of Asia and is of interest to culturists in the United States and in the Caribbean region. Interest in pompano (*Trachinotus carolinus*) culture has redeveloped after a hiatus since the 1970s when research was conducted in Florida. There had been interest in the culture of dolphin (*Coryphaena hippurus*, which is a finfish, not a mammal, and is marketed as Mahi Mahi) in the past, but there appears to be little activity associated with its culture at present.

Milkfish (*Chanos chanos*) are cultured in Southeast Asia (particularly in the Philippines) and mullet (primarily grey or striped mullet, *Mugil cephalus*) are cultured in many parts of the world, including the Mediterranean region and Southeast Asia.

With the collapse of the cod (*Gadus morhua*) fishery in the western north Atlantic and severe declines in the eastern north Atlantic, cod aquaculture has become the focus of commercial ventures in Norway, Canada, and Maine, in the Unites States.

A variety of native species in Brazil are currently being cultured to at least some extent, while many others may be suitable candidates for fish culture and have yet to be evaluated in that context. The same is undoubtedly true for other species in Latin America.

Finally, Australia is a nation that is developing a variety of cultured foodfish species. One that is quite popular is the barramundi (*Lates calcarifer*).

In Hawaii, local finfish species are sold under their Native Hawaiian names. Two fishes, kahala (amberjack or Hawiian yellowtail, *Seriola* sp.) and moi (Pacific threadfin, *Polydactylus sexfilis*), are being cultured and are available in restaurants and in processed form for tourists interested in taking some back to from whence they came.

Unlike domestic livestock production, which today and for centuries, has relied on a handful of species, the number of fish species under culture has been growing for a few decades and there is no end in sight. Many of the ones of particular interest are marine species with extremely small eggs and larvae, making them difficult to feed after yolk sac absorption. Atlantic and Pacific halibut, which can reach sizes of 200 kg or more, are among the species with tiny eggs. Not only are the eggs small, they are extremely fragile. It is testimony to the resourcefulness and tenacity of those who have developed industries around such species that they have achieved the level of success that is seen today. Other species elude the talents and technology to bring them into commercialization, but breakthroughs will be made and new species can be expected to appear on the tables of restaurants and in the supermarkets and fish markets of the world bearing a label indicating that they were cultured.

2.6 Shrimp culture

Shrimp aquaculture has its origins in Southeast Asia, where for centuries farmers raised incidental crops of wild shrimp in tidal fish ponds, but the shrimp were not considered of value. The earliest record of Penaeidae is in Chinese history and traces back to the eighth century BPE. Japanese literature referred to penaeids in 730 BPE. The first scientific record of a penaeid was in 1759, when Seba of Amsterdam named and sketched a penaeid. In 1815, Rafinesque recognized that penaeids were a distinct group within the Decapoda.

Artisanal capture of freshwater shrimp was probably incidental to pond fish culture in some locations. Spawning, larval rearing, and growout of freshwater shrimp was not developed until the twentieth century.

2.6.1 Marine shrimp

Documentable shrimp culture began in 1933, when Japanese biologists, including Motosaku Fujinaga (who also published as Hudinaga), first studied

(Hudinaga 1935) the artificial propagation and culture of the Kuruma shrimp, *Penaeus japonicus,* now referred to as *Marsupenaeus japonicus.* A graduate of Tokyo University, Fujinaga entered the Hayatomo Fisheries Institute of the Kyodo Gyogyo Company at Oyanoshima, Amakusa (later called the Nippon Suisan Company), and worked as a biologist (Tseng 1987). There were only three people assigned to conduct the work under very primitive conditions in a small lab with no electricity and only kerosene lamps for light.

In the first year, Fujinaga used a plankton net to collect juvenile shrimp and studied their morphology, taxonomy, and life cycle. However, he was not able to obtain eggs using that method of collection. He then placed mature spawners in aquaria to obtain eggs. Nauplii were obtained but died when they were not fed properly and failed to metamorphose into zoea. A series of feeding trials were conducted over the next several years. It was not until 1939 that Fujinaga successfully fed shrimp larvae the diatom *Skeletomema costatum,* which came from a mass culture technique developed by Matsue of Tokyo University.

Fujinaga (Hudinaga) published from 1935 to 1969 and conducted pioneering shrimp research for more than forty years (Hudinaga 1935, 1941; Fujinaga 1969). He set up a pilot commercial hatchery in 1959 that was instrumental in transferring the technology to the commercial sector, and in the early 1960s the first commercial farms were built along the Seto Inland Sea of Japan. By 1967 some twenty operations using his methods produced 4,000 metric tonnes of shrimp from 8,500 ha of water. Dr. Fujinaga was given the name "Founder of Kuruma Shrimp Culture in Japan." The next real breakthrough in mass culture of shrimp larvae came later, when Fujinaga was credited with feeding *Artemia* nauplii to postlarval shrimp and increasing their survival and health (Tseng 1987).

While the Asian (Japanese and later Taiwanese) approach to larval shrimp culture was considered the green water method (most feeds such as algae were grown in the same large volume, deep tanks with the shrimp larvae), another method of shrimp culture was developed in the Western hemisphere. The Galveston method, or clear water method, was developed in Galveston, Texas. Shrimp larval feeds were grown in separate tanks and fed when needed. This was generally synonymous with smaller, conical tanks with heavy aeration to keep everything suspended, clearer water in batch culture operation, higher stocking densities, higher larval survivals, and higher labor requirements.

Limited studies of penaeid shrimp had been conducted in the United States in the 1930s, and even though none were aquaculture oriented, they added to the knowledge base on penaeids. In 1935 J.C. Pearson described the eggs of some species of penaeid shrimp, and he described the life histories of some American penaeids (Pearson 1939). In 1953, Paul E. Heegaard attempted to spawn the white shrimp, *P. setiferus,* in Port Aransas, Texas, and Gunter and Hildebrand worked with the wild shrimp life cycle in 1954 (Gunter & Hildebrand 1954). About that time the Bureau of Commercial Fisheries, later named the National Marine Fisheries Service, began to work with the biology of commercial shrimp in the Gulf of Mexico. According to Harry Cook (Treece 1993) the Director of the Laboratory, Milton J. Lindner, was successful in obtaining the funding necessary to develop the methodology for a prototype shrimp hatchery system at the Galveston Laboratory. In 1959 Harry Cook was a biologist for the Bureau of Commercial Fisheries and the following account is from an oral interview with Harry Cook by Bob Rosenberry of Shrimp News International (2007). Papers related to these developments include Cook (1965) and Cook and Murphy (1971).

Three other people were hired to study the life history of Gulf of Mexico shrimp. Wooden troughs about 0.75 m by 1 m in diameter were used as rearing containers. When the research started they had trouble finding gravid brown shrimp *(Penaeus aztecus)* so they worked with *Xiphopenaeus* and published a paper that described the nauplius stage. Later the researchers were able to spawn brown shrimp and they sketched the various larval stages so that they could identify them in seawater samples. They published the life history of the brown shrimp and a key to the genus. There were researchers in the red tide section of their lab who knew how to culture algae. They relied heavily on the work of Fujinaga but their initial attempts were unsuccessful. The algae and debris would settle and the larvae would concentrate in the corners. The investigators were growing larval feeds in inverted carboys and saw how the water circulation kept everything suspended and the debris from clumping. They decided to rear the larval shrimp in fiberglass tanks with conical bottoms with an air stone at the bottom of the cone and around the sides of the tank. They also developed a filter system that allowed water exchange without damaging the larvae. Cook had visited an oyster hatchery on Long Island, New York, that was using airlifts in its algae tanks. When Cook returned to Texas he put airlifts in the tanks and that basically became the Galveston method: tanks with conical bottoms, airlifts for aeration, a daily batch exchange of water and the addition of live feed, either a mixture of *Skeletonema and Monochrysis* for zoea or brine shrimp for mysis and postlaravae (Cook 1965; Cook & Murphy 1971). Although the conical tanks are rarely seen now, this general method is still in use today in many Western hemisphere hatcheries. The Galveston method generally used a number of different algal feeds but eventually evolved into a combination of small cell algae and larger cell algae (*Chaetoceros* sp. and *Tetraselmis chui*) as beginning feeds for shrimp larvae.

At that same time in Florida, Tom Costello and Don Allen at the Miami Lab of the Bureau of Commercial Fisheries used fluorescent light bulbs to grow algae indoors. At the Galveston laboratory, researchers were growing algae in outdoor greenhouses, but eventually fluorescent bulbs and indoor culture became part of the Galveston method (Rosenberry 2007). Also as part of the development of the Galveston method, a variety of food types, such as rotifers and nematods were tried, but they eventually were dropped and replaced by *Artemia* as food for late stage larval shrimp.

The water in the closed seawater system at the Galveston laboratory had a mineral imbalance that became a major problem. In England, a researcher reported better production when using EDTA in algal cultures so the Galveston group tried it and made a major breakthrough. EDTA is a chelator that helps improve water quality by helping the minerals and metals stay free and unbound. With the success of EDTA in algae culture they decided to put it in the shrimp tanks and that's when they started getting better larval survival. The Galveston group did not count the algae; they used a spectrometer to get an estimate of algal density.

Another procedure developed in the Galveston method involved replacing the volume removed from the algae cultures to feed the larval shrimp each day with sterilized water and nutrients, so that the algae would grow back to optimum density by the next day. This kept a continuous, clean source of food available for the shrimp with minimal effort and expense.

Although the Japanese research had some influence on the research in Texas, Fujinaga's and Mitsutake Miyamura's visit to the Galveston lab in 1963 was not for the purpose of information transfer and was not a turning or beginning point for shrimp culture in Texas as we were led to believe for many years by various historical accounts. According to Harry Cook (personal communication, 1992) the purpose for the Japanese visit (which only lasted a few hours) was to find a place for shrimp growout in the United States The Japanese wanted to lease East Matagorda Bay, Texas, for that purpose but ended up in Florida and in 1967 established Marifarms, Inc., which operated from 1968 to 1982. A full account of those activities was published by John Chesire (2005). Storm damage and other problems including environmental concerns led to moving from the United State to Ecuador (Treece 1993).

J. J. Ewald worked on the laboratory rearing of pink shrimp *P. duorarum* in 1965 (Ewald 1965). Harry Cook described the rearing and identification of Gulf of Mexico native shrimp larvae in 1965, published about a dozen other papers on the subject from 1966 to 1970, and Cook and Murphy described the developmental stages of native shrimp larvae in 1971. It wasn't until after Cook left the Galveston lab that other researchers further developed maturation techniques for shrimp (Rosenberry 2007, interview with Harry Cook). A newcomer (Cornelius Mock) came to the National Marine Fisheries Service laboratory about the time Harry Cook was leaving and Mock continued to publish on the Galveston Method.

Generally speaking, in the past the Western hemisphere practiced the Galveston method in company-owned hatcheries that supplied large farms. At the same time, culturists in the Eastern hemisphere practiced the green water or Taiwanese method in smaller, but more numerous, usually family or government-owned hatcheries. However, in recent years the two major methods of larval rearing have merged. The small backyard hatcheries in Asia further blurred the distinction of the two methods. In Asia small and medium-scale hatcheries continue to be most popular but worldwide, the distinction between the two methods or styles of larval culture continues to be more blurred as new operations are built, borrowing the best from both methods.

Commercial shrimp growout attempts were made in Ecuador in the 1960s and in the United States starting in the late 1960s and early 1970s. The Ecuadorian industry was based upon *L. vannamei* and *L. stylirostris*, and was started by accident when a broken dike on a banana farm allowed shrimp to enter the farm. By the time the farmer repaired the dike a crop of shrimp had been produced. Expansion of the Ecuadorian industry was made possible by an abundance of wild postlarval shrimp. After the industry matured and could not rely entirely upon the seasonally wild-caught postlarvae, brood collection stations were developed along the coast, which captured and spawned wild-mated females and provided an important nauplii source to meet the growing hatchery demand. The Ecuadorian industry became more dependent upon hatcheries as it grew and larvae from hatcheries became stronger when new hatchery techniques and combination diets were developed. The initial United States industry followed that lead but was based on native species of white, brown, and pink shrimp (*Litopenaeus setiferus, Farfantepenaeus aztecus,* and *F. duorarum,* respectively). When researchers in the United States grew exotic shrimp from the Pacific coast of Central and South America, those species proved to be easier to culture and more productive in the ponds. Gradually commercial producers concentrated on exotics such as *L. vannamei,* now the most popular farm-raised shrimp in the world.

Once shrimp hatcheries began supplying large quantities of shrimp to farmers, the production of farm-raised shrimp expanded rapidly worldwide. The explosion of the industry continued into the early 1990s until problems began with disease outbreaks and water quality deterioration, which slowed worldwide production for a few years. Many groups from different countries worked on shrimp culture research and development. SEAFDEC (the Philippines), TASPAC/MEPEDA (India), AQUACOP (France and Tahiti), NOAA/Sea Grant and USDA/USMSFP (US), Ralston Purina, DOW Chemical, Coca Cola, W. R. Grace, Prince, King James Shrimp, (US companies), Marifarms, Inc. (Japanese-American owned in Panama City Florida and later Ecuador), Sea Farms (Honduras), and United Nations/FAO (Rome, Italy) are just a few groups that made progress on shrimp culture through research.

In addition to Japan, Taiwan became a very successful shrimp producing country in Asia. In 1969, Taiwan began to culture a number of economically important species of shrimp (Lien & Ting 1969). In 1970, with financial aid from the Rockefeller Foundation the Tung Kang Marine Lab was established and led the way to establishing *P. monodon* in Taiwan and other parts of Asia. I-Chiu Liao, a follower of Fujinaga, worked many years at the Tung Kang Marine Lab with *P. monodon* and is known as the "Father of *monodon*" in Taiwan. By 1978 most of the farmers in Taiwan had mastered the culture techniques and in 1979 the production in Taiwan reached a historical peak of 600 metric tonnes. By 1986, the production was up to 45,000 metric tonnes in Taiwan alone. Production crashed in 1988, when farms experienced severe disease outbreaks of Monodon Baculovirus (MBV) caused by their own activities that had led to environmental degradation. The problem caused a near collapse of the industry in Taiwan and the high production levels of 1986 were never again reached.

However, the success in Taiwan with *P. monodon* spread to other areas in Asia. The Philippines, Hong Kong, Malaysia, Singapore, Sabah, Borneo, Southern China, Thailand, and Australian shrimp industries developed rapidly. In 1985 there were 148,000 metric tonnes of shrimp produced in Asia (Scura 1987).

Shrimp culture became an organized industry on a worldwide scale and similar culture techniques were applied to various species in different regions. For example, *P. orientalis* (later named *Fenneropenaeus chinensis,* a colder water species) was grown in China; *P. japonicus* in Japan; *P. monodon* in Taiwan and the Philippines; *P. merguinsis* in Thailand, Indonesia, and parts of China; *Metapenaeus ensis* in Hong Kong and other parts of China and *Macrobrachium rosenbergii* in Malaysia, other parts of Asia and Hawaii; and *P. vannamei* and *P. stylirostris* in Ecuador, Mexico, the United States, and other Western hemisphere countries.

In 1997, Isabel Perez Farfante and Brian Kensley (Farfante & Kensley 1997) proposed some changes in the way scientists referred to the popular farmed shrimp species. Except for three species (*Penaeus monodon, P. esculentus,* and *P. semisulcatus*), the genus names were changed for the other penaeids to *Litopenaeus, Fenneropenaeus, Marsupenaeus,* or were based on a reexamination of the phylogenetic relationships among species in the family Penaeidae.

The majority of the shrimp produced for human consumption have historically been grown in Asia (80% Eastern hemisphere—*P. monodon, P. chinensis,* and others—and 20% Western hemisphere—*P. vannamei, P. stylirostris,* and others). Diseases caused a collapse of the industry in both hemispheres (IHHN virus, Monodon Baculovirus, Yellowhead virus, Taura Syndrome Virus, White Spot Syndrome Virus, and others) first in the Eastern hemisphere and later the Western hemisphere. The estimated economic loss in Taiwan (1987–1988) was US$420 million; in China (1993) US$1 billion; and in Thailand (1991) US$180 million (NACA 1994–1995). World Bank estimates of shrimp disease losses for all of Asia from 1994 to 1996 ranged from US$1 billion/year to $3 billion/year. In 1999 to 2000, Ecuador, the Western hemisphere's largest shrimp producer at the time, lost US$300 million to $500 million due to viral diseases (Wickins & Lee 2002).

Disease resistant *P. vannamei* replaced diseased stocks once biosecurity techniques were developed and genetic selection of shrimp began. Genetic selection for disease resistant penaeids started on a fast track in the United States in the late 1980s and early 1990s, thanks to the US Department of Agriculture's Marine Shrimp Farming Program and the technology and practice spread to other countries in the mid- to late-1990s. Since disease resistant stocks have been developed and *L. vannamei* culture has grown in both hemispheres, *L. vannamei* has replaced most of the species that were once grown. Once disease resistant stocks of *P. monodon* are more widely distributed, that species may come back as well. A number of companies are now working to make that happen with one US company now promoting and marketing ninth-generation selected animals.

Coupled with disease resistant strains, low salinity stocks were also developed that allowed *P. vannamei* farming in freshwater starting in the mid-1990s. The technology matured in China and freshwater culture of the species was a major driving force behind China's shrimp farming boom in the early 2000s.

According to Lem (2006), shrimp (both freshwater and marine and both harvested and farm grown) is by far the most important commodity by value in the international fish trade. Yearly exports worldwide of shrimp and shrimp products exceeded US$10 billion in 2003 and represented 20% of world total exports of fish and fishery products (FAO 2004). Marine shrimp farming grew into a US$6 billion industry but diseases and lack of biosecurity continue to give the shrimp-industry problems, even with disease-resistant stocks. Yet each country that experienced a crash in production has started a comeback. Most countries have shifted to genetically improved, domesticated stocks of *L. vannamei* or plan to do so in the near future.

China is the largest marine shrimp farming country in the world and its 490,000 metric tonnes in 2003 accounted for 27% of total world production (Yuan *et al.* 2006). China's shrimp farming industry has continued to grow since 1997. The initial success of *L. vannamei* attracted more and more farmers, resulting in more than a doubling of production in only four years from 218,000 metric tonnes in 2000 to 500,000 metric tonnes in 2004. The US shrimp farming industry started to decline in 2004 and continues a downward trend. To make matters worse, after being funded for twenty-five years, the US Congress terminated the funding of the USDA US Marine Shrimp Farming Program in 2011.

2.6.2 Freshwater shrimp

The Malaysian freshwater shrimp or giant river prawn (*Macrobrachium rosenbergii*) became the subject of aquaculture research and commercial enterprise less than fifty years ago. In 1961 a major breakthrough was achieved at the Marine Fisheries Research Institute, in Glugor, Penang, Malaysia, when it was discovered that a certain amount of salinity was an important basic requirement for the survival, growth, and development of the larval stages of the species (Ling & Merican 1961). The discovery quickly led to other breakthroughs on the food and environmental requirements of the larvae. Techniques for rearing hatchlings through all their larval stages were successfully developed and the first laboratory-reared juvenile was produced in June 1962 (Ling 1977). Techniques for spawning and other hatchery operations were soon developed, and in the spring of 1963 sufficient numbers of juveniles were produced for conducting growout experiments in ponds. Other refinements were reported by Ling (1967a, b). The results of growout experiments were encouraging and news of the achievements spread rapidly to other countries (Tham 1968; Liao & Huang 1972; Racek 1972). According to Ling (1977), growers in Hawaii asked the Marine Fisheries Research Institute for specimens in 1969 and soon became a major producer of *M. rosenbergii*.

Mistakidis (1969) published an excellent biological account of the freshwater shrimp with line drawings of eggs and larval stages. Fujimura & Okamoto (1970) at the University of Hawaii further refined the mass production of juveniles, and thereafter, freshwater shrimp culture began to spread to many other areas. Bardach *et al.* (1972) gave an early account of progress with the species. The growout technology spread to areas such as Mauritius, French Polynesia (Aquacop 1977), Israel, and in the US state of Florida. The Weyerhaeuser Company in Florida started a research and development program in 1974. Culture then began in the Unites States in Texas (1974); Puerto Rico (1975); Martinique, French West Indies (1977); Jamaica and the Dominican Republic (1978); Central America (1979); and Brazil (1981).

Liao *et al.* (1977) in Asia and Johnson (1977) in the Western hemisphere described diseases found in freshwater shrimp. Durwood Dugger assisted Sun Oil Company from 1974 to 1976 with a shrimp project in Texas and he formed a new company, Commercial Shrimp Culture International (CSCI). When CSCI started, Dugger and some CSCI small investors built a freshwater shrimp hatchery called Aquaprawns, Inc. Dugger built, developed, and ran the Aquaprawns hatchery for Sun Oil in Port Isabel, Texas. Aquaprawns had some very small experimental ponds in Brownsville where they were the first company to harvest and transfer freshwater shrimp with a fish pump and weren't the first company to fail to overwinter *M. rosenbergii* in open ponds even with a large heating system in spite of having some of Sun Oil's best engineers working with them. Aquaprawns conducted one of the first large scale *M. rosenbergii* restaurant marketing studies on mainland United States (5,500 mt), and was one of the first to look at aeration and various device efficiencies. After hurricane Allan destroyed the project in August 1980, CSCI received a US Department of Commerce Small Business Administration (SBA) disaster loan for US$300,000 and moved to a 162 ha site in Los Fresnos, Texas, where they rebuilt a multispecies hatchery and grew *M. rosenbergii, P. vannamei,* and *P. setiferus* in Rio Grande irrigation water in a two-phase, twenty-pond project consisting of two hectare ponds and four twelve-meter by sixty-five-meter greenhouse nurseries. CSCI was involved with one of the first commercial uses of head-start shrimp nurseries, and was one of the first groups to grow *L. vannamei* and *P. setiferus* in freshwater in commercial quantities, seeing that *L. vannamei* could adapt better to freshwater than *P. setiferus,* and at the same time produce more kilograms per hectare than with *M. rosenbergii.*

During this time CSCI was selected by the US Army Corp of Engineers (COE) to develop a 121 ha dredge containment area for a marine shrimp farming demonstration on the Brownsville ship channel. However, for political reasons, this COE project was not funded for another two years. Later, from 1988 to 1992 the project was completed under a new name and new group (MariQuest, Inc.) and tried to overwinter cold-tolerant shrimp species from Asia.

According to Durwood Dugger, CSCI commercially marketed the largest quantities of cultured *M. rosenbergii* in the United States to date. That marketing effort proved a number of things about *M. rosenbergii* as a premium seafood product—mainly that the cost of sales in serving the premium seafood market offset any price premiums that might be received on a large-scale commercial basis. After proving they could produce *L. vannamei* in freshwater commercially and that took almost any of the advantages away from *M. rosenbergii* that might be realized, this was Dugger's last effort to try to commercialize *M. rosenbergii* in the United States. The freshwater shrimp operation in Los Fresnos was purchased first by a local businessman (Ted Hollin) and later by Marshall Snider from New York, who operated it a for a few years until an early winter storm killed the shrimp in November 1991. Snider moved his operation to Puerto Rico and continued to supply the New York market.

Durwood Dugger completed a feasibility trial on aquaculture in far west Texas (Pecos County) and Texas A&M University set up a pilot culture system near Imperial, Texas. Seven farms built commercial operations for brackishwater shrimp in the following years, but only one has survived (Bart Reid's Permian Sea Organics farm in Imperial, a 26 ha facility that cultures *L. vannamei*).

In 1983, Aquaculture Enterprises, Inc., acquired an unsuccessful prawn farm in Puerto Rico (Shrimps Unlimited, Inc.) and John Glude restarted the farm. It experienced a large debt service and construction delays for five years before it was considered economically viable in 1988. Failure to obtain projected production levels led to a change in strategy that was developed and tested in 1989 and 1990 when production of 3,000 kg/ha/yr was achieved. White postlarvae (PL) disease caused by *Rickettsia* hit the company while a recession in the United States caused a drop in demand for the product. Production was put on hold in 1992 (WAS 1994), by which time inexpensive Taiwanese frozen shrimp had appeared in the world market at US$10/kg. This created fierce competition in the industry. For many producers, their production costs were higher than the prices that shrimp were bringing on the market.

Large freshwater shrimp similar to *M. rosenbergii* are found in practically all tropical and subtropical regions. *M. rude* was cultured along with *M. rosenbergii* in India for awhile. The monsoon river prawn, *M. malcolmsonii*, was grown in Pakistan and India; the painted river prawn, *M. carcinus*, in Barbados; *Cryphiops caementarius* in Peru; the oriental river prawn, *M. nipponense*, in China; the African river prawn, *M. vollenhovenii*, in Africa; and species native to the United States (*M. acanthurus*, *M. carcinus*, *M. ohione*, and *M. olfersii*) have been grown but none of the species listed species has been as successfully or as widely produced in aquaculture as *M. rosenbergii*. However, in China, culture of *M. nipponense* has recently experienced very significant expansion.

Polyculture of Chinese and common carp with *Macrobrachium* has been tried (Malecha *et al.* 1981) and early Taiwanese research demonstrated that polyculture of milkfish and grey mullet led to efficient utilization of pond resources (Liao & Chao 1982). Researchers in the Philippines found that *M. lanchesteri* could be polycultured with rice (Guerrero *et al.* 1982). Culture of freshwater shrimp in combination with fingerling catfish was successfully demonstrated under small-scale experimental conditions and appeared commercially feasible (D'Abramo & Brunson 1996). A scheme for intercropping of freshwater shrimp and red swamp crawfish was developed and evaluated as well. Numerous intercropping scenarios involving such species as bait minnows, tilapia, and other fish species have also been evaluated but the monoculture of *M. rosenbergii* is the sustainable method.

There has been a resurgence in the culture of *M. rosenbergii* in the United States in recent years due to researchers in Kentucky and Tennessee finding that cooler temperatures are actually beneficial to growth. Apparently, in warmer climates such as Mississippi and Texas, the animals put more energy into reproduction than growth. There have been several new hatcheries built in the last few years to accommodate that increased growth in the southeastern states. The economics and management of freshwater shrimp culture in the Western Hemisphere were addressed by Valenti & Tidwell (2006).

The Malaysian prawn was introduced into China in 1976 and has become a major freshwater species there with a 20% annual growth rate during the late 1990s. Total farming area exceeds 30,000 ha in China, creating a yield of 97,420 metric tonnes in 2000. Under competition from freshwater *L. vannamei*, farming of *M. rosenbergii* was reduced to 87,252 metric tonnes in 2003, but farming activities still exist in over twenty provinces of China (Yuan *et al.* 2006). China is by far the largest freshwater shrimp producing country. Annual world production of *M. rosenbergii* has reached a volume greater than 300,000 metric tonnes valued at more than US$1.1 billion.

2.7 Mollusk culture

The Greeks apparently were involved in some type of oyster culture by about 350 BPE (Gosling 2003). The Romans reportedly constructed ponds to stockpile European oysters (*Ostrea edulis*) and placed wood fascines (bundles of sticks) on the ponds to collect spat (Herál & Deslous-Paoli 1991). According to those authors, human population increases in Europe led to overfishing of the oyster beds in France, Ireland, and England in the eighteenth century and regulation of the fishery failed to deter the practice. The French refined the Italian fascine technique in about 1860 and also employed other materials such as slate as cultch. Bottom and rack culture were then developed. The French began importing Japanese oyster (*Crassostrea gigas*) spat and adults in 1968 and terminated the process in 1975 after reproduction of the species had become well established. In Japan, oyster farming dates back to 1624 (Gosling 2003).

Oyster culture in the United States ranges from just scattering oyster shell as cultch to spawning and setting spat in a hatchery for distribution to appropriate beds. The oysters may be collected using tongs or dredges subtidally or by hand intertidally. Some tray culture is also practiced. The American or eastern oyster *Crassostrea virginica* is cultured along the Atlantic and Gulf of Mexico coasts and *C. gigas* was first imported from Japan to the Pacific Northwest early in the twentieth century. Annual imports continued until World War II, after which hatchery technology was developed to ensure a constant supply of spat.

Mussel culture in Europe is supposed to have been accidentally initiated in 1225 by a sailor who was shipwrecked along the coast of France (Gosling 2003). He drove poles into the substrate and tied netting around them to attract birds. Mussels soon colonized the poles and were of better quality than those on the bottom. What became the bouchot technique continues to be conducted in France. In Spain, mussels are cultured in large quantities on lines suspended from rafts. Longline culture is another common culture method. In the United States, mussel consumption was low except in places where it was a part of the culture of one ethnic group that settled there. The popularity of mussels has increased dramatically over the past two or three decades and mussel culture has developed to help supply the demand.

Some form of clam culture has been conducted in China for several hundred years but few details appear to be available (Pillay & Kutty 2005). Native hard clams, *Mercenaria mercenaria* and *M. campechiensis,* are cultured on the east coast of the United States and the Manila clam, *Ruditapes philippinarum,* is cultured in the Pacific Northwest where it was introduced. Clams are cultured on the bottom, and in the Pacific Northwest, intertidal clam beds are covered with netting to reduce predation by benthic predators. The clams are harvested at low tide with mechanical harvesting devices.

While successful spat collection method was developed thirty years earlier, culture of the yesso scallop *Patinopecten yessoensis* began in Japan in the 1960s. Other popular species are *Chlamys* spp., *Argopcten* spp., and *Pecten maximus.* The major scallop-producing nations are Japan and China (Gosling 2003). Culture of at least one species produced in China is said to have been initiated when a shipment of about fifty individuals was sent from the United States several years ago. Much of the export market for those scallops from China is back to the United States.

2.8   Controversy

The history of aquaculture would not be complete without mention of the controversy that swirls around the practice, particularly with respect to marine fishes (including anadromous salmon) produced in coastal or offshore facilities. Opposition to fish culture in the commons first appeared in the United States in the 1980s. It seems to have initially been largely an issue with upland homeowners in the Puget Sound region of Washington State who objected to what they called visual pollution from net pen salmon farms. Interestingly, while fish culture operations in Washington were seen as eyesores, their counterparts in Japan were considered amenities because they are producing highly desirable food items. In the United States, Canada, and part of Europe, opposition to aquaculture sprang up wherever fish farming was initiated in public waters. Over time, a laundry list of issues was developed. Even in areas where rigorous permitting processes were put into place, opponents were often able to tie up the permitting process for months or even years by raising objection after objection, often to the point that the prospective fish farmer spent all the resources allotted for establishing the fish farm on lawyer fees.

While the controversy has slowed marine fish culture development significantly, at least some of the issues raised by opponents do need to be addressed. It is certainly the case that there have been instances where fish farms have been stocked too heavily in a given area and the carrying capacity has been exceeded. This has caused considerable negative impact on the local environment. That happened, for example, in Japan during the 1970s. For approximately the last two decades a considerable amount of research has been conducted to determine which objections have merit so best management practices or new approaches could be developed to address the problems. Technology has also been advanced in terms of cage designs for offshore systems that will better withstand storms and prevent escapes.

A major issue has been the use of fish to feed fish. To provide the proper amino acids and fatty acids, fish meal and fish oil have long been employed as ingredients in fish feeds. In recent years a considerable amount of work has been conducted to find alternative protein sources to replace fish meal and the percentage of fish meal in many feeds has been reduced. Plant sources of n-3 fatty acids, including sources from genetically modified plants, are alternatives to fish oil in fish feeds and may ultimately completely replace fish oil. Fish nutritionists have also been looking for ways to improve phosphorus digestibility in fish feeds to reduce the potential for eutrophication.

Proper siting of aquaculture facilities in public waters is critical for avoiding issues associated with eutrophication and the creation of anoxic areas under cages due to buildup of feces and uneaten feed. Maintaining appropriate densities of fish in cages can also help ensure water quality as well as reducing stress on the fish and possibly reducing the potential for diseases.

Adoption of new technologies and research results by the industry not only lead to more increased profitability, but also improvement in the sustainability of marine fish culture enterprises. It is certainly not in the best interest of the fish farmer to cause environmental damage because it will also be the farmer that suffers.

2.9   References

Anonymous (Undated) *History of Aquaculture.* www7.taosnet.com/platinum/data/whatis/history.html. Accessed April 2008.

Anonymous (2008) Status and outlook of United States catfish industry. *Aquaculture Magazine* **January/February**:21–3.

Aquacop (1977) *Macrobrachium rosenbergii* (DeMan) culture in Polynesia: Progress in developing a mass intensive larval rearing technique in clear water. *Proceedings of the World Mariculture Society* 8:311–9.

Arnold, C.R., Bailey, W.H., Williams, T.D., Johnson A. & Lasswell, J.L. (1977) Laboratory spawning and larval rearing of red drum and southern flounder. *Proceedings of the Southeastern Association of Fish and Wildlife Agencies* 31:437–40.

Atlantic Marine Aquaculture Center (2007a) http://ooa.unh.edu/finfish/finfish_flounder.html. Accessed May 2008.

Atlantic Marine Aquaculture Center (2007b) http://ooa.unh.edu/finfish/finfish_halibut.html. Accessed May 2008.

Bardach, J.E., Ryther, J.H. & McLarney, W.O. (1972) *Aquaculture: The Farming and Husbandry of Freshwater and Marine Organisms.* Wiley-Interscience, New York.

Beveridge, M.C.M. & Little, D.C. (2002) The history of aquaculture in traditional societies. In *Ecological Aquaculture: The Evolution of the Blue Revolution* (Ed. by B.A. Costa-Pierce), pp. 4–29. Blackwell Scientific, Oxford.

Borgese, E.M. (1980) *Seafarm: The Story of Aquaculture.* H. N. Abrams, New York.

Bowers, M. (1900) *A Manual of Fish-Culture, Based on the Methods of the United States Commission of Fish and Fisheries.* US Government Printing Office, Washington, DC.

Chimits, P. (1957) Tilapia in ancient Egypt. *FAO Fisheries Bulletin* 10:211–5.

Cook, H.L. (1965) Rearing and identifying shrimp larvae. *US Fish and Wildlife Service Circular No. 230.*

Cook, H.L. & Murphy, M.A. (1971) Early developmental stages of the brown shrimp reared in the NMFS Galveston lab. *Fisheries Bulletin* 69(1):223–39.

Costa-Pierce, B.A. (2002) The *Ahupua'a* aquaculture ecosystems in Hawaii. In *Ecological Aquaculture: The Evolution of the Blue Revolution* (Ed. by B.A. Costa-Pierce), pp. 30–43. Blackwell Scientific, Oxford.

Costa-Pierce, B.A. & Rakocy, J.E. (Eds.) (1997) *Tilapia Aquaculture in the Americas*, Volume 1. World Aquaculture Society, Baton Rouge.

Costa-Pierce, B.A. & Rakocy, J.E. (Eds.) (2000) *Tilapia Aquaculture in the Americas*, Volume 2. World Aquaculture Society, Baton Rouge.

Cowey, C.B., Andron, J. & Blair, A. (1970a) Studies on the nutrition of marine flatfish: The essential amino acid requirements of plaice and sole. *Journal of the Marine Biology Association of the United Kingdom* 50:87–95.

Cowey, C.B., Adron, J.W., Blair, A. & Pope, F. (1970b) The growth of O-group plaice on artificial diets containing different levels of protein. *Helgolander Wissenshaftliche Meersuntersuchungen* 20:602–9.

Cowey, C.B., Pope, J.A., Adron, J.W. & Blair, A. (1971) Studies on the nutrition of marine flatfish. Growth of the plaice, *Pleuronectes platessa*, on diets containing proteins derived from plants and other sources. *Marine Biology* 10:145–53.

D'Abramo, L.R. & Brunson, M.W. (1996) Biology and life history of freshwater prawns. *Southern Regional Aquaculture Center (SRAC) Pub. 483*. Mississippi State University.

Dannevig, H. (1897) On the rearing of the larval stages of the plaice and other flatfishes. *Fisheries Board of Scotland, 15th Annual Report* Part 3:175–93.

Dannevig, H. (1898) Report on the operations at the Dundee Marine Hatchery for the period July 1896 to December 1897 with notes on the rearing of flat fishes. *Fisheries Board of Scotland 16th Annual Report* 3(7):219–25.

Davis, H.S. (1956) *Culture and Diseases of Game Fishes*. University of California Press, Berkeley.

Dawes, B. (1930) Growth and maintenance in the plaice (*Pleuronectes platessa* L.). *Journal of the Marine Biological Association of the United Kingdom* 17:949–75.

Drews, R.H. (1951) The cultivation of food fish in China and Japan: A study disclosing, contrasting national patterns for rearing fish consistent with the differing cultural histories of China and Japan. Doctoral dissertation, University of Michigan.

Dushkina, L.A. (1994) Farming of salmonids in Russia. *Aquaculture Research* 25: 121–6.

Ewald, J.J. (1965) The laboratory rearing of pink shrimp, *P. duorarum*. *Bulletin of Marine Science in the Gulf and Caribbean* 15:436–49.

FAO (2004) *Status and important recent events concerning international trade in fishery products*. COFI-FT/IX/2004/2. FAO, Rome.

FAO (2006) *The State of World Fisheries and Aquaculture 2007*. FAO, Rome.

Fabre-Domergue, P. & Bietrix, E. (1905) *Developpment de la sole (Solea vulgaris)*. Paris: Travail du Laboratoire de Zoologie Maritime de Concarneau. Buibert et Nony.

Federation of European Aquaculture Producers (2006) *Aquamedia European Production*. http://aquamedia.org/Production/euproduction/euproduction_en.asp. Accessed May 2008.

Farfante, I.P. & Kensley, B. (1997) *Penaeoid and Sergestoid Shrimps and Shrimp of the World. Keys and Diagnoses for the Families and Genera*. Memories Du Museum National D'Historie Naturelle, Paris.

Fujimura, T. & Okamoto, H. (1970) *Notes on progress made in developing a mass culturing technique for Macrobrachium rosenbergii in Hawaii*. FAO Indo-Pacific Fisheries Council, 14th Session, Bangkok, Thailand.

Fujinaga (Hudinaga), M. (1969) Kuruma shrimp (*Penaeus japonicus*) cultivation in Japan. *FAO Fisheries Report* 3(57):811–32.

Gosling, E. (2003) *Bivalve Molluscs.* Fishing News Books, Oxford.

Guerrero, L.A., Circa, A.V. & Guerrero III, R.D. (1982) A preliminary study on the culture of *Macrobrachium lanchesteri* (de Man) in paddy fields with and without rice. In *Giant Prawn Farming, Developments in Aquaculture and Fisheries Science*, Volume 10 (Ed. by M.B. New), pp. 203–06. Elsevier, Amsterdam.

Guerrero III, R.D. (1975) Use of androgens for the production of all-male *Tilapia aurea* (Steindachner). *Transactions of the American Fisheries Society* 104:342–8.

Gunter, G. & Hildebrand, H. (1954) The relation of total rainfall and catch of the marine shrimp (*P. setiferus*). *Bulletin of Marine Science in the Gulf and Caribbean* 4(2):95–103.

Hargreaves, J.A. & Tucker, C.S. (2004) Industry development. In *Biology and Culture of Channel Catfish* (Ed. by J.A. Hargreaves & C.S. Tucker), pp. 1–14. Elsevier, Amsterdam.

Herál, M. & Deslous-Paoli, J.M. (1991) www.ifremer.fr/docelec/doc/1991/publication-3038.pdf.

Hickling, C.F. (1962) *Fish Culture.* Faber and Faber, London.

Hickling, C.F. (1968) *The Farming of Fish.* Pergamon, London.

Hudinaga, M. (1935) The study of *Penaeus.* 1. The development of *Penaeus japonicus* Bate. *Report from Hayatomo Fisheries Research Laboratory* 1(1):1–51

Hudinaga, M. (1941) Reproduction, development and rearing of *Penaeus japonicus* Bate. *Japanese Journal of Zoology* 10:305–93.

Hudinaga, M. & Kittaka, J. (1967) The large-scale production of the young Kuruma prawn, *Penaeus japonicus* Bate. *Bulletin of the Plankton Society of Japan* December:35–67.

Jessé, G.J. & Casey, A.A. (Undated) Study of the chronological dates in world aquaculture (fish farming) history from 2800 BC. www.thehobb.tv/wow/water_culture_origins.html. Accessed April 2008.

Johnson, S.K. (1977) Crawfish and freshwater shrimp diseases. *Publication #TAMU-SG-77-605.* Texas A&M University, Sea Grant College Program.

Kirk, R. (1987) *A History of Marine Fish Culture in Europe and North America.* Fishing News (Books), Farnham.

Kirk, R.G. & Howell, B.R. (1972) Growth rates and food conversion in young plaice (*Pleuronectes platessa* L.) fed on artificial and natural diets. *Aquaculture* 1:29–34.

Kurata, H. (1956) On the rearing of larvae of the flatfish, *Liopsetta obscura,* in small aquaria. *Bulletin of the Hokkaido Regional Fisheries Research Laboratory* 13:20–9.

Laird, L.M. (1996) History and applications of salmonid culture. In *Developments in Aquaculture and Fisheries Science*, Volume 29 (Ed. by W. Pennell & B.A. Barton), pp. 1–28. Elsevier, Amsterdam.

Lem, A. (2006) An overview of global shrimp markets and trade. In *Shrimp Culture—Economics, Market And Trade* (Ed. by P.S. Leung & C. Engle), pp. 3–10. Blackwell Publishing, Oxford.

Liao, I.C. & Huang, T.L. (1972) Experiments on propagation and culture of prawns in Taiwan. In *Coastal Aquaculture in the Indo-Pacific Region* (Ed. by T.V.R. Pillay), pp. 213–43. Fishing News (Books) Ltd., London.

Liao, I.C. & Chao, N.H. (1982) Progress of *Macrobrachium* farming and its extension in Taiwan. In *Giant Prawn Farming, Developments in Aquaculture and Fisheries Science* Volume 10, (Ed. by M.B. New), pp. 357–79. Elsevier, Amsterdam.

Liao, I.C., Yang, F.R. & Lou, S.W. (1977) Preliminary report on some diseases of cultured prawn and their control methods. *JCRR Fisheries Series* **29**:28–33 (Chinese/English Summary).

Lien, G.K. & Ting, Y.Y. (1969) Artificial propagation of edible shrimps in Taiwan. *Bulletin of the Taiwan Fisheries Research Institute* **15**:127–34. (In Chinese).

Ling, S.W. (1977) *Aquaculture in Southeast Asia: A Historical Overview*. University of Washington Press, Seattle.

Ling, S.W. & Merican, A.B.O. (1961) Notes on the life and habits of the adults and larval stages of *Macrobrachium rosenbergii* (de Man). *FAO Indo-Pacific Fisheries Council* **9**(2):55–60.

Ling, S.W. (1967a) Methods of rearing and culturing *Macrobrachium rosenbergii* (de Man). *FAO World Conference on the Biology and Culture of Shrimps and Prawns. Mexico City.* FR: BCSP/67/E/31.

Ling, S.W. (1967b) *The General Biology and Development of* Macrobrachium rosenbergii (*de Man*). *FAO World Conference on the Biology and Culture of Shrimps and Prawns. Mexico City.* FR: BCSP/67/E/30.

Maar, A., Mortimer, M.A.E. & Van der Lingen, I. (1966) *Fish Culture in Central East Africa*. FAO, Rome.

Malard, A.E. (1899) Sur le development et la pisciculture du turbot. *Les Comptes Rendus de l'Académie des Sciences* **129**:181–3.

Malecha, S.R., Buck, D.H., Bur, R.J. & Onizuka, D.R. (1981) Polyculture of the freshwater prawn, *M. rosenbergii*, Chinese & common carps in ponds enriched with swine manure. *Aquaculture* **25**:101–16.

May, R.C. (1971) An annotated bibliography of attempts to rear the larvae of marine fishes in the laboratory. *National Marine Fisheries Service Special Scientific Report—Fisheries Series No. 632.* Washington, DC.

Mistakidis, M.N. (Ed.) (1969) The freshwater prawn, *Macrobrachium rosenbergii*. *FAO Fisheries Report* **5**(3).

McHugh, J.L. & Walker, B.W. (1948) Rearing marine fishes in the laboratory. *Journal of California Fish and Game* **34**:37–8.

NACA (1994-95) *Fish Health Management in Asia*. Programme proposal to the Office Internationale des Epizooties by Network of Aquaculture Centres in Asia-Pacific.

Nash, C.E. (1968) Power stations as sea farms. *New Scientist* **14**:366–9.

NASS (National Agricultural Statistics Service) (2008) *Catfish Processing*. http://www .usda.gov/nass/PUBS/TODAYRPT/catf1208.txt.

Pearson, J.C. (1939). The early life histories of some American penaeidae, chiefly the commercial shrimp, *Penaeus setiferus* (Linn). *Bulletin of the U.S. Bureau of Fisheries* **30**(49).

Perez, K.R. (2006) *Fishing for Gold: The Story of Alabama's Catfish Industry*. University of Alabama Press, Tuscaloosa.

Philippart, J-Cl. & Ruwet, J-Cl. (1982) Ecology and distribution of tilapias. In *The Biology and Culture of Tilapias*, ICLARM Conference Proceedings 7 (Ed. by R.S.V. Pullin & R.H. Lowe-McConnell), pp. 15–59. International Center for Living Aquatic Resources Management, Manila.

Pillay, T.V.R. & Kutty, M.N. (2005) *Aquaculture Principles and Practices*. Blackwell Publishing, Oxford.

Rabanal, H.R. (1988) History of aquaculture. FAO, Rome. www.fao.org/docrep/field/009/ag158e/ AG158E00/htm. Accessed April 2008.

Racek, A.A. (1972) Indo-West Pacific penaeid prawns of commercial importance. In *Coastal Aquaculture in the Indo-Pacific Region* (Ed. by T.V.R. Pillay), pp. 152–72. Fishing News (Books) Ltd., London.

Rad, F. (2007) Evaluation of the sea bass and sea bream industry in the Mediterranean with emphasis on Turkey. In *Species and System Selection for Sustainable Aquaculture* (Ed. by P. Leung, C-S. Lee & P. O'Bryen), pp. 445–59. Blackwell Publishing, Oxford.

Riley, J.D. & Thacker, G.T. (1963) Marine fish culture in Britain. III. Plaice (*Pleuronectes platessa* L.) rearing in closed circulation at Lowestoft, 1961. *Journal du Conseil International Exploration Mer* 28:80–90.

Rollefsen, G. (1939) Artificial rearing of fry of sea water fish. Preiminary communication. *Rapport Conseil Exploration Mer* 109:133–4.

Rosenberry, B. (2007) *An Oral History of Shrimp Farming in the Western Hemisphere.* Shrimp News International website http://www.shrimpnews.com/HistoryFolder/HistoryIndex.html.

Sarig, S. (1989) Introduction and the state of art of aquaculture. In *Fish Culture in Warm Water Systems: Problems and Trends* (Ed. by M. Shilo & S. Sarig), pp. 1–19. CRC Press, Boca Raton.

Scura, E. (1987) *Review of World Prawn Farming.* Seminars on Prawn Farming for Profit. SCP Fisheries Consultants Australia Pty. Ltd.

Shelbourne, J.E. (1963a) Marine fish culture in Britain. II. A plaice rearing experiment at Port Erin, Isle of Main, during 1960, in open sea water circulation. *Journal du Conseil International Exploration Mer* 28:70–80.

Shelbourne, J.E. (1963b) Marine fish culture in Britain. IV. High survivals of metamorphosed plaice during salinity experiments in open circulation at Port Erin, Isle of Man, 1961. *Journal du Conseil International Exploration Mer* 28:246–61.

Shelbourne, J.E. (1963c) A marine fish-rearing experiment using antibiotics. *Nature* 198:74–5.

Shelbourne, J.E., Riley, J.D. & Thacker, G.T. (1963) Marine fish culture in Britain. I. Plaice rearing in closed circulation at Lowestoft 1957–1960. *Journal du Conseil International Exploration Mer* 28:50–69.

Shelbourne, J.E., Riley, J.D. & Thacker, G.T. (1964) The artificial propagation of marine fish. *Advances in Marine Biology* 2:1–83.

Shira, A.F. (1917) Notes on the rearing, growth and food of the channel catfish, *Ictalurus punctatus. Transactions of the American Fisheries Society* 46:77–88.

Stickney, R.R. (1989) *Flagship: A History of Fisheries at the University of Washington.* Kendall/Hunt, Dubuque.

Stickney, R.R. (1996a) *Aquaculture in the United States: A Historical Survey.* Wiley-Interscience, New York.

Stickney, R.R. (1996b) Fish production and distribution in the United States, Part 1. Government initiatives. *World Aquaculture* 27(1):31–41.

Stickney, R.R. (1996c) Fish production & distribution in the United States, Part 2. Pacific salmon distribution prior to 1900—McCloud River salmon activities. *World Aquaculture* 27(2):32–8.

Stickney, R.R. (1996d) Fish production & distribution in the United States, Part 3. Trout distribution begins. *World Aquaculture* 27(3):33–40.

Stickney, R.R. (1996e) Fish production & distribution in the United States, Part 4. Rainbow and cutthroat trout to 1900. *World Aquaculture* 27(4):62–6.

Stickney, R.R. (1997a) Fish production & distribution in the United States, Part 5. Early activities with nonsalmonids. *World Aquaculture* 28(1):53–8.

Stickney, R.R. (1997b) Fish production & distribution in the United States, Part 6. Salmon introductions from 1901 to 1940. *World Aquaculture* 28(2):55–8.

Stickney, R.R. (1997c) Fish production & distribution in the United States, Part 7. Trout distributions from 1901 to 1940. *World Aquaculture* 28(4):64–7.

Stickney, R.R. (1997d) *Annotated Bibliography of Flatfish Culture.* Texas Sea Grant College Program, College Station.

Stickney, R.R. (2005) Atlantic and Pacific Halibut. In *Aquaculture in the 21st Century* (Ed. by A. Kelly & J. Silverstine), pp. 471–89. American Fisheries Society, Bethesda.

Stickney, R.R. & White, D.B. (1973) Effects of salinity on the growth of *Paralichthys lethostigma* postlarvae reared under aquaculture conditions. *Proceedings of the Southeastern Association of Game and Fish Commissioners* 27:532–40.

Stickney, R.R. & White, D.B. (1974) Lymphocystis in tank cultured flounder. *Aquaculture* 4:307–8.

Stickney, R.R. & White, D.B. (1975) Ambicoloration in tank cultured flounders, *Paralichthys dentatus. Transactions of the American Fisheries Society* 104:158–60.

Stickney, R.R. & Seawright, D. (1993) A bibliography on Atlantic Halibut (*Hippoglossus hippoglossus*) and Pacific Halibut (*Hippoglossus stenolepis*) culture, with abstracts. In *Technical Report No. 29* (Ed. by R.R. Stickney & D. Seawright), pp. 27–57. International Pacific Halibut Commission, Seattle.

Stickney, R.R., McVey, J.P., Treece, G., George, N. & Hulata, G. (Eds.) (1999) Flatfish culture. *Aquaculture* 176:1–188.

Swingle, H.S. (1957) Preliminary results on the commercial production of channel catfish in ponds. *Proceedings of the Southeastern Association of Game and Fish Commissioners* 10:160–2.

Swingle, H.S. (1958) Experiments on growing fingerling channel catfish to marketable size in ponds. *Proceedings of the Southeastern Association of Game and Fish Commissioners* 12:63–72.

Szücs, I., Stundi, L. & Varadi, L. (2007) Carp farming in central and eastern Europe and a case study in multifunctional aquaculture. In *Species and System Selection for Sustainable Aquaculture* (Ed. by P. Leung, C-S. Lee & P. O'Bryen), pp. 389–413. Blackwell Publishing, Oxford.

Tham, A-K. (1968) Prawn culture in Singapore. *FAO Fish Report* 57(2):85–93.

Tilseth, S., Hansen, T. & Moller, D. (1991) Historical development of salmon culture. *Aquaculture* 98:1–9.

Treece, G.D. (1993) Texas aquaculture, history and growth potential for the 1990s. *Pub.TAMU-SG-103.* Texas A&M University, Sea Grant College Program. National Sea Grant Library, http://nsgl.gso.uri.edu.

Tseng, W. (1987) *Shrimp Mariculture—A Practical Manual.* Chien Cheng Publisher, Kaohsiung, Rep. of China.

US Department of Commerce (2007) *Fisheries of the United States 2006.* US Government Printing Office, Washington, DC.

Valenti, W.C. & Tidwell, J.H. (2006) Economics and management of freshwater prawn culture in Western Hemisphere. In *Shrimp Culture—Economics, Market, and Trade* (Ed. by P.S. Leung & C. Engle), pp. 261–76. Blackwell Publishing, London.

WAS (1994) Special Report. *Journal of the World Aquaculture Society* 25:5–17.

White, D.B. & Stickney, R.R. (1973a) A bibliography of flatfish (Pleuronectiformes) research with partial annotation. *Georgia Marine Science Center Technical Report Series Number 73-6.* Savannah.

White, D.B. & Stickney, R.R. (1973b) A manual of flatfish rearing. *Georgia Marine Science Center Technical Report Series No. 73-7*. Savannah.

Wickins, J.F. & Lee, D. O'C. (2002) *Crustacean Farming—Ranching and Culture*, 2nd Edition. Blackwell Science, London.

Yuan, Y., Cai, J. & Leung, P.S. (2006) An overview of China's cultured shrimp industry. In *Shrimp Culture—Economics, Market, and Trade* (Ed. by P.S. Leung & C. Engle), pp. 197–220. Blackwell Publishing, Oxford.

Chapter 3

Functions and Characteristics of All Aquaculture Systems

James H. Tidwell

Aquaculture is an extremely diverse enterprise. We work in very different environments (freshwater, brackish water, saltwater), which can represent extremely different physiological challenges to the animal being raised. We also work with many different species, some estimates exceed 400, with the number growing each year. To make things even more complicated, most of these species have not even been domesticated. How different from terrestrial livestock is it working with aquatic animals? *Very*. Here are a few examples to make a point.

3.1 Differences in aquatic and terrestrial livestock

(1) Many of the finfish we raise are carnivores. Why? Because this is what consumers want to buy. However, it creates many difficulties when it comes to formulating diets and also in terms of production systems. Chicken houses would look a lot different if the chickens were prone to *eat* each other.

(2) Many of our culture animals live suspended or swimming in the water column (termed pelagic). I don't know of any terrestrial livestock that floats or flies in mid-air. For aquaculture this can be an advantage as we can utilize all three dimensions of the culture system. However, it can also be a disadvantage in that many fishes do not readily feed at the surface or at the bottom.

(3) For some aquaculture species, the opposite is the case. The animals are completely sessile and attached to bottom (such as oysters). In terrestrial

agriculture this is true of plants, but not the animals! Because of this, these animals are unable to avoid issues of poor water quality or move away to avoid predators. In fact, siltation can be a problem. Imagine poultry producers that had to worry about their flock being suffocated by a dust storm.

(4) Many aquacultured animals are filter feeders. Again, can you think of a terrestrial livestock species that gets its food by catching dust blowing in the wind?

(5) Most aquaculture animals are r-strategist while most terrestrial livestock are K-strategists. In simple terms r-strategist animals produce a lot of offspring but invest little care or energy into each. In contrast, K-strategists have fewer offspring with more investment in each. Fecundity and potential number of offspring are often far higher in aquatic than in terrestrial animals. A cow will produce only a maximum of two calves in a year or twenty in a lifetime; a pig maybe eight piglets per litter and fifty in a lifetime; and a chicken may lay 325 eggs per year over a maximum of eight years with a possible total of 1,800 offspring. However, a single adult female oyster can produce 20 to 30 million eggs at once (Galtsoff 1964) and the Atlantic cod (*Gadus morhua*) produces 10 million eggs (Williams 1975).

(6) For many aquaculture species the offspring are tiny! Can you picture the poor swine producer whose piglets could pass through a window screen? What and how do you feed them? Well, that is the problem that producers of marine fish and shellfish species face. Newly hatched larvae may only be 100 to 200 μm in length and require unicellular algae or rotifers (<200 μm) as their early diets.

(7) Osmoregulation. For terrestrial animals air is pretty much air but for aquatic animals freshwater and saltwater represent very different environments and physiological challenges. Virtually all aquatic animals have to work against their external environment to maintain their internal environment. Most fish require an internal osmotic concentration of 250 to 500 m Osmol/kg. However, freshwater is <0.1 m Osmol/kg and seawater is approximately 1,000 m Osmol/kg (Evans & Claiborne 2008). That means that in marine fish, water is constantly trying to leave the fish and in freshwater species, water is constantly trying to move into the fish. Some species, known as euryhaline species, have mechanisms that allow them to adapt to a range of salinities. Others known as stenohaline are strictly confined to one environment or the other.

We also raise these organisms many different ways. Production systems range from being basically natural systems with a few extra animals added and little management intervention (for example reservoir ranching, chapter 8), all the way up to super-intensive recirculating systems with the animals basically being on life support and the aquaculturist having to "play God" around the clock (see recycle systems, chapter 11). However, despite all of these differences, there are also characteristics and functions that all aquaculture systems share.

Even more than terrestrial farm animals, aquatic animals are captives of their environment. All terrestrial livestock are homeotherms (warm-blooded). That means they are able to regulate their body temperatures to stay within the narrow range needed for proper function. However, aquacultured animals are primarily poikilothermic (cold-blooded), so their body temperature is basically the temperature of their environment.

Terrestrial animals live in a relatively high oxygen gaseous environment (>21% by weight) and use lungs for gas exchange. Aquatic animals have evolved diverse structures to utilize the oxygen that is dissolved in the water. However, oxygen is much scarcer in water (0.00001% by weight) making an adequate oxygen supply a bigger consideration for the aquatic animal. Also, to complicate matters even more, the effects of temperature on the fish and the availability of oxygen in the water operate in conflict. As water temperatures increase, the metabolism of the animal (and its oxygen demand) *increases*. However, as temperature increases the solubility of oxygen in the water *decreases*, meaning less is available just when the animal needs it most. Nature played a cruel joke on the aquaculturist (Timmons *et al.* 2002).

Another factor is the way that gills function. The water the fish lives in and the fluid of the animals' blood are separated by only a few cells. Whatever is in the water passes through the gills and into the fish's blood very readily. Also, the skin of most fish is very permeable making absorption of compounds passive and selective exclusion difficult or impossible. This issue is compounded by the fact that, through diffusion, anything added to a body of water rapidly disperses to all parts. If a noxious compound gets into the water it can spread throughout the system and quickly be absorbed into the aquacultured animal's body. No escape.

3.2   Ecological services provided by aquaculture production systems

As this discussion shows, aquatic animals are very much "captives" of their environment. As aquaculturists we do not so much manage the animals as much as we manage their environment. That is also largely the function of the different aquaculture systems—to manage the animals' environment. Not only must the environment be maintained to support life but in the case of aquaculture, it needs to be maintained in such a way as to support maximum growth rate, with maximum efficiency, and a minimum of waste.

As described earlier, many factors affect the survival and growth of aquatic animals. However, a few environmental variables are fundamental and a discussion of the ways that aquaculture systems control them is the unifying theme of this chapter and this book. Throughout the different chapters illustrations will be given on how the specific system being examined provides the cultured animal with the (1) proper temperature for growth, (2) sufficient oxygen to breathe, (3) removal of inevitable waste products, and in some cases (4) some or all of the animal's food needs.

3.3   Diversity of aquaculture animals

Terrestrial animal agriculture relies on relatively few species. In cattle, milk and meat production utilize one (*Bos taurus*) and (maybe) a second species

(*B. indicus*). In pigs, all commercial production is based on one species (*Sus domestica*). In poultry we have hundreds of varieties of chickens but they are all actually one species (*Gallus gallus*), and we also have the turkey (*Meleagris ocellata*). These animals are all warm-blooded and differ at the genus or class level. However, in aquaculture we raise well over 400 species (Duarte *et al.* 2009), all are cold blooded, and many differ at class or even phylum level. So what determines what their environmental needs and tolerances are, and what conditions must the chosen production system provide to raise that animal?

Basically, the environmental requirements of an animal are determined by its evolutionary adaptations for the environment it evolved in. In fish, we have broad categories that, based on the characteristics of their "natural" or natal environment, often tell us much about what they need, what they can tolerate, and even their nutritional requirements. These requirements and tolerances are often represented as a set of minima and maxima with a range in between. For some variables there will also be an optimum at which the animal operates most efficiently.

3.4   Temperature classifications of aquacultured animals

One way of classifying fish is based on water temperature (table 3.1). Fish are often characterized as coldwater, coolwater or warmwater species. Some descriptions add a fourth category of tropical species. The characteristics, requirements, and especially tolerances of the fish within these groups are largely controlled by enzyme functions and efficiencies. Many enzymes only operate within a limited temperature band, which, in the case of a poikilothermic animal, means that the animal also operates efficiently within a narrow temperature band (Somero & Hochachka 1971). In homeothermic animals such as cows, pigs, and chickens, they are able to burn calories to maintain their internal environment within a very narrow range, ensuring that the enzyme systems essential to many metabolic functions continue to work efficiently. In poikilotherms, the internal environmental temperature is controlled by the external environment, so it is important that the culture system provide that proper temperature.

Table 3.1 Generalized characteristics and trends among fishes characterized according to temperature range and salinity range.

Group	D.O. Requirement	Ammonia Tolerance	Protein Requirement	n-3 Fatty acids
Coldwater	>5 mg/L	Low	High	Required
Coolwater	>5 mg/L	Low	Moderate	Not required, but beneficial
Warmwater	>2 mg/L	Moderate	Moderate	Not required
Tropical	>1 mg/L	High	Low	Not required
Marine	Higher		Higher	Required
Freshwater	Lower		Lower	Not required

As stated earlier, the environment the species evolved to live in determined what temperature range it would be best adapted for. The conditions of that natal environment also affected other aspects of the animals' requirements and tolerances in terms of nutrition and water quality tolerances. Common generalities of fishes within the different temperature classifications are described below.

3.4.1   Coldwater species

Finfish and invertebrates whose thermal optimum for growth is below 20°C are classified as coldwater species. Examples of commercially important aquaculture species within this group include the marine Atlantic salmon (*Salmo salar*), the freshwater rainbow trout (*Oncorhynchus mykiss*), and the marine Pacific oyster (*Crassostrea gigas*). Rainbow trout are thought of as freshwater fish but are actually close relatives (same genus) of the Pacific salmons. Their optimum temperature for growth is about 10 to 16°C. They require relatively high oxygen levels (>5 mg/l) and tolerate only low levels of ammonia (<0.0125 mg/l unionized). This probably relates to their evolution in a coldwater environment where oxygen is abundant (the solubility of oxygen in water is inversely related to temperature) and accumulated nitrogenous wastes are relatively rare and of low toxicity (ammonia is less toxic at low temperatures and quickly flushed away in mountain streams). In the United States over 80% of commercial trout production occurs in the Hagarman Valley of Idaho (Hardy 1989) where huge volumes of groundwater break out of springs at approximately 15°C. Trout are very efficient converters of feed to flesh with about a 1:1 conversion ratio (Hardy 2002).

3.4.2   Coolwater species

Species whose optimum temperature is around 20°C are considered coolwater species. Currently there are fewer commercially important aquaculture species in this category. The striped bass (*Morone saxatilis*), yellow perch (*Perca flavescens*), and European perch (*Perca flaviatilis*) are examples. For the striped bass the optimal temperature is reported to be 15 to 17°C (Kohler 2000). However, much of the commercial production involves *Morone* hybrids whose optimum temperatures may be higher (25 to 30°C; Webster & Lim 2002) and would more properly be considered as warmwater fishes. The yellow perch and their cousin the European perch are coolwater species who are both being cultured commercially. It is estimated that approximately 226,000 kg/year of yellow perch were produced in 2005 (Hart *et al.* 2006). The European perch production in 2005 was estimated to be 315,000 kg (FAO 2007). The optimal temperature for yellow perch is 22 to 24°C with an upper lethal limit of 30°C (Hart *et al.* 2006).

3.4.3 Warmwater species

Many important aquaculture species are considered warmwater species, with an optimum temperature around 30°C. Among crustaceans, they would include the Pacific white shrimp (*Litopenaeus vannamei*), the tiger shrimp (*Penaeus mondon*), and the freshwater prawn (*Macrobrachium rosenbergii*). Among mollusks are the American oyster (*Crassostrea virginica*), Northern quahog (*Mercenaria mercenaria*), and blue mussel (*Mytilus edulis*). Among finfish we would include the common carp (*Cyprinus carpio*), channel catfish (*Ictalurus punctatus*), sea bass (*Dicentrachus labrax*), gilthead sea bream (*Sparus aurata*), and yellowtail (*Seriola quinqueradiata*).

Compared to coldwater species, warmwater species in general tend to have a greater tolerance for lower dissolved oxygen (DO) levels. This is logical as warmwater will hold less oxygen so the fish would be more likely to evolve mechanisms (mechanical and biochemical) to deal with low DO environments. They also tend to tolerate higher levels of un-ionized ammonia. Again, because of the dynamic equilibrium whereby higher temperature (and high pH) shift more total ammonia into the un-ionized form, warmwater species are more likely to need to evolve mechanisms to deal with high concentrations.

3.4.4 Tropical species

As stated earlier, some classifications add a fourth category. Tropical species would be those whose optimum temperature would be >30°C. You may also add a characteristic of having a minimal lethal temperature of ≤10 to 15°C. Examples would be tilapia, with an optimum temperature of 29 to 31°C (Popma & Masser 1999). To demonstrate the impact of temperatures, the growth rate of tilapia at 30°C is three times greater than at 22°C. However, growth is not the only variable that is affected. When tilapia are exposed to temperatures <18°C they handle poorly and get sick easily as their immunocompetence is severely compromised at temperatures below their optimal range. At temperatures of <10 to 12°C they normally die within a few days as enzyme systems cease to function. However, at temperatures >25°C, tilapia are very tolerant of handling low oxygen levels and high ammonia levels (Stickney 2000). The tilapia species preferred in aquaculture are native to the Middle East and Africa. In much of their home range tilapia have evolved to live in small water bodies that can shrink during the dry season and have low oxygen levels and poor water quality. Tolerance of these conditions was a strong selective pressure in their evolutionary history.

3.5 Temperature control in aquaculture systems

As stated earlier, temperature permeates all aspects of an aquacultured animal's growth, health, and even nutritional requirements. Different production systems approach temperature control (or a lack of it) in different ways (reviewed in chapter 4). However, even *within* the aquaculture animals' acceptable temperature range, rapid changes can be stressful or even lethal. Why? Again, the animal cannot compensate and control its internal environment. Because of the high specific heat of water, compared to air, the aquatic environment tends to be a very stable environment. Under most natural conditions, aquatic animals would rarely be exposed to rapid temperature changes and have not developed mechanisms to allow them to deal with it. Aquatic animals can adapt to temperature changes within their range of tolerances, but only very slowly. A rule of thumb is that fish should only be lowered about 5°C per hour (Stickney 1979). Rapid changes represent stressful or even lethal conditions for the fish. Because fish cannot control their internal temperature, and the "operating range" of enzymes is normally very narrow, fish must actually induce (turn on) the production of new enzymes to be able to function at these new temperatures.

In some aquaculture systems, such as open ponds, water temperatures are controlled by ambient air temperatures and solar radiation. Most pond production occurs using warmwater animals in tropical or semi-tropical climates where water temperatures vary only a few degrees seasonally. However, even a 2 to 3°C seasonal variation can significantly impact production during the tropical winter and summer seasons. When warmwater fish are raised in temperate ponds, such as catfish production in the continental United States, fish growth may stop entirely for several months during the winter. An even more pronounced affect occurs when tropical animals such as tilapia or freshwater prawns are raised in temperate ponds. Water temperatures can reach the animal's thermal minimum by late fall and the animals will die if not harvested before temperatures drop too low (Tidwell & D'Abramo 2010). This also creates a need to overwinter broodstock indoors in heated tanks and requires that a hatchery and/or nursery be operated seasonally, increasing costs and the potential of missing an entire year of production should hatchery or nursery problems occur.

Some systems, such as trout raceways, often utilize groundwater resources. Water temperatures in these systems usually do not vary over 1 to 2°C season to season. In this case, we can't change the water temperature so we must choose a species that performs well at *that* given temperature. The majority of rainbow trout production in North America is conducted in this type of system and the optimum temperature range of the animal (10 to 15°C) closely matches the temperature of the groundwater, which breaks out of the ground as springs. While there are no heating costs to the water, there is conversely little flexibility in siting. The production system must be brought to the water resource and commercial scale springs (minimum water flow of >1,900 Lpm) can be difficult to find. That is why approximately 80% of the trout production in the United States is located along the Hagerman Valley of Idaho where there are many large springs (Hardy 1989).

The opposite extreme is found with the use of the recirculating aquaculture systems. Most of this type of production is conducted indoors and systems can be designed to operate at almost any temperature in almost any climate. This offers tremendous advantages in terms of what species will be raised and where. We can raise a tropical species, such as tilapia, in far northerly latitudes or coldwater species, such as trout, in southerly latitudes. We can also site production near, or even in, urban settings with their large market potentials. However, energy costs are major considerations in these systems and the greater the temperature differential that must be maintained between outdoor temperatures and culture system temperatures, the more it will cost.

3.6 Providing oxygen in aquaculture systems

Oxygen is our second consideration. As discussed previously, compared to the atmosphere, when we move into an aquatic environment oxygen is much less abundant. While air is approximately 21% oxygen by weight, water, which is saturated with oxygen, contains <0.0005% oxygen by weight. To dissolve the oxygen in water it has to be squeezed in between the water molecules and there just isn't as much room there as there is in widely spread gas molecules.

3.6.1 Oxygen in open systems

In open systems, such as shellfish or sea cages, the production system relies on the natural environment's processes (primarily algal photosynthesis and diffusion) to provide the oxygen needed for the animals. The aquaculturist also relies on wind, currents, or tides to move the water and oxygen to and through the system. For those systems, proper siting is a major consideration.

3.6.2 Oxygen in semi-closed systems

In semi-closed systems, such as ponds, we rely on similar processes (photosynthesis and diffusion) to provide oxygen. However, in these smaller static water systems algal photosynthesis becomes much more important. In fact, on a calm day diffusion plays only a minor role. On a molecular scale, diffusion can move oxygen through the water column only relatively slowly. Supplied by photosynthesis and atmospheric diffusion, an area of saturation develops at the air/water interface, but oxygen does not efficiently move deeper down the water column. In pond systems, availability of oxygen is normally the first limiting factor in production intensification. The carrying capacity of an un-aerated pond is approximately 1,500 kg/ha. However, once supplemental aeration is provided, this figure increases threefold. Supplemental aeration is basically the use of mechanical devices to increase diffusion by increasing the contact between atmospheric oxygen and the water. This can be done either by injecting bubbles into the water column so that the oxygen inside the bubble can easily diffuse out into the water, or by pumping the water up into the air and breaking it into small droplets, which readily pickup atmospheric oxygen. The efficiencies of both types of aeration are greatly increased when they also provide circulation to move low oxygen water into and highly oxygenated water out of the zone of the aerator. They also function to circulate the oxygenated water down through the water column.

3.6.3 Oxygen in closed systems

When we move into closed recycle systems (RAS) we now have to take over the entire oxygen demand of all of the components of the system. Just as with a pond, the cultured animals are not the only source of oxygen demand in this system. The bacteria colonizing the biofilter (see chapter 11) can also represent a significant demand. In a RAS, the water coming out of the biofilter should always be ≥ 2.0 mg/L (Timmons *et al.* 2002), ensuring that the oxygen demands of the nitrifying bacteria in the biofilter are being adequately met.

3.7 Waste control in aquaculture systems

3.7.1 Aquaculture produces less waste than terrestrial systems

The third function of all aquaculture systems is to remove waste products. All animals produce wastes and compared to terrestrial animals, aquatic animals produce relatively smaller amounts, for several reasons. One is the fact that they are poikilothermic. This means they burn no calories when maintaining an internal body temperature higher than the environmental temperature. The main reason a cattle farmer feeds his cows all winter, especially on the coldest days, is so they can metabolize the food to maintain their internal body temperature. Another reason aquatic animals produce fewer waste products is that most secrete ammonia passively, directly from the blood into the water through their gills. They don't have to expend any energy converting ammonia to less toxic forms such as urea (cows) or uric acid (chickens). Terrestrial animals also expend energy removing water from the wastes and storing those wastes for later excretion. These reduced metabolic costs represent significant energy savings and less wastes produced. Also, aquatic animals live in, what is for them, a basically weightless environment. Most can control buoyancy so they expend little or no energy fighting gravity. All of these energy savings mean it takes less food to grow an aquatic animal. For example, it takes 7 to 8 kg of feed to produce a kilogram of weight gain in cows, 3 to 4 kg of feed in swine, and 2 to 3 kg in poultry. However, for fish it takes only about 1.5 kg of feed. Because of this relatively efficient feed conversion, less feed in means less waste out. However, even in fish there is always some waste.

3.7.2 Ammonia production by aquacultured animals

One of the most important waste products produced by fish is the nitrogenous waste product ammonia. It is largely the breakdown product of protein consumed in the feed. As stated earlier, it is primarily passively diffused through the gills. This is the classic "double-edged sword," though (and there are a lot them in aquaculture!). While passive diffusion saves energy compared to the other forms of nitrogenous waste excretion, it is the passive part that can create problems in production aquaculture. To be profitable, production aquaculture means we usually stock and feed fish at relatively high rates. Despite their efficiency relative to other farmed animals, approximately 87% of the nitrogen that fish take in as feed protein is released back into the water (Boyd 1979). This means that ammonia will accumulate in the water if not removed relatively rapidly and efficiently. As we pointed out earlier, aquatic animals readily take up compounds dissolved in the water through both their gills and skin, so toxic levels of ammonia can readily accumulate in the blood if allowed to build up in the environment. There must be some process to continually remove or detoxify the ammonia being excreted by the cultured animals into its surrounding.

3.7.3   Ammonia as a limiting factor in intensification

Different aquaculture production systems have several different methods to prevent the accumulation of ammonia produced by the culture animals. The rate and efficiency of the removal method is often the second limiting factor in system intensification (oxygen supply being first). In open systems, such as shellfish production, waste loads are low and natural diffusion can be sufficient to remove them. In other open system technologies, such as sea cages, natural water movement like tides or currents can augment diffusion to move the waste products away from the animals where they are processed or assimilated by natural processes (i.e., algal or bacterial). Even within semi-closed systems, raceways rely on a constant flow of new water to flush away the wastes. In raceways arranged in series and tiered in elevation, water dropping from one raceway to the next is reoxygenated with each drop, so the factor that limits the number of times it can be reused is the accumulation of ammonia. In semi-closed system ponds, mechanical aeration can provide for the oxygen budget of biomass densities beyond the current 5,000/kg/ha of many pond-based industries. It is the ability of the algal and bacterial populations in the ponds to process the ammonia loads that currently restrain additional intensification.

3.7.4   Ammonia control in closed systems—chemoautotrophic bacteria

As we move to recycle systems, the processes to remove (really convert) ammonia must be intensified. In most recycle systems this means pumping the water into specialized vessels where nitrifying bacteria are cultured at very high densities. Since these particular nitrifying bacteria grow best attached to a surface, these vessels are filled with specialized materials (such as pellets or strands), which have a lot of surface area for the bacteria to grow on and space for the water to freely flow and deliver the ammonia to the bacteria. These vessels are known as biofilters because they use living organisms (bacteria) to "filter-out" the waste products. Their functionality and efficiency can often be improved by prefiltering with a mechanical filter. The removal of solids prevents clogging and reduces ammonia load and oxygen demand associated with the decomposition of solid wastes (such as feces and uneaten feed).

3.7.5 Ammonia control by heterotrophic bacteria

In recent years another type of closed recycle system has been in development. As opposed to more traditional recycle systems, which rely on autotrophic nitrifying bacteria confined in a biofilter, these systems are based on heterotrophic bacteria unconfined in the systems. These systems do not rely on highly efficient solids removal. In fact, significant amounts of solids are left in the system and suspended by heavy aeration. These suspend particles, known as bioflocs, not only support the growth of the heterotrophic organisms, which quickly and efficiently convert ammonia into bacterial biomass, but also can be consumed by the animal being cultured in the system and represent a source of nutrition and nutrient recycling.

3.8 Aquaculture systems as providers of natural foods

A fourth function provided by some systems is to also provide some or all of the food for the culture animals. In the culture of bivalves in open systems (chapter 5) essentially all of the food is provided by the culture environment. This is also true in reservoir ranching as described in chapter 8. In semi-closed system ponds there is a range of importance for the system being a provider of natural foods. For what are known as extensive ponds, the animals consume only natural foods and production is relatively low with a carrying capacity (CC) of approximately 150 kg/ha (table 3.2.) However, if we add organic or inorganic fertilizers, primary productivity is now increased. The culture animals still rely on natural foods but the CC is increased threefold to 500 kg/ha. If we

Table 3.2 Effect of fertilization, feeding, and aeration on estimated carrying capacity of a pond and the limiting factor to further intensification.

Management Input	Carry Capacity (kg/ha)	Limiting Factor
No inputs	150	Availability of natural foods
Organic fertilization	500	Availability of nutrients
Supplemental feeds	1,500	Low morning dissolved oxygen
Complete feeds with aeration	5,000	Ability to process nitrogen waste products

supplement the fertilized system with cereal grains for the animals to consume directly as "supplemental feed," CC is increased to 2,000 kg/ha (Tidwell *et al.* 1997). If we move to high-quality "complete" diets (diets containing all of the macro- and micro-nutrients needed by the species for growth and reproduction, CC can now be increased to >5,000 kg/ha. However, in aquaculture everything is interrelated. In pond systems, once we exceed 1,500 to 2,000 kg/ha, oxygen availability becomes limiting and above 4,000 to 5,000 kg/ha the ability to process nitrogenous waste products becomes limiting.

As you can see, while the basic functions of all aquaculture systems are quite similar, the path taken to accomplish these functions can be quite different. The theme of this book will be a more in-depth look at the major categories of aquaculture production systems with an explanation of how each one performs these basic functions or ecological services. Then we will examine some hybrid systems that combine positive aspects of different types of systems. Then we will look down the road to see what technologies and methodologies might await us to accomplish these tasks more rapidly, efficiently, or cost effectively in the future.

3.9   References

Boyd, C.E. (1979) *Water Quality in Warmwater Fish Ponds.* Auburn University Agricultural Experiment Station, Auburn.

Duarte, C.M., Holmer, M., Olsen, Y., Soto, D., Marba, N., Guiu, J., Black, K. & Karakassis, I. (2009) Will the oceans help feed humanity? *BioScience* 59(11):967–76.

Evans, D.H. & Claiborne, J.B. (2008) Osmotic and ionic regulation in fishes. In *Osmotic and Ionic Regulation: Cells and Animals* (Ed. by D.H. Eraus), pp. 295–366. CRC Press, Boca Raton.

FAO (2007) *State of the World Fisheries and Aquaculture.* FAO, Rome.

Galtsoff, P.S. (1964) The American Oyster *Crassotrea virginica* Gmelin. *Fishery Bulletin* 170(1):11–28.

Hardy, R.W. (1989) Practical feeding—salmon and trout. In *Nutrition and Feeding of Fish* (Ed. by T. Lovell), pp. 185–99. Van Nostrand Reinhold, New York.

Hardy, R.W. & Tacon, A.G.J. (2002) Fish meal historical uses, production trends, and future outlook for sustainable supplies. In *Responsible Marine Aquaculture* (Ed. by R.R. Stickney & J.P. McVey), pp. 311–25. CABI, New York.

Hart, S.D., Garling, D.L. & Malison, J.A. (2006) Yellow perch (*Perca flavescens*) culture guide. *North Central Regional Aquaculture Center (NCRAC) Culture Series Number 103.* Iowa State University, Ames.

Kohler, C.C. (2000) Striped bass and hybrid striped bass culture. In *Encyclopedia of Aquaculture* (Ed. by R.R. Stickney), pp. 898–907. John Wiley & Sons, Inc., New York.

Popma, T. & Masser, M. (1999) Tilapia: life history and biology. *Southern Regional Aquaculture Center Publication No. 283.* Texas A&M University, College Station.

Somero, G.N. & Hochachka, P.W. (1971) Biochemical adaptation to the environment. *American Zoologist* 11(1):159–67.

Stickney, R.R. (1979) *Principles of Warmwater Aquaculture.* John Wiley & Sons, Inc., New York.

Stickney, R.R. (2000) *Encyclopedia of Aquaculture*. John Wiley & Sons, Inc., New York.

Tidwell, J.H., Coyle, S.D., Webster, C.D., Sedlacek, J.D., Weston, P.A., Knight, W.L., Hill, S.J., D'Abramo, L.R., Daniels, W.H. & Fuller, M.J. (1997) Relative prawn production and benthic macroinvertebrate densities in unfed, organically fertilised, and fed pond systems. *Aquaculture* **149**:227–42.

Tidwell, J.H. & D'Abramo, L.R. (2010) Growout systems—culture in temperate zones. In *Freshwater Prawns: Biology and Farming* (Ed by M.B. New, W.C. Valenti, J.H. Tidwell, L.R. D'Abramo & M.N. Kutty), pp. 180–94. Wiley-Blackwell, Oxford.

Timmons, M.B., Ebeling, J.M., Wheaton, F.W., Summerfelt, S.T. & Vinci, B.J. (2002) *Recirculating Aquaculture Systems*. Cayuga Aqua Ventures, Ithaca.

Webster, C.D. & Lim, C. (2002) *Nutrient Requirements and Feeding of Finfish for Aquaculture*. CABI, New York.

Williams, G.C. (1975) *Sex and Evolution*. Princeton University Press, Princeton.

Chapter 4

Characterization and Categories of Aquaculture Production Systems

James H. Tidwell

The basic progression of this chapter, and this book, is from less intensive production systems to more intensive production systems. For aquaculture, intensification implies a number of things. One factor is how densely the animals are stocked into the system. For reservoir ranching systems, there may be 10 kg of animals in a hectare (10 kg/ha) while in intensive recirculating heterotrophic systems, there may be 10 kg of animals in *one square meter* (or 100,000 kg/ha).

Intensification also implies human intervention and outside inputs into the production system. For example, in some shellfish systems the primary intervention is to add hatchery-reared juveniles into the natural ocean habitat or to add some additional substrate for the shellfish to grow on. This is in contrast to an intensive recirculating system where the aquaculturist must provide for almost every biological function—including the oxygen for the animals to breathe.

Quite often, intensification also implies energy inputs. As we move up the scale of intensification the amount of energy invested in each kilogram of production also tends to increase. First we add aeration, then we add pumps and filters, and then we begin to add heat. In times of instability in the energy markets, these costs could become even greater considerations. A comparison of the same fish raised in two different systems can illustrate this. Carp raised in ponds with fertilization and limited feeding require 11 gigajoules (GJ) of energy input per ton of edible protein produced, while the same fish raised in a recycle system

Aquaculture Production Systems, First Edition. Edited by James Tidwell.
© 2012 John Wiley & Sons, Inc. Published 2012 by John Wiley & Sons, Inc.

requires 56 GJ of energy input per ton of edible protein produced (De Silva & Soto 2009).

Within this continuum of increasing intensity we have three major categories or classifications for aquaculture production systems. These groupings are primarily based on the amount of control or intervention the aquaculturist provides in terms of the three basic functions or ecosystem services that each system must provide: proper temperature, adequate oxygen, and waste removal. Again, with these classifications we progress from lower inputs and outputs to greater inputs and outputs. However, the demarcations between these systems are not always clear and distinct. These are general categories and there are gradations and overlaps between categories. There are even hybrids among and between these categories.

4.1   Open systems

Production systems within this category rely entirely on natural ecological processes to address the three major functions. These systems are normally natural bodies of water that are now being stocked for commercial production. Many of these systems could be considered stock enhancement rather than aquaculture. Biomass densities are usually low enough that natural processes can provide sufficient oxygen for the biomass being supported. The oxygen can be sourced from diffusion, photosynthesis by natural algal communities, or both. Waste products are also removed by natural processes within the systems operating at natural rates. Bacterial breakdown of solid wastes is by heterotrophic bacteria and fungi. Nitrogenous waste products, such as ammonia excreted by the animal, are either flushed away or processed into less toxic forms by the chemoautotrophic components of the natural nitrogen cycle (nitrification) or assimilated by algae. Water temperatures in these systems are ambient. In open systems, site selection is the major control factor the producer has for all environmental services. Because of this, GIS technologies have in recent years become major tools in identifying and evaluating suitable production locations.

Many production methods in open systems rely on natural water movement from tides, currents, or wind action to move waste products away from the animals and bring new, clean, and highly oxygenated water to the animals. To some extent, stocking rate, and sometimes added substrate, are the only management inputs. Production methods that function in an open system environment include shellfish systems, cages, net pens, ocean ranching, and reservoir ranching, each of which will be described in much more detail further on.

Open systems are probably the oldest aquaculture systems. Early examples include the enhanced oyster systems of ancient Rome (fig. 4.1) and early examples of cage culture. Open systems (like all systems) have both positive and negative attributes. Because these systems use the natural environment to produce their crops, initial investments can be relatively low. It may be just a matter of releasing hatchery fish or shellfish, letting them grow for a period of time, and then

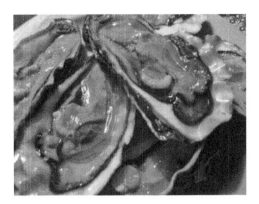

Figure 4.1 Oysters were raised or at least enhanced by the early Romans.

coming back and harvesting the crop. Therefore, management and input costs are also relatively low. However, the large water bodies needed for open system culture often involve public waters or waters that are surrounded by multiple land owners. This can bring up questions of ownership of the culture system and even ownership of the crop (Pillay & Kutty 2005). Many shellfish systems rely on leases from government entities, which can be strongly influenced by public opinion (where concerns may or may not be based on hard science). There is a strong suspicion, often fueled by sensationalist reporting, that all aquaculture systems are somehow polluting and harmful (De Silva & Soto 2009). A balanced review demonstrates that relative to the great majority of the other food commodities, the environmental impact of aquaculture is low (De Silva & Soto 2009). In fact, shellfish systems can actually be significant agents for positive environmental change by filtering out excess nutrients and productivity sourced from terrestrial livestock production, lawn fertilization, and other factors (Subasinghe 2007).

A negative of the low management inputs in aquaculture is that we also have little oversight or control. Poaching can be a problem. Many people feel they have every right to catch or harvest any aquatic animal from a natural or public body of water. For enforcement agencies responsible for public waters, ownership issues are often ill defined and oversight and enforcement responsibilities remain largely unresolved. The agencies with oversight over public waters are usually guided and empowered by environmental and wildlife regulations, not theft or livestock rustling statutes. Even if ownership is defined (by a lease agreement) who then has the right or obligation of enforcement—the owner or the public agency? I once visited a net-pen operation in Corsica. They kept an armed guard out on the water twenty-four hours a day, 365 days a year to protect their crop from poachers (fig. 4.2). You would have to consider this expense as a significant management input.

Poaching is not the only significant source of losses in open systems; predation can also be a problem. In the 1970s ocean ranching showed great promise as a method for producing salmon. However, in reality, losses during the two to three

Figure 4.2 Guarding sea cages from poachers.

years of ocean growth were so high that they made the method economically unfeasible (Arnason 2001). In shellfish systems constructed to get the animals up off the bottom, these steps are taken to get them away from predatory starfish and snails (such as the oyster drill), which cannot swim (Stickney 1979).

4.1.1 Bivalve culture: floats, trays, and rafts

Within the category of open systems we have a number of production methods. For bivalves these include floats, trays, and rafts. By placing the animals in the containers, these normally benthic animals can be suspended off of the bottom. This has the advantage of not only reducing predation (as previously mentioned), but also opening up all three dimensions of the water column to production. It also allows them to be suspended at depths where maximum phytoplankton (their primary food source) densities are found.

4.1.2 Cages and net pens

Cage culture is another open system production method. Cage culture is primarily used for finfish and occasionally for crustaceans. It basically represents a "fencing-off" of a portion of the natural aquatic habitat. Some cages are literally fenced compounds in shallow water. The bottom of the cage is the mud bottom of the bay or lake. In others, the cages have net bottoms and are suspended off of the bottom by flotation. These can be small cages (1 to 4 m^3) floating in shallow freshwater ponds. Larger cages used in marine environments are usually referred to as net pens. They were initially used in protected waters such as fjords and bays. In recent years they have been scaled up to sizes of 20,000 to 40,000 and even 60,000 cubic meters and engineered to take the abuse of unprotected offshore waters. Some are now fully enclosed structures that can be moved if needed, or (as discussed later) may evolve self-powered mobility (Cohen 2009).

4.2   Semi-closed systems

Within this category we still rely largely on nature to provide the three basic ecological services of proper temperature, sufficient oxygen, and waste removal. However, within the semi-closed category the production units themselves are now largely manmade. Production methods within semi-closed systems include ponds and raceways. Within the production units we now have the ability to add or remove water. There is more management input in these systems, and the first steps toward supplementing or enhancing natural processes exist in these systems, as well.

In semi-closed systems water is taken from a natural source such as rainfall, springs, streams, or rivers. The water is then gravity-flowed or pumped into specially designed and constructed production units. The water can be used once and discharged or constantly cleaned and reoxygenated by natural processes. Compared to open systems, semi-closed systems have several advantages. One is much higher production rates, as much as 1,000 times the productivity of an open system. This is due to the greater control and inputs into these systems and the fact that their physical parameters can be maximized for greater productivity.

As an illustration, compared to a deep reservoir used for cage culture, an aquaculture pond is shallow, allowing it to warm quickly by the sun. This also reduces the possibility of turnover, as there is little thermal stratification. It is also easier to monitor dissolved oxygen levels and aerate if needed. Because of these factors, while open-system cage culture in a large pond or reservoir is normally limited to a maximum production of approximately 2,000 kg/ha, in a semi-closed system pond culture production is increased to greater than 5,000 kg/ha.

Advantages of semi-closed systems over open systems include easier and more efficient use of prepared feeds, control over water depth or water replacement, practical and cost-effective mechanical aeration, more easily controlled poaching and predation, elimination of competitors and predators, effective detection and rectification of water quality deterioration and diseases, and potential temperature control. However, there are also negatives. Construction and equipment costs can be significant, there are more management demands for monitoring and intervention, energy and feed inputs are higher, and there is a greater likelihood that water quality issues and diseases will occur.

Figure 4.3 A flow-through raceway. Photo courtesy of Charles Weibel.

4.2.1 Raceways

Raceways are basically large manmade earthen or concrete troughs (fig. 4.3). A typical length:width:depth ratio in linear raceways is 30:3:1 (Stickney 2009). High-quality water flows into and through the trough bringing in needed oxygen and flushing away wastes. Water sources are usually ground waters coming to the surface in the form of springs or surface water from snow melt or rain runoff from higher elevations. The water can often be reused several times as the water flows through multiple raceways in series. Inputs come in the form of high-quality feeds, simple aeration between raceways, cleaning of raceways, size grading of the animals, and easy observation of the fish for disease problems and efficient feed utilization. Raceway production is very intensive in terms of land use. On a per hectare basis, an excess of 300,000 kg of fish can be produced per year. However these systems require *a lot* of water. To produce 1 kg of trout in a raceway requires 98,000 liters of water, compared to 1,250 to 1,750 liters to produce a kg of catfish in a leveed pond (Fornshell & Hinshaw 2008). Because of this extremely high water demand (>1,500 Lpm), the siting of commercial raceway operations is almost completely dictated by availability of suitable water resources. A suitable source must provide sufficient volumes of water at correct temperatures constantly, year around.

In raceways the oxygen is provided by incoming water. It must come into the raceway saturated with oxygen. If ground water is used, it must be coming from unconfined underground spaces where it has been tightly exposed to air. If the water has been confined between strata, it can have low levels of oxygen and be supersaturated with certain undesirable gases.

Wastes produced in these systems are passed on for processing further downstream in the receiving waters, or onsite in designed treatment units. Temperatures in raceway systems reflect their water source. Retention time is low so temperature changes little within the system. Raceways using ground water have water temperatures the same as the region's groundwater, which is directly correlated to proximity to the equator. Exceptions include raceways utilizing surface waters or deep source geothermal waters. Raceways are covered in more detail in chapter 9.

4.2.2 Ponds

There are several types of ponds. The simplest and easiest to construct is a watershed or impoundment pond. These are constructed by building a dam across a natural waterway to retain the rain runoff at a level set by the dam. It is essential when building this dam that it be properly cored or keyed with an impermeable material to prevent seepage. They are usually triangular in shape (fig. 4.4) and can be relatively inexpensive to construct. However, it can be difficult to control how much water goes into a watershed pond during rain events. In areas with rolling topography, these ponds do not have consistent depths as they tend to be deep at the dam end and shallow on the far end. The pond's shape is largely dictated by the land's topography. This type of pond needs to be constructed with a proper water-control structure, such as a spillway or overflow pipe, to ensure that excess water can leave without cutting the dam. However, this also allows nutrients and even fish to be washed out of the production unit.

A pond specifically designed and built for aquaculture is usually a leveed pond (fig. 4.5). They commonly have a 2:1 length to width ratio. They can be efficiently

Figure 4.4 A watershed type pond. Photo courtesy of Charles Weibel.

Figure 4.5 Leveed style ponds used in catfish production.

constructed by "cut and fill." By this method if 1 meter of soil is excavated then moved to the perimeter of the planned pond and used there to build a levee, this 1-meter cut can result in 2 meters of water depth. By using this method, large ponds (>10 ha) can be constructed with relatively small construction equipment (i.e., tractors with dirt pans). These ponds are not normally constructed to depths deeper than 1.5 meters because (1) the cost of pond construction is the cost of moving dirt; and (2) shallow ponds tend to stay well mixed, reducing the chance of crop loss due to pond turnover.

The type of soil used for pond construction is important to minimize leakage. In general, clay content of ≥20% is desirable. If these conditions are not available, pond liners are widely available. However, for most applications in medium- to large-scale commercial aquaculture, they can be cost prohibitive. Ponds used in commercial aquaculture production vary widely in size. Ponds used in freshwater prawn production in the United States are often 0.1 to 0.2 ha in size while catfish production ponds are often ≥8 ha in size. In early years, catfish production ponds were sometimes built as large as 30 ha. In ponds, most of the oxygen budget is based on oxygen production by photosynthetic phytoplankton. In the past this was the limiting factor for production within these systems. Without supplemental feeding the carrying capacity of a pond is around 500 kg/ha. With supplemental feeding this can be increased to about 1,500 kg/ha. However, at this biomass density and the accompanying feeding rate of 30 to 40 kg/ha/day, the chance of low oxygen periods during the night or early morning starts to become unacceptably high (Boyd 1979). In most commercial-scale pond production systems man has intervened by providing mechanical aeration. With this change, feed rates can be increased to about 100 kg/ha/day and production can be increased over threefold to more than 4,500/kg/ha.

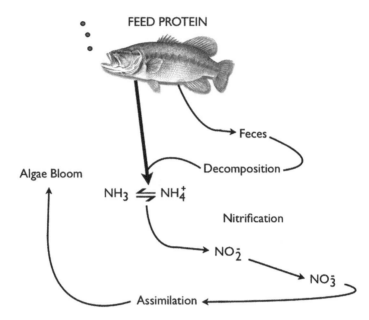

FEED PROTEIN

Feces

Algae Bloom   Decomposition

$NH_3 \rightleftharpoons NH_4^+$

Nitrification

NO_2^-

NO_3^-

Assimilation

Figure 4.6 Nitrogen cycle in ponds.

Ponds still rely on natural processes to remove waste products. Again, solid wastes are broken down primarily by heterotrophic bacteria and detritivores on the pond bottom. Ammonia (NH_4^+) excreted by fish or shrimp is directly assimilated by algae or converted to less toxic nitrite (NO_2^-) by *Nitrosamonas* bacteria then on to nitrate (NO_3^-) by *Nitrobacter* bacteria. The nitrate form can then be assimilated by algae (fig. 4.6). The efficiency of this nitrogen removal system is now the primary bottleneck in further pond production intensification. Maybe man can again intervene by increasing the rate of algal assimilation (see Partitioned Aquaculture Systems, chapter 13) or by increasing the rate of bacterial nitrification. Pond systems are discussed more thoroughly in chapter 10.

Water temperatures in these systems are basically ambient. They lag behind, but reflect, the mean air temperature in the region. Near the equator there is little seasonal fluctuation. As you move further from the equator seasonal fluctuations become more pronounced and can either affect growth or even produce mortality. There has been some work in using waste heat from power plants and other industries to warm pond waters. However, fish production is usually a secondary consideration. The needs of the primary industry usually take priority and have created problems in the past during shut down for repairs or maintenance.

In ponds, a significant portion of the food for the culture organism can also be generated internally. If this is the primary food for the system it is said to be an extensive pond system. The natural carry capacity of an unfed pond is in the range of 250 kg/ha. This carrying capacity can be increased by adding nutrients to the system. In most extensive or low-input systems these nutrients are supplied in the form agricultural byproducts, or animal (or human) waste. This is known as organic fertilization. If needed nutrients are supplied in their purely chemical form (often derived from petrochemicals) they are knows as inorganic fertilizers. Again, there are positives and negatives to each.

Positives of organic fertilizers include low costs, slow release of nutrients, and sustainability aspects of reuse. Negatives include the need to handle bulk and sometimes wet (and heavy) materials for small amounts of nutrients. To be made available to the phytoplankton and food web these materials must first be decomposed by microbes, which can be fairly slow and is an oxygen consuming process. These products can also directly deteriorate water quality (for example, by yielding ammonia) if misapplied. Inorganic fertilizers have the positive aspects of acting quickly, without deteriorating water quality through added nitrogenous wastes. However, inorganics can be more expensive and can actually work "too well" if misapplied. A slight over application of phosphorous can cause of a rapid phytoplankton bloom, which can die off just as rapidly. This can result in oxygen depletion as the phytoplankton decomposes.

4.3   Closed systems

In closed systems water is reused within a manmade culture system (fig. 4.7). Also, there is human intervention of some type and at some level in *all* of the basic processes. The major advantage of closed systems is that they provide the operator complete control over all of the environmental variables in the culture

Figure 4.7 A recirculating aquaculture system. Photo courtesy of the Conservation Fund's Freshwater Institute.

system. The major *disadvantage* of closed systems is that the operator now has complete *responsibility* for all aspects of the animals' environment.

In closed systems water temperature can be maintained very near the optimum growing temperature for the cultured animal. This can have a tremendous positive impact on not only growth rate but also efficiency, both of which are highly important in these systems. Because of this temperature control, we can now raise tropical animals in temperate zones, if that is where the markets are. Waste heat from industrial processes can provide economic advantages if the schedules and proximities of the systems are compatible.

With closed systems, water can be constantly disinfected with ultraviolet (UV) lights or ozone to crop down pathogenic organisms. Predators and poachers can be completely eliminated. External environmental events like floods or cold snaps are no longer a problem. Feed can be efficiently administered and consumption and conversion accurately monitored. Water supply volumes become less of a concern. However, for large systems, their loss of more than 5% of their total water volume per day (for maintenance reasons) can still become substantial.

4.3.1   Biofilter-based systems

Recirculating aquaculture systems (RAS) are also known as closed loop systems, recycle systems, and intensive recycle systems. As these names imply, as opposed to trout raceways that have a constant inflow of new water, these systems use the same water over and over. They do this by constantly adding air or oxygen to the water and removing the waste products produced by the fish. If aeration is used, production is limited to about 40 kg/m^3 (0.33 lb/gal). With the use of pure oxygen (oxygenation) production can be increased to approximately 120 kg/m^3 (1.0 lb/gal; Timmons *et al.* 2002). To remove waste products most systems rely on mechanical filters to remove solid wastes. Then, nitrogenous wastes such as ammonias are detoxified to nitrite and then nitrate, using the same nitrifying bacteria discussed regarding ponds. However, these bacteria are now cultured at very high densities inside containers known as biofilters. The nitrifying bacteria in the biofilter need a surface to attach to so special materials, known as media, are packed or suspended inside of the biofilter vessel to provide extra surface area for large numbers of bacteria to grow on. However, it is also important for the biofilter media to have sufficient open areas for the water to flow through and wash over the bacteria so that they can "consume" the inorganic waste and excrete less toxic versions.

4.3.2   Heterotrophic or biofloc-based systems

In recent years another type of closed or recycle system has been developed. Instead of relying on chemoautotrophic nitrifying bacteria, which utilize inorganic compounds such as ammonia for energy, these systems are colonized with heterotrophic bacteria that consume the organic wastes. These bacteria are not confined in biofilters but live suspended in the culture vessel along with the animal being cultured. Once this bacterial population is established and stabilized, very high production rates can be achieved (>5 kg/m^3). However, these systems also have very high oxygen demand and relatively little research has been conducted on their complex microbial ecology. These systems can quickly remove nitrogenous waste products from the animals by directly converting it to bacterial biomass. In traditional RAS, quick removal of solids is important to their function. In heterotrophic or biofloc systems these solids are largely retained in the culture tanks and become colonized with heterotrophic bacteria, fungi, and protozoans into suspended particles called bioflocs. The bioflocs also recycle the wastes as they can be grazed by the animals and directly consumed as high-protein forage. Heterotrophic/biofloc systems are covered in more detail in chapter 12.

4.4 Hybrid systems

Many new approaches in recent years are blurring the lines between the different production systems and even major categories (open, semi-closed, and closed). They take aspects of different systems and combine them in new ways to overcome the shortcomings of one, capitalize on the positives of another, or break a system into its functional components so they can be individually manipulated.

4.4.1 Aquaponics

In a system called aquaponics, the basic components of a recirculating system are utilized. However, the biofilter has been replaced by plants that assimilate the nitrogenous waste products and then turn it into saleable plant products (fig. 4.8). Aquaponics systems are increasingly of interest in areas where water availability is limited. Much of their early development has occurred on small islands where the supply of fresh fish, freshwater, and fresh vegetables are all limited. There is increasing interest in evaluating these systems in arid countries, such as those in the Middle East. Now there are also efforts to apply these technologies to generating fresh fish and vegetables in or near major urban areas. These are meant to address the phenomenon known as "food deserts." These are urban areas where city dwellers do not have ready access to healthy foods at reasonable prices (Ford & Dzewaltowski 2008). This makes these populations, especially certain ethnic groups, even more susceptible to health problems such as type 2 diabetes and obesity. Urban aquaponics might represent an "oasis" of healthy foods in or near the food deserts.

Figure 4.8 An aquaponics system at the University of the Virgin Islands.

4.4.2   In-pond raceways

In-pond raceways are similar to cages floating in open system waters or semi-closed system ponds. However, they take on the characteristics of a raceway by constantly flowing water through cages by mechanical means. In-pond raceways allow the fish to be confined in small systems so they can be more efficiently fed and monitored (like a raceway). They also give at least the potential of capturing waste products for removal, again like a raceway. However, unlike regular raceways, these systems are not confined to locations with large groundwater springs or flowing surface water resources. These "raceways" can be located wherever large ponds or reservoirs exist. The temperature and waste removal functions of these systems are still handled by the same pond processes. The oxygen needs of these systems are also primarily addressed by the pond system. However, since the fish are confined at much higher densities, dissolved oxygen supplies are usually supplemented by mechanical aeration, and at times by addition of pure oxygen. This system is covered in more detail in chapter 15.

4.4.3   Partitioned aquaculture system

With the partitioned aquaculture system (PAS), the concept of a pond as an algae-driven system is modified to address one of a traditional pond's limiting factors. In heavily fed ponds dense plankton blooms develop and light cannot penetrate deeply, so only the upper level of the water column is fully functional (i.e., has sufficient light for photosynthesis). By taking the water and circulating it through channels that are only 40 to 60 cm deep, essentially all of the water volume gets sufficient sunlight to be productive, allowing threefold to fourfold production increases in the same area and volume. The PAS system is covered in more detail in chapter 13.

The PAS is not so much a hybrid system as it is a "deconstruction" of a pond-based system into its functional components so that each component can be modeled and its efficiency maximized. It is still basically a pond system. However, the primary culture fish are now confined in a cage similar to an in-pond raceway. Again, there is at least the potential of removing solid wastes from the system. Water is moved very efficiently using low rpm hydraulic paddlewheels that push the water out into the shallow channels. The combination of water movement and shallow water depths results in conditions where light penetration is no longer a limiting factor and algae fixation rates are no longer light limited. To keep algal cells in a rapid growth phase, and again maximize their efficiency, an algae grazing fish is confined in another cage within the water flow. Their feeding crops the algae, keeping cell age down and their metabolism (and oxygen production and ammonia removal) high.

As you can see, aquaculture is an extremely diverse enterprise not only in the number of species we raise, but also in how we raise them. While some methods are millennia old, some are only a few decades old at most. As the demands placed on aquaculture continue to increase, there will undoubtedly continue to be changes in the way we raise aquaculture species. Some of these changes will be revolutionary. Most will be evolutionary. However, change and improvement *is* inevitable.

4.5   References

Arnason, R. (2001) The economics of ocean ranching: Experience, outlook, and theory. *FAO Fisheries Technical Paper No. 413.* FAO, Rome.

Boyd, C.E. (1979) *Water Quality in Warmwater Fish Ponds.* Agricultural Experiment Station, Auburn University, Auburn.

Cohen, A. (2009) MIT tests self-propelled cage for fish farm. *MIT News.* http://web.mit.edu/newsoffice/2008/aquaculture-6902. Accessed November 19, 2009.

De Silva, S.S. & Soto, D. (2009) Climate change and aquaculture: Potential impacts, adaptations and mitigation. In *Climate Change Implications for Fisheries and Aquaculture: Overview of Current Scientific Knowledge* (Ed. by K. Cochrane, C. De Young, D. Soto & T. Bahri), pp. 151–212. FAO Fisheries and Aquaculture Technical Paper No. 530. FAO, Rome.

Ford, P.B. & Dzewaltowski, D.A. (2008) Disparities in obesity prevalence due to variation in the retail food environment: three testable hypotheses. *Nutrition Reviews* 66(4):216–28.

Fornshell, G. & Hinshaw, J.M. (2008) Better management practices for flow-through systems. In *Environmental Best Management Practices for Aquaculture* (Ed. by C.S. Tucker, & J.A. Hargreaves), pp. 331–88. Wiley-Blackwell, Oxford.

Pillay, T.V.R. & Kutty, M.N. (2005) *Aquaculture: Principles and Practices.* Blackwell Publishing, Oxford.

Stickney, R.R. (1979) *Principles of Warmwater Aquaculture.* John Wiley & Sons, Inc., New York.

Stickney, R.R. (2009) *Aquaculture: An Introductory Text*, 2nd Edition. CABI, Oxfordshire.

Subasinghe, R.P. (2007) Aquaculture statues and prospects. *FAO Aquaculture Newsletter* 38:4–7.

Timmons, M.B., Ebeling, J.M., Wheaton, F.W., Summerfelt, S.T. & Vinci, B.J. (2002) *Recirculating Aquaculture Systems.* Cayuga Aqua Ventures, New York.

Chapter 5
Shellfish Aquaculture

Robert Rheault

In the scheme of global aquaculture production, freshwater finfish dominate, followed by aquatic plants, mollusks, crustaceans, and marine fish. Among marine species in culture, shellfish are the dominant group with annual harvests exceeding 13.1 million tonnes worldwide, compared with 5.3 million tonnes of crustaceans, and 3.4 million tonnes of diadromous and marine fish. Crustaceans (mostly shrimp) lead in terms of the dollar value of production at over US$24.1 billion per year, but molluscan shellfish are not far behind with annual harvests just under US$12.9 billion (FAO 2009). Production of cultured mollusks continues to expand at a remarkable rate with global landings increasing 42% in the ten-year span from 1999 to 2009.

The overwhelming majority of shellfish culture occurs in China and the Asian Pacific (80% of the value and over 90% of the biomass). For the past twenty years Asia's production of oysters has doubled about every ten years while their clam production has nearly tripled every ten years. A plot of international production makes Asia's dominance in aquaculture readily apparent (fig. 5.1). Western Europe, North America, and Oceania (Australia, New Zealand, and the Islands of the South Pacific) are all significant shellfish producers, but Africa, India, and the Middle East so far have relatively small shellfish culture industries.

Aquaculture Production Systems, First Edition. Edited by James Tidwell.
© 2012 John Wiley & Sons, Inc. Published 2012 by John Wiley & Sons, Inc.

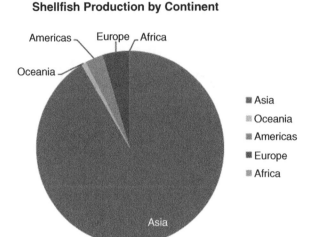

Figure 5.1 Shellfish production by continent (FAO 2009).

5.1   Major species in culture (oysters, clams, scallops, mussels)

The FAO (2008) subdivides molluscan shellfish production into seven categories: oysters, clams (including cockles and arc shells), scallops, mussels, "miscellaneous mollusks," freshwater mollusks, and abalones (including winkles and conchs). The world harvest of cultured oysters in 2006 totaled 4.7 million tonnes. The FAO lists fourteen species of oysters in culture, but the most significant contributor by far is the Pacific or Japanese oyster, *Crassostrea gigas*, which alone accounts for 97% of the world oyster production.

The American or Eastern oyster, *Crassostrea virginica*, ranks second, accounting for 1.4% of world production; followed by the Suminoe oyster, *Crassostrea ariakensis*, with 0.5%; and the European flat oyster, *Ostrea edulis*, and the Sydney rock oyster, *Saccostrea commercialis*, each with 0.1%. These are the principle oyster species in culture by volume, but there are dozens more oyster species in culture around the world.

World clam culture produced 4.3 million tonnes in 2006 (FAO 2008). There are relatively few species of clams currently being produced in culture, the most significant by far being the Manila or carpet clam, *Ruditapes phillipinarium*. China's production of this one species exceeds 3 million tonnes a year. In the United States in the 1930s, Manila clams were accidentally introduced in the Pacific Northwest with a shipment of Pacific oyster seed and the Manila clam is now a major aquaculture species from California to British Columbia with regional annual harvests of 6,000 tonnes valued at over US$25 million. In Europe, overfishing and irregular yields of the native (European) grooved carpet shell, *Ruditapes decussatus*, led to imports of *R. philippinarum* into European waters. European countries now grow 58,000 tonnes of Manila clams, slightly eclipsing their wild harvest fishery.

The constricted tagelus, *Sinonovacula constricta,* supports annual harvests of 680,000 tonnes in Asia making it the second most popular cultured clam species. The northern quahog or hard clam, *Mercenaria mercenaria* is grown in the United States and Canada where annual harvests exceed 38,000 tonnes. Most of this production has developed in the past ten years with Virginia and Florida being the biggest producers. Also of note because of its impressive size is the geoduck, *Panopea abrupta.* It is the largest known burrowing bivalve weighing as much as 3.2 kg, with a shell measuring up to 18- to 23-cm long and siphons that extend up to 1.3 m. These clams are native from Baja California to Alaska, burrowing as deep as a meter or more in intertidal or subtidal sands. Growers are trying to expand production of this species to meet the demand from lucrative Asian markets. The geoduck can live over 100 years, but farmers typically harvest them after only four to seven years.

World scallop harvests totaled 1.4 million tonnes in 2006 (FAO 2008). Many species of scallops have been used in aquaculture because of their high fecundity, rapid growth, and the high value of their meats. Scallops are typically grown in lantern nets or by ear-hanging, but production from these labor-intensive methods is dwarfed in localities where growers are able to "ranch" the scallops by dispersing seed on the bottom. Countries with inexpensive labor, such as China, continue to employ massive suspended culture operations, accounting for over 81% of the world scallop production—primarily the Japanese scallop, *Patinopecten yessoensis,* and the northern bay scallop, *Argopecten irradians.* There is also substantial scallop production in Japan, Korea, and the Russian Federation. Lesser scallop producers include Peru and Chile (*Argopecten purpuratus*), Canada (sea scallop, *Placopecten magellanicus*), and the United Kingdom (Queen scallop, *Chlamys opercularis*).

The world mussel harvest totals 1.9 million tonnes and is spread around the globe (FAO 2008). China produces 39% (primarily the green shelled mussel, *Perna viridis*), Thailand 14% (*Perna canaliculus*), New Zealand 5% (*P. canaliculus*), and Spain 12% (Mediterranean mussel, *Mytilus galloprovincialis*). The blue mussel, *Mytilus edulis,* representing 8% of world harvest, is raised primarily in western European nations with contributions from Canada and the United States.

The FAO's category of "miscellaneous mollusks" lists harvests of 1.3 million tonnes. There are dozens more bivalve species being grown in significant numbers, with many more being studied and evaluated for aquaculture. Annual harvests of abalones, winkles, and conchs total 367,000 tonnes. Freshwater mollusks are a minor player on the world stage with production totaling only 154,000 tonnes, but that number has grown tenfold in the past five years.

5.2   History

The history of the development of shellfish aquaculture reveals that culture efforts only become significant when natural stocks become depleted from overfishing. China and Japan are generally believed to be the first nations to practiced have shellfish culture. In China the first efforts are thought to have started 2,000 years ago during the Han Dynasty (Nie 1982). Early growers simply drove wooden stakes or provided piles of rock or shell to attract sets of oysters in areas that were otherwise too muddy to grow oysters. These relatively primitive techniques remained virtually unchanged for hundreds of years, and production didn't really take off until 1950s. Subsequent production in China doubled every ten years until the 1980s when it really exploded. Between 1980 and 1990 production increased fourfold to 1.8 million tonnes, and in the next decade production again increased fourfold, reaching 7.5 million tonnes by 2000 (FAO 2009). Japan was one of the first nations to develop intensive shellfish culture techniques. Spat collection efforts were recorded as early as the seventeenth century (Lou 1991). Early efforts involved fencing off intertidal flats with woven bamboo to eliminate predators, while inside these enclosures, bamboo stakes, rocks, and shell were planted in rows during spawning season to provide suitable substrate for larval oysters to attach. Once the spat were of suitable size the screens could be removed, and after two to three growing seasons the oysters were harvested.

In the 1920s Japanese growers became some of the first to experiment with hanging culture techniques (Nie 1982). Oyster spat were still collected on intertidal racks, but they were moved to deeper water in various enclosures that were suspended under rafts or on longlines. Suspended culture, was shown to increase growth rates and eliminate mortality from benthic predators while allowing growers to use areas that were not suitable for bottom culture (Imai 1978). Yields increased dramatically with suspended culture and time to harvest could be cut by nearly a full year. Scallop growers using hanging culture in Hiroshima Bay claim some of the highest production densities on record, with up to 20,000 kg/ha of meat per year (Lou 1991). Japanese scallop culture really took hold with refined methods for spat collection and the development of hanging culture methods (ear hanging and lantern nets) in the late 1960s. In ten years' time cultured harvest eclipsed wild landings, propelling Japan into the position of top scallop producer in the world. Successful spat collection efforts also led to huge increases in bottom plantings with commensurate increases in yields (Lou 1991).

Around the same time wild scallop landings started to decline in China. By borrowing spat collection technology and lantern net design from the Japanese, scallop culture was initiated in China in the late 1970s. Seed supply issues were resolved with massive investments in hatcheries and improvements in spat collection, and the industry took off (Lou 1991). By 1985 production was up to 10,000 tonnes and by 1992 China surpassed Japan as the number-one producer of scallops, with harvests of 338,000 tonnes. In 1996 China reported cultured landings totaling 81% of the world's scallop production (FAO 2007).

Mussels became important in Spain when farmers began culturing them at the beginning of the twentieth century. The first mussel culture developed near the Iberian Peninsula in the early 1900s by placing poles in the sediment to catch spat (FAO 2006). Raft culture of mussels was introduced in the Galician region in 1946 and within a few years production had increased sharply (FAO 2006).

French and British oyster culture efforts began in the 1800s. In 1866 the famous British biologist Thomas Huxley was commissioned to investigate the causes of the 30 year decline of the oyster fishery. He declared that the preposterous regulations hampering the various fisheries should be abolished because the supply was "inexhaustible." Huxley reasoned that regulations such as "closed seasons" were ineffective at controlling over-dredging because they did not control the fishing effort when the restrictions were lifted. In his 1883 inaugural address to the London Fisheries Exhibition he conceded that oysters "may be exhaustible" and recommended that the "State can grant a property in the beds to corporations or to individuals whose interest it will become to protect them efficiently. And this I think is the only method by which fisheries can be preserved" (Blinderman 2008). At the time, oyster culture consisted simply of holding oysters that had been dredged, but were too small for sale, in wooden "barks" that suspended the oysters in protected tidal areas.

In France in 1869 a boatload of Portuguese oysters, *Crassostrea angulata,* was inadvertently dumped near the mouth of the Gironde River. They grew well and were found to be more durable than the native *Ostrea edulis.* They became the foundation of much of France's oyster production for the next 100 years, until the decision to introduce *C. gigas.*

In the United States oysters were a huge source of easily accessible protein from the time of the early settlers until the late 1800s when harvests began to peak. The completion of the Transcontinental Railroad in 1869 and the invention of steam-powered boats in the 1880s led to rapid depletion of vast beds of oysters in the mid-Atlantic and New York waters. In the 1860s oyster growers started taking schooners full of oysters and seed from these populations to be bedded on thousands of acres of private leases in Long Island Sound and Narragansett Bay, where they would fatten and grow quickly (Kurlansky 2006).

In 1884 the "Oyster Panic" hit New York when cholera outbreaks (a disease attributed to filthy living conditions in Manhattan's slums) resulted in a number of deaths among rich oyster eaters. That same year Koch proved sewage-related bacteria was the cause of typhoid and by 1890 Pasteur's "germ theory" had become accepted doctrine (Kurlansky 2006). At the time, every major city was dumping millions of gallons of untreated sewage and horse manure into estuaries, and deforestation was leading to siltation of many prime oyster beds. The advent of running water and the invention of the flush toilet at the turn of the century made a severe problem much worse. The first oil refinery was built on the East River in 1872 and shortly thereafter oil spills and discharges made oysters and fish taken from estuaries near many major cities taste like oil. None of this was good for the oysters or their delicate larvae.

In 1925 the surgeon general mandated the formation of the National Shellfish Sanitation Program in response to the decades of outbreaks of typhoid fever related to the consumption of raw oysters (Yuhas 2002). Shellfish-related illness continued to plague the industry for decades, but advances in sewage treatment, monitoring, and restrictions on ocean dumping have greatly diminished the risk.

In the United States, overharvesting, habitat destruction, and the oyster diseases MSX and Dermo that hit the mid-Atlantic states in the late 1950s and

1990s, respectively, decimated Eastern oyster populations. Natural populations were reduced to a tiny fraction of historical levels. More recently, interest in oyster aquaculture has been stimulated by a renaissance in oyster consumption at trendy raw-bars across the country.

Development of shellfish hatchery techniques became the focus of research when natural oyster recruitment along the Eastern Seaboard became unpredictable in the early 1900s. Joe Glancy working at Blue Points Company on Long Island in the 1940s did pioneering work that was later refined by Victor Loosanoff and Harold Davis at the National Marine Fisheries Service Lab in Milford, Conn. (Carricker 2004; Matthiessen 2001). A prerequisite for the development of successful hatchery production was the refinement of techniques to produce microalgal food. Glancy (who was building on the work of W. F. Wells) described a simple method of coarsely filtering seawater down to 5 or 10 μm and introducing some fertilizer to stimulate algal growth. After a few days of sunlight, a rich broth of mixed-algal species was often the result. The so-called Wells-Glancy method was the primary way to produce micoalgal shellfish food for decades. Unfortunately, the quality of the algae produced using the Wells-Glancy method is unpredictable and prone to contamination (Carricker 2004). Robert Guillard later developed techniques to isolate and grow axenic, bacteria-free, single-species cultures and developed a collection of hundreds of species from around the world. Cultures from his collection, and others that have followed, are used by hatcheries around the world.

5.3   Biology

The vast majority of shellfish in culture are bivalves from the phylum Mollusca. These are relatively primitive organisms that feed by passing large volumes of water across a fine ciliated gill to filter out microscopic plants and organic particles. Like most filter-feeders, they are omnivorous consumers of small particles, including microalgae (diatoms and flagellates), organic detritus, bacteria, viruses, and small protists, (collectively referred to as "organic seston"). Mollusks are typically infaunal or epifaunal and are mostly sedentary. Scallops and razor clams have been known to swim significant distances in search of food or to avoid predators, but clams rarely move more than a few centimeters a day. Oysters have evolved to spend their entire life cemented in one place.

Shellfish are classic r-strategists, meaning that each adult is capable of producing large numbers of larvae (up to several million), ranging from 60 to 250 μm in diameter (Mackie 1984). Such high fecundity means shellfish are able to maintain stable populations, even if only one in a million survives to reproductive age; and it also means that under optimal conditions they can colonize new habitats opportunistically, sometimes at densities so great that they smother each other.

Shellfish larvae are free-swimming plankton for the first two to three weeks of their lives. Under a microscope, most shellfish larvae look like a typical clam,

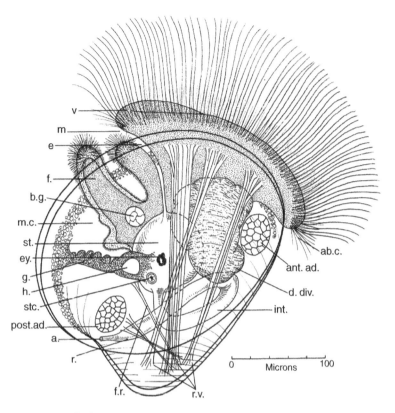

Figure 5.2 Diagram of veliger larvae. Abbreviations in diagram are (a.) anus; (ab.c.) aboral circle of cilia; (ant.ad.) anterior adductor muscle; (b.g.) byssus gland; (d.div.) digestive diverticula; (e.) esophagus; (ey.) eye; (f.) foot; (f.r.) foot tractor muscles; (g.) gill rudiment; (h.) heart; (int.) intestine; (m.) mouth; (m.c.) mantle cavity; (post.ad.) posterior adductor muscle; (r.) rectum; (r.v.) velar retractor muscles; (st.) stomach; (stc.) statocysts; (v.) veulum. Figure from Paul Galtsoff (1964).

but with a transparent shell and ciliated vellum that is a combination feeding structure and propulsion mechanism (fig. 5.2). Capable of swimming only a few centimeters per minute, they ride the tides and can be dispersed for many kilometers before settling.

After the larvae develop, they seek an appropriate place to settle and spend the rest of their lives. Infaunal species such as clams seek sand or muddy bottom while oysters look for firm substrate (preferably shell) to which they can cement themselves. Mussel, clang and scallop larvae secrete an epoxy-like byssal thread that hardens on contact with seawater, allowing larvae to fasten themselves to firm structures, usually off the bottom and away from predators such as crabs and starfish.

Once the larvae find an appropriate place to settle they undergo a complex metamorphosis and take on the adult morphology. Under proper conditions they will grow rapidly from about 250 μm at settlement to several millimeters in a few weeks. Under ideal conditions fast growing species can reach 7 cm or more in the first growing season and will be able to reproduce the following year, with some species living only two years (e.g., *Argopecten irradians*).

5.4   Culture basics

Shellfish aquaculture is much like other agriculture in that it involves planting and nurturing seed; protecting the crops from predators, disease, or storms; harvesting the crop; and marketing the perishable product. There are, however, unique challenges tied to working on the water that add several degrees of difficulty. Monitoring the crop condition, growth rate, and survival through the lens of several meters of water usually requires hauling cages or donning scuba gear. Aquaculture also requires the prospective grower to obtain exclusive use of a public resource that has many other user groups vying to protect their interests. In today's world, all this must be done under the watchful eye of environmental groups that are increasingly wary of any new potential perturbations or stresses on the fragile, inshore marine ecosystem. Unlike terrestrial agriculture, which has developed over centuries, shellfish aquaculture is relatively new, and the scientific underpinnings that guide growers' actions have mostly been developed in the last 50 to 100 years. Growers are still refining techniques to surmount the many challenges that arise. Given the tremendous spatial and temporal variability of the marine environment, it is common that techniques that work well in one area are ineffective a short distance away.

In its essence, shellfish aquaculture involves two, sometimes conflicting, goals: (1) protecting the animals from predators while (2) ensuring an ample supply of food. Shellfish farmers have developed dozens of different ways to grow shellfish while balancing these two objectives. Because each species has slightly different requirements and each growing area poses a unique set of challenges, no single culture method will work in every growing area.

There is an amazing array of potential predators that consume shellfish, especially when the shellfish are small. Consequently, if a grower leaves seed unprotected it is not uncommon for predators to consume most of the crop. Predators can include birds, starfish, crabs, fish, and predatory gastropods such as whelks or oyster drills. A flock of diving ducks can pick a mussel raft clean in a few days, while roving populations of starfish can annihilate whole shellfish beds in a matter of weeks. Many species of crabs feed voraciously on shellfish and juveniles are especially vulnerable. The diminutive sand shrimp (*Crangon* sp.) can feast on newly set shellfish, while a single blue crab (*Callinectes sapidus*) can consume hundreds of clam seed in an hour (Arnold 1984). In response, growers have devised a wide array of protective enclosures.

The ideal environment for shellfish is one that has ample available food and protection from predators. Whereas growth in a solitary bivalve is primarily determined by the concentration of particulate food in the water, a dense population of shellfish will rapidly consume all of the food in the immediate area unless there is a good current to replenish the food supply. For filter-feeders being grown at commercial densities on an aquaculture farm, current speed is

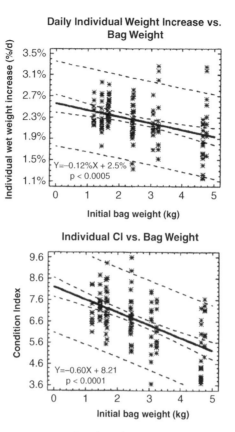

Figure 5.3 Stocking density versus growth and condition index in individually tagged oysters after six weeks. Growth rates (upper graph; percent increase in weight per day) and condition index (lower graph) are plotted against initial stocking density of the bag. Linear regression formulae are shown with each line plotted with 95% confidence limits (inner dashed lines) and 95% prediction limits (outer dashed lines). From Rheault and Rice (1996).

usually just as critical as food concentration in determining food availability (Peterson *et al.* 1984).

The product of current speed and food concentration is called "seston flux" and this number determines how densely a grower can stock the shellfish before food competition starts to limit growth (Grizzle & Lutz 1989). In any given location growers must experiment with different stocking densities to determine the optimal production density. At higher stocking densities growth rates will decline because the shellfish are experiencing local food depletion and suboptimal food availability (Rheault & Rice 1996; fig. 5.3). Lowering the planting density will reduce stress, improve the condition index (the ratio of meat weight to size), speed growth, and reduce the time to harvest. Shortening the time to harvest in turn reduces the amount of gear needed for multiple crops and lowers the risk of mortality.

Growers are often faced with an economic dilemma. At lower stocking densities, growers must deploy and maintain much more gear in order to produce the same biomass, which dramatically increases unit production costs, so there is a clear economic trade-off. At some point, all growers need to calculate the optimal production density for their farm by balancing the increased costs of more gear stocked at lower densities with the benefits of low stocking densities (reduced stress, lower mortality, faster growth, shorter time to harvest, and better meat yield). This exercise needs to be repeated for every growing area, for every species, and for different stages of growth because the curves will have different shapes and inflexion points for each (Rheault & Rice 1996). The curves will also be displaced by changes in food quality, by monthly changes in tidal velocity, and seasonal changes in temperature. In intensive production systems the cost of buying and maintaining additional gear to maintain low densities will have a profound impact on the economics of production. All the benefits of lowering densities described above may be lost if the grower is unable to properly maintain the additional gear.

5.5   Extensive versus intensive culture

In their efforts to protect shellfish from predators, growers have resorted to various predator barriers such as cages, mesh bags, or netting. These techniques are considered "intensive culture." By growing clams under netting, or oysters in mesh bags or cages, farmers are often able to reduce predation losses to near zero. These approaches, however, require increased capital investment and involve tremendous amounts of labor to maintain and to clean the gear. The mesh of the netting (and any firm substrate left in the marine environment) quickly becomes fouled by a remarkable variety of fouling organisms. Fouling restricts the flow of water to the crop, which in turn reduces the availability of food for the shellfish, resulting in slower growth and poor condition.

The alternative to using predator barriers is to take a more natural approach such as free-planting seed on the bottom. These extensive culture techniques accept a certain level of losses due to predation, compensating with sheer numbers and maintaining profitability by keeping costs low and minimizing inputs of energy, capital, and labor.

Sometimes it is difficult to draw a line between a well-managed fishery and extensive aquaculture. The efforts of extensive farmers may be limited to enhancing natural spat collection or rudimentary predator control. Conversely, several wild shellfisheries are augmenting their catch by utilizing culture techniques such as hatcheries, spat collection, planting, and rotational harvesting (Booth & Cox 2003). The line is often drawn based on the fact that growers lease the bottom from which they harvest, but this distinction is subtle enough to confuse both regulators and statisticians.

For example, there is significant debate as to whether the substantial oyster landings of Connecticut, Louisiana, and Mississippi should be considered cultured or wild harvested. These growers use techniques that are little changed since the late 1800s. Typically, either the growers or the state managers (or both) spread tons of clean shell, called "cultch," in areas (often public) where they know larvae will be abundant. Once the young oyster spat are established on the cultch, they are harvested and spread on leased beds for growout until the oysters reach market size. Growers move the seed to different beds depending on the density, size, or growing needs of the oysters. Some leaseholders may engage in starfish "mopping" or crab trapping to reduce predation losses. Because the level of husbandry varies greatly from grower to grower, the federal government continues to struggle with the legal definition of "cultured" in these multimillion-dollar fisheries.

In reality, no clear lines can be drawn separating sustainably managed wild shellfisheries, extensive culture, and intensive culture. Instead there is a broad spectrum of approaches, each designed to take advantage of local conditions, the biology of the species, the various predators, labor costs, materials costs, and the economics of production. Intensive culture usually involves hatchery production, nursery structures such as upwellers or raceways, and growout strategies involving predator barriers such as netting, cages, racks, and bags. Extensive culture typically relies on wild set (sometimes enhanced using spat collectors); growout efforts are limited to spreading the seed on the bottom and very little is spent on culture gear or predator control. The lines become blurred when intensive culture systems rely on wild spat collection or when extensive systems use hatcheries or nurseries. Such hybrid approaches are common.

5.6 Spat collection: hatchery, nursery, growout

Shellfish aquaculture can be subdivided into three components based on the life cycle needs of the species in culture and the culture techniques used by the grower: hatchery (or spat collection), nursery, and growout. The lines between these three steps can be blurred, depending on the species and the culture techniques. Some growers will move seed as small as 1 mm or less from a hatchery into their nursery systems, but most will wait until the seed are 5 to 10 mm. Nursery culture techniques such as upwellers, raceways, and spat bags can theoretically be used for larger animals, but at some point growers typically transition to a more economical growout method. Growout methods vary widely from extensive bottom planting to intensive methods involving rafts, lantern nets, or various types of racks and bags.

5.6.1 Spat collection

The first step in any culture effort is the acquisition of juveniles. Most early shellfish culture efforts concentrated on improving spat collection and the survival of tiny, post-set juveniles. Shellfish larvae are microscopic and free-swimming for the first weeks of their life and they are a terrific food for larval fish, crustaceans, and gelatinous zooplankton. Natural mortality rates for larvae and post-set juveniles are astronomical, so anything that can be done to attract the settlement of larvae and improve the survival of these delicate early life stages can have a dramatic impact on survival rates and harvest numbers.

Late-stage larvae will begin to seek the bottom and sample the substrate looking for ideal conditions in which to spend the rest of their lives. Some larvae can delay settlement for days while their instincts drive them to seek the right combination of environmental cues that give them the best chance to survive long enough to make their own contribution to the gene pool. Some species have evolved to seek out brackish intertidal waters to avoid certain predators, while some require subtidal or deeper waters. Clam larvae typically seek out sandy or muddy bottom, while oysters typically seek shell or other firm substrates to which they can cement themselves; scallops typically seek grasses or other off-bottom substrates.

Many larval and juvenile shellfish species possess a byssal gland that secretes tiny adhesive threads allowing them to adhere tenaciously to various objects. In the hard clam (*M. mercenaria*) this ability is lost once the seed grow beyond a few millimeters, while most scallop species still use byssal threads until they are a centimeter or more in length. Mussels never lose the ability to secrete strong byssal threads.

The earliest aquaculture efforts consisted of attracting larvae by simply providing the proper conditions to enhance natural settlement. Early naturalists observed oysters setting on almost any firm substrates in protected waters, and found that by placing ceramic tiles, shell, or even sticks of wood in the appropriate waters they could collect oyster seed in great quantities. When natural populations of shellfish dwindled due to overharvesting and disease, fishermen and resource managers began to experiment with various spat collectors to enhance the natural set. By testing different materials placed at different times of year they were able to determine that oysters prefer clean shell placed in the water shortly before peak settlement. Substrates placed in the waters too early became fouled with other organisms. They also discovered they could greatly enhance oyster sets by coating twigs or ceramic tiles with cement because the larvae detect the lime, which resembles shell.

Similarly, naturalists observed that scallop larvae seek eelgrass or similar vertical structures to which they can attach themselves via a byssal thread. It was discovered that mesh bags stuffed with burlap, polypropylene, or old monofilament line and suspended in various locations could reliably catch scallop sets. Further experiments have resulted in refinements in the types of fibers used, as well as in the ideal timing and location for the placement of spat bags so that yields of several hundred spat per bag are not uncommon. Most sea scallop farms still rely on spat collector bags, while nearly all of the mussels cultured around the world are the product of spat collectors, which consist of specially designed fuzzy ropes. Some mussel spat are still scraped from rocks to fill growout mesh tubes called "socks," and, remarkably, the substantial New Zealand green mussel industry still relies on wild spat collected from seaweed that predictably washes ashore on a certain beach each year.

Unfortunately, wild spat sets are not always reliable. In the late 1800s, oyster growers in Connecticut and Long Island had built a multimillion-dollar industry based on the placement of millions of bushels of shell in established setting grounds in estuaries along the coast. When spat falls failed for four consecutive years starting in 1912, the industry nearly collapsed. Eventually, researchers developed hatchery techniques in an effort to revive the industry and stabilize production. Historically, investments in hatcheries have been sporadic because hatchery production is easily dwarfed when environmental conditions align to favor a strong natural set.

5.6.2   Hatchery methods

Hatchery techniques were initially developed for shellfish in the 1940s by Joe Glancy working at Blue Points Company on Long Island and were later refined by Victor Loosanoff, Harold Davis, and Robert Guillard at the National Marine Fisheries Service Lab in Milford, Conn. (Carriker 2004). A prerequisite for the development of successful hatchery production was the refinement of techniques to produce significant amounts of single-celled algal food for the broodstock and subsequently for the larvae and post-set spat. Glancy (who was building on the work of W. F. Wells) described a simple method of coarsely filtering seawater down to 5 or 10 μm and introducing some fertilizer to stimulate algal growth. After a few days of sunlight, a rich broth of mixed algal species was often the result. The so-called Wells-Glancy method was the primary way to produce micoalgal shellfish food for decades (Carriker 2004).

Unfortunately, the quality of the algae produced using the Wells-Glancy method is unpredictable because there is no control over which species dominate the brew. Not all algal species are good shellfish foods because some species are indigestible, noxious, or even toxic. Also the Wells-Glancy method is prone to contamination by various flagellates and protists that can multiply almost as fast as the algae and can eat most of the desirable small cells.

5.7   Cultured algae

Robert Guillard developed techniques to isolate and grow axenic, single-species cultures and developed a collection of hundreds of species from around the world (Guillard 1975). Several single-celled algae species are good bivalve foods, easily digested, and provide a balance of critical essential amino acids, proteins, fatty acids, and carbohydrates needed to promote growth (Webb & Chu 1983; Wikfors *et al.* 1992). Hatcheries sterilize large volumes of seawater using heat, chlorine, or microfiltration so they can reliably produce the large volumes of algae needed to commercially produce shellfish. These can be produced in large batches or semi-continuously by harvesting a percentage and replenishing the sterile media daily.

Modern hatcheries typically will try to have several algal species in culture so they can offer a mixed diet and have different cell sizes and nutritional profiles available to meet the dietary needs of different life stages of larvae, post-set and broodstock. More often than not, the key to good quality egg production and good larval survival is related to the quality, variety, and volume of algal food provided.

The economics of microalgae production remains one of the roadblocks to the development of land-based culture systems for filter feeders. Microalgal production costs can be as high as US$200 per dry kilo because of the costs of labor and the expense of sterilizing seawater (Persoone & Claus 1980). More recently, systems have been developed to automate continuous-flow systems (SeaCAPS, UK), and microalgae can be purchased in paste form (Instant Algae, Reed Mariculture, Campbell, CA). Efforts to supplement or replace microalgae with less expensive microparticulate foods such as yeasts or rice starch have not been very successful (Epifanio 1981). The nutritional profiles and hydrodynamic properties of the particles are critical to their acceptance and utilization by filter feeders (Webb & Chu 1983; Wikfors *et al.* 1992). The food demands of shellfish increase exponentially as they grow in length, so it becomes imperative to start feeding the spat with natural algae as early as possible.

5.8   Spawning

Bivalve spawning is usually simple to induce if the brood stock are properly conditioned and fully ripe. Most species can be induced to ripen by simulating late spring water temperatures and providing ample algal food for four to six weeks. Many bivalve species are hermaphroditic, producing both eggs and sperm, but typically they do not release both at the same time because survival of self-fertilized eggs is poor. Many bivalve species are protandric hermaphrodites, starting life as a male, but developing ovaries as they get larger; however, in lean years the sex ratio is likely to favor males because it takes more energy to produce eggs than sperm.

Depending on the species and size of the individual, it is common for a ripe female to produce 3 to 30 million eggs in a single spawn. In subtropical environments they can do this several times a year. As much as 30% of the body weight of bivalves is gonad, and spawning is so energetically costly that it leaves the adults thin, weak, and prone to mortalities.

Many temperate populations only spawn once a year and have evolved strategies to synchronize the release of gametes to optimize their breeding success. While several species of bivalves remain active in winter temperatures (e.g., *Mytilus*), some are no longer able to feed when water temperatures drop below 7 to 10°C (e.g., *Crassostrea* and *Mercenaria*). Their digestive enzymes no longer function and filter-feeding ceases for several months a year (Brock *et al.* 1986). Their metabolism slows and they only need to open their valves briefly for respiration. When waters warm in the spring most of their energy is devoted to building up their gonad, and following a string of warm days the entire population will spawn at once.

Most shellfish release their eggs and sperm into the water, where they mix and fertilize. There are a few species of "brooders" (*Ostrea* sp.) that retain the larvae in the mantle cavity, allowing the larvae to swarm around the gill for the first weeks of development (Chaparo *et al.* 1993). Most temperate species can be induced to release eggs and sperm with thermal shock by raising and lowering the water temperature by 5 to 10°C. If this fails, spawning can often be induced chemically by adding a few drops of sperm to the water. Spawning has also been induced with injections of neurotransmitters such as serotonin. Sperm and eggs can also be obtained by "strip spawning," which involves macerating the gonad with a razor blade and filtering out the chunks of tissue. Typically, naturally released eggs will have much better survival rates than those obtained by these more drastic measures.

When fertilizing batches of eggs, hatchery operators need to exercise care. Shortly after a single sperm penetrates the egg, a change on the cell surface makes it refractory to additional sperm; however, in certain species of shellfish this transformation takes longer than in higher organisms. If sperm cells are too concentrated, several may penetrate a single egg at once (polyspermy). When this happens, abnormal development and mortality can be expected. Proper dilution and judicious additions of sperm are required to ensure that a good percentage of the eggs are fertilized, but not too many are fertilized more than once (Bricelj 1979).

5.9   Larval development

Fertilization takes place quickly, and at proper temperatures cell division begins in under an hour. In 12–18 hours the embryos are called trochophore larvae, which appear to be ciliated balls, swimming in lazy spirals. After 24 hours "D-stage" larvae are visible. These larvae look like a microscopic clam, starting out at about 60 to 110 μm in diameter, developing in two to three weeks from veligers to pediveligers (fig. 5.2), and increasing to 150 to 250 μm or larger at the time of settlement.

For the first few weeks of their lives, most bivalve larvae are free-swimming and planktonic. The shell is transparent, revealing the heart pumping and algal cells being processed in the gut. Veliger larvae swim using the vellum, which is a ciliated organ that also captures algal cells. Veligers feed on single-celled algae, typically flagellates and diatoms of 1 to 3 μm in size. Algal concentrations need to be maintained at around 10,000 cells/ml, as veligers will starve if densities are too low and will not thrive if concentrations are too high.

Larvae are typically reared in batches of 1,000 to 100,000 liters or more at densities from 5 to 25 per ml. Operators usually drain the tanks through a very fine-mesh sieve every other day and refill the tank with filtered seawater and algae. More recent experiments have shown that larvae can be grown in much higher densities in flow-through systems by placing a fine filter over the drain. Gentle aeration is provided to stir the tank, otherwise the larvae tend to swarm together. Growers generally try to produce seed in early spring so they can be moved out of the hatchery in early summer to take advantage of the entire growing season. This means incoming water needs to be heated to maintain good growth.

5.10   Setting

After two to three weeks, the larvae develop a foot and start investigating the bottom of the vessel in preparation for settlement. Hatchery operators will present these pediveliger larvae with an appropriate substrate (e.g., for oysters this is usually crushed shell). Clams and scallops will often set on the bottom or sides of the vessel. Once the larvae are no longer swimming in the water column, they need to be spread out horizontally so they are not overcrowded. Setting trays are the best way to help the larvae survive the metamorphosis from free-swimming larvae to sessile spat. Setting trays consist of a fine-mesh screen that is stretched taut and floated or suspended in a large tank. Water (and algae) is gently pumped into the top of the tray and it flows out through the screen. To keep oysters from setting on the sides of the setting trays, growers will often coat the sides with a thin layer of wax or Vaseline.

Post-set mussels, scallops, and clams will start to produce byssal threads and they will glue themselves to the tank, each other, or almost any other firm substrates offered by the grower. They can release the byssal thread if they decide they need to move. Negatively geotactic scallops will sometimes climb up to the water surface where they need to be gently brushed back into the water. These "escape" attempts can be prevented by offering scallops an eelgrass mimic such as weighted tufts of nylon, polypropylene, or burlap fiber. Post-set clams will attach their threads to each other, forming a mat that must periodically be sieved gently to break up clumps so the clams don't smother each other.

Post-set shellfish (referred to as "spat") are highly susceptible to bacterial and protozoan infections, predators, and fouling organisms. Their shells are fragile and they need to be handled delicately. Given proper food and growing conditions they will grow rapidly, but getting good larval survival through metamorphosis and the early nursery stages can be difficult. Young shellfish are ravenous feeders whose food demands grow exponentially as they increase in size. Feeding millions of spat once they get to be a few millimeters in length requires hundreds of liters of microalgae. The food needs to be the right species, harvested in good condition, and offered at the right concentrations ($\sim$25,000 cells/ml).

As soon as possible, hatchery operators try to move away from static systems (draining the tank and changing water every other day) to continuous flow systems in an attempt to flush away waste and maintain good water quality. Unfortunately, this brings a new set of problems as precious food gets flushed down the drain and large volumes of water must be heated and filtered. Some of the cultured algal food can be supplemented with natural food by pumping raw seawater, but in early spring ambient phytoplankton concentrations are still typically low and the water needs to be heated to maintain good growth. Also, raw seawater needs to be coarsely filtered to remove potential predators,

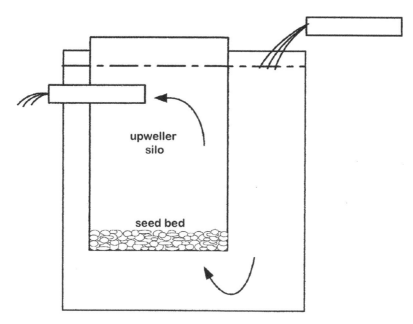

upweller
silo

seed bed

Figure 5.4 Diagram of an upweller. Figure courtesy of Robert Rheaulf.

competitors, and fouling organisms, as well as detritus and algal cells that are too large to eat. Hatchery operators often use 5 μm or 10 μm bag filters, which require regular cleaning.

The most common way to rear post-set juveniles is to use upwellers and downwellers. A screen fine enough to retain the spat is stretched taut and floated or suspended in a large tank while water is drawn upward or downward through the screen (fig. 5.4). Each screen-bottom vessel is referred to as a "silo" because it is often fashioned out of PVC pipe, but square silos are also common. Smaller seed are usually grown in downwellers because the seed are so light that the upflow current in upwellers can carry them up and out the drain. Once seed grow over 1 mm, they have enough mass to be placed in upwellers, but it is still prudent to place a screen over the drain. In upwellers the water current carries away feces and fluidizes the bed of seed with a gentle tumbling motion. As seed grow larger they can be packed into upwellers in greater volumes in a bed several centimeters thick. As the seed grow, the flow rate to each silo needs to be increased to provide adequate food and to keep the seed bed fluidized.

Screens on the wellers need to be regularly cleaned and checked for damage. The finer mesh screens on downwellers are especially delicate, prone to fouling, and will clog quickly with silt and feces. A small imperfection in a screen can allow thousands of spat to escape. As the animals grow they can be moved onto larger mesh screens, which are sturdier and less prone to fouling.

One approach to oyster culture is called "remote setting." Late-stage larvae are placed in large tanks that have been filled with bags of shell. After a few days, several larvae have set on each shell and the bags are moved into field nurseries. After a few weeks, the spat are well established and the shell is spread on the growing grounds. Since several spat are growing on each shell, they often grow into each other, forming large, irregular clumps. Those that grow into irregular clumps are acceptable for shucking, but few would be considered suitable for the more lucrative half-shell raw-bar trade, where single, uniform, cupped oysters are more desirable. Hatchery operators have learned that single oysters can be produced by presenting oyster larvae with "micro-cultch," tiny chips of ground-up shell about 150 μm in diameter, only big enough to allow one or two larvae to attach to each fragment.

5.11   Nursery and growout scale considerations

To better understand the economic factors driving the various growout options, it is useful to consider the scale of operations. Data for the following example are gleaned from a New England oyster hatchery and growout operation (Rheault & Rice 1995). This is simply an example; growth rates will vary greatly depending on temperature, flow rate, stocking density, food quality, and shellfish species.

One million larvae can be grown in as little as 100 liters of water and subsist on a few liters of cultured algae a day. One million, 1-mm seed have a total volume of about two liters. At this size they can be held in a few small upwellers and fed with a few hundred liters of algae a day. In two weeks the same crop can grow to an average size of 3 mm, with a packed volume of about 12 liters, while food demands can exceed 1,000 liters of cultured algae a day. At this point (or earlier) it becomes imperative to wean the animals from cultured algae and offer them raw seawater so they can feed on natural microalgae and other organic particles.

This grower moves his seed to floating upwellers in early summer at a size of 1 mm. Each upweller silo is 60 cm square (360 cm^2), and food-rich water is pumped up at 200 to 400 Lpm. At this size, the oyster seed line the bottom of each silo in a layer that is 30-cm deep, tripling in volume each week. Over the next three weeks the oysters grow quickly in size and volume, growing to an average of 10 mm with a total crop volume of about 500 liters, overflowing 10 upweller silos. During this time, the seed need to be stirred daily, restocked every five to seven days, and graded by size every other week.

In the following three weeks growth slows somewhat, mean size reaches about 15 mm, and the total crop volume increases to about 3,600 L. This grower uses a modified rack-and-bag growout system with bags measuring 0.6 m by 0.6 m. Over this period the largest oysters are stocked into over 1,000 growout bags at two liters per bag, while the slower growers are retained in upwellers.

By the end of the first growing season the mean size approaches 35 mm, and the entire crop fills over 4,000 growout bags stocked at four liters per bag. Each bag is constantly getting fouled and needs to be cleaned and divided every one to two months. This grower field-plants his oysters on the bottom the following summer, freeing the bags that are needed for next year's crop. In the fall, he starts to harvest about one ton each week of eighteen-month-old, market-size animals (>75 mm). Harvest continues throughout the year, but feeding and growth cease once water temperatures dip below 10°C (mid-November to May at this location). Slower growing animals may take another season to reach market size.

Production-scale intensive shellfish culture is essentially a massive materials-handling exercise, necessitating the handling of tons of live animals on a regular basis. Where labor costs are high, automation and mechanization are essential to keep operating costs down.

5.12   Nursery methods

The nursery phase of shellfish culture is arguably the most challenging. There is no best time to move seed from the hatchery to the nursery, and different growers will do this at different sizes. Hatcheries typically set the larvae inside and grow the spat up to several millimeters before they are moved out to nurseries. Because filtering seawater and feeding cultured algae is expensive, hatcheries try to shift over to raw seawater as soon as possible. If ambient temperatures allow it, the spat can be switched to raw seawater once they reach a few millimeters. This can be done on land in upwellers or raceways, or in the water in various types of containers.

5.12.1   Nursery culture: static systems, pearl nets, and rack-and-bag systems

The simplest approach to growing shellfish seed is to disperse them in the wild, but mortality rates for small, unprotected shellfish usually approach 100%. To protect the small spat, they are usually held in fine-mesh containers and suspended where natural water currents will keep them well fed. The Japanese were probably the first to attempt this at any significant scale using pearl nets and lantern nets. A pearl net is a pyramid-shaped container with a flat bottom and sloped walls made of a fine-mesh material. Spat are introduced through a slit on one side, the slit is sewn shut, and the pearl nets are suspended in food-rich waters until the spat are large enough to be moved to a larger-mesh container. Suspended and strung together in long chains, pearl nets look somewhat like a string of pearls. Many growers still use pearl nets to this day, especially for early life stages of scallops, but most growers have moved to other systems because pearl nets are labor intensive to load, sew shut, and later unload.

The next step after pearl nets is often the Japanese lantern net, so-called because it resembles a paper lantern (fig. 5.5). Lantern nets hold several tiers of shellfish on horizontal shelves up to 0.5 m in diameter, all protected by a cylinder of netting. Similar to pearl nets, shellfish are stocked into lantern nets through a slit on the side that is sewn shut. Novel lantern net designs use trap doors or Velcro to eliminate the labor-intensive sewing step.

Suspending culture systems from the surface from rafts or floats (typically on anchored long-lines) allows growers to utilize areas where the bottom is

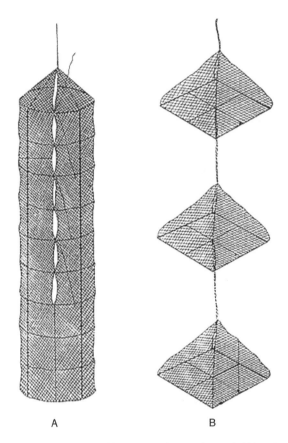

A   B

Figure 5.5 Lantern net. Photo courtesy of http://www.fao.org/docrep/field/003/AB739E/
AB739E03.htm

not suitable or where benthic predators are too numerous. Surface waters are
also usually warmer, with stronger currents and more plentiful microalgae—all
conditions conducive to faster growth. Another advantage is that by utilizing the
vertical dimension of the water column, growers can hold much greater numbers
of animals per square meter of bottom.

A number of contraptions have been developed to hold and grow shellfish
seed. Early efforts used wood, natural-fiber netting, or galvanized wire, but
with the invention of synthetic fibers, plastic netting, PVC, and injection-molded
plastics came a wide array of novel designs. Trays, pillow-shaped bags, and
many other containers have been tried. Most of these can be lumped into the
category of modified "rack-and-bag" systems. Bags can be floated at the surface,
suspended from rafts, placed directly on the bottom or held off the bottom by
trestles, cages, or racks (fig. 5.6). The mesh of the container needs to be fine
enough to retain the seed, yet sturdy enough to resist the attack of crabs. The
bottom of the container should be held flat so the seed don't all slide down into a
pile. This would negate the advantage of having a container with a large surface
area since the animals in the middle of the pile might starve or suffocate.

Figure 5.6 Bags on trestles. Photo courtesy of Robert Rheault.

Each of these nursery culture methods requires regular maintenance. Fine-mesh containers needed for smaller seed need the most frequent attention because the fine mesh will clog quickly with silt and fouling organisms, restricting the flow of water and food. As the spat grow they continually need to be thinned so that competition for food doesn't limit growth. Periodic sieving and grading is recommended because larger animals will tend to consume most of the food, and small seed will never grow to their full potential unless they are held separately.

Small seed can double or triple (or more) in volume each week, so the increase in total crop volume is geometric. As the seed get larger, their metabolism and growth rate slows so the stocking density (measured by biomass) can be increased, while the number of individuals per unit area drops. Larger-mesh containers have more open area so it takes longer for fouling organisms to colonize and occlude the openings; therefore the time between cleaning and restocking can be extended. But with every restocking, the number of containers needed to hold the crop increases geometrically.

5.12.2 Nursery culture: pumped systems, upwellers, and raceways

Raceways are a popular method of nursery culture. These are long, gently sloped, shallow tanks with water pumped into one end and flowing out at the other. Raceways need to be drained regularly to wash out any accumulated silt and feces. The stocking density is limited by total surface area of the tank, as the animals must be grown in a single layer. Corrugated plastic can be used in raceways to grow scallops or abalone, increasing the surface area for attachment.

In raceways, the animals at the front end will grow faster because they get more food than the animals near the drain.

Upwellers are one of the most effective nursery culture methods, but as the crop expands in volume, the silos need to be substantially larger and flow rates increased to ensure that the spat receive adequate food. The flow demands of a specific system will vary depending on the ambient food concentration, the species, and the biomass of the spat. Pumping costs quickly become a major economic concern.

The energy required to pump water is directly proportional to the "head pressure," which is determined by the length and diameter of the pipe and the vertical height that the water must be lifted. To reduce pumping costs growers move their nursery systems as close to the water as possible. Floating the upweller in the water reduces the head pressure to near zero, cutting pumping costs dramatically. Floating upwellers (FLUPSYs) have the additional advantage of not needing any expensive waterfront property.

5.13   Growout methods

As mentioned earlier, there is no clear division between nursery culture and growout. In theory, all of the same methods used for nursery culture of seed can be used to grow shellfish all the way to market size. From a practical standpoint, however, most of these systems become too expensive as the total crop volume expands.

Generally, growers try to limit the use of gear-intensive and labor-intensive culture methods to the nursery phase; once the seed reaches a size where it is resistant to most predators, seed can be broadcast on the bottom for growout to market size. There is theoretically a "size of refuge" that needs to be determined experimentally in each location by free-planting a range of seed sizes and monitoring survival. Larger seed will have a higher survival rate, but in most areas there is no size that guarantees shellfish will be predator proof. Each grower needs to determine at what size predation losses decline to acceptable levels. This is often referred to as "field-plantable" size, and it varies with species and with the predator assemblage at each location.

5.13.1   In-bottom or on-bottom growout

The simplest growout method involves broadcasting the seed on the bottom and letting it grow to market size. For clams and oysters, planting densities can vary from 300 to 1,500 per m^2, but optimal densities need to be determined for any given site. In cases where seed are inexpensive and mortality rates acceptable, this approach is economical and can be highly productive; however, most sites will suffer staggering predation mortality.

Intertidal clam farmers have learned that most predators can be thwarted with light plastic bird-netting rolled out over the seed at low tide and weighed down around the periphery. The netting comes in rolls that are 4.5-m wide with mesh sizes of 6 mm and 12 mm being the most commonly used. Maintenance consists of periodic inspections for damage to the net and removal of fouling during the two- to three-year growout period.

5.13.2   Rack and bag (subtidal or intertidal)

Various rack-and-bag systems are commonly used for both nursery culture and growout, with the only real distinction being that larger shellfish can be grown in larger-mesh bags. Again, these bags can be placed on the bottom or held off the bottom on trestles or in cages and racks; they can be used in subtidal or intertidal locations. Anything the grower can do to improve the flow of water through the mesh will increase the seston flux, resulting in faster growth, lower stress, and better condition index. Larger mesh takes longer to become fouled and will have less resistance to flow, so growers typically use the largest size mesh that will retain the seed and protect it from predators. Water will take the path of least resistance, with most of it tending to flow around the bag as opposed to through it. Shellfish held in a clean, 12-mm mesh Vexar (Conwed Plastics, LLC, Minneapolis, MN) bag will experience a 90% reduction in flow compared to animals just outside the bag (Rheault unpub. data).

Since current speed drops to zero right at the bottom, most growers try to keep their bags at least a few centimeters off the sediment. This allows better water circulation and keeps feces and silt from accumulating in the bags.

5.13.3   Suspended gear (fixed or floating): Taylor floats and lantern nets

As stated earlier, moving the gear higher up in the water column usually ensures better current flow, higher phytoplankton concentrations, and warmer water—all conducive to faster growth. Many growers use trestles or tables to hold their bags off the bottom in the intertidal zone where they are exposed for part of the tidal cycle. Since animals suspended from floats or rafts are able to feed continuously, floating gear is perhaps the best of all worlds for shellfish. Suspended gear allows growers to use areas where benthic conditions might not be suitable for bottom culture while keeping shellfish high in the water column and allowing them to feed continuously.

Growers have had success putting floats inside bags, attaching floats to the sides of bags, and fabricating all manner of floating rafts. Using huge rafts to hang long ropes of mussels has proven to be a successful approach for hundreds of years. The Japanese use rafts and longlines to grow scallops using a

Figure 5.7 Taylor float. Photo courtesy of Daniel J. Grosse.

technique called "ear hanging." By drilling a small hole in the shell of the scallop near the hinge (the ear), each scallop can be individually fastened to a larger rope. Recently, growers have developed machines to mechanize this otherwise laborious process.

A float design that has gained popularity in the mid-Atlantic region is the Taylor float (fig. 5.7). Constructed out of a ring of PVC pipe to provide floatation with a wire basket hung below, Taylor floats come in a variety of shapes and sizes and can hold several Vexar bags.

Floating gear can be susceptible to storm damage and is best suited for sheltered waters, which can pose permitting issues in areas where aesthetic considerations or navigation are important. In deep water another option is to hang gear near the surface from horizontal longlines. This is the preferred technique for growing mussels and scallops. State-of-the-art mussel farms have developed paired-backbone longlines, 100 to 200 m long, with large buoys capable of supporting a tonne or more, with 8-tonne anchors at either end (fig. 5.8). Mussel seed are injected into tubular mesh called "socks," which are suspended at regular intervals from the backbone lines extending 6 to 15 m below the surface. A typical New Zealand longline is about 110 meters long, holds around 3,000 meters of culture sock and produces around 25 tonnes of mussels in an 18-month rotation.

Scallop farmers use similar longlines, but instead of mussel socks they hang lantern nets. As the animals grow and the weight of fouling organisms on the

Figure 5.8 Mussel longlines. Photo courtesy of Richard Langan.

gear increases, the growers must either remove the fouling or add additional floatation to keep the whole operation from sinking. As scallops are sensitive to excessive movement, growers in sites exposed to excessive wave action will suspend the backbone lines below the surface so that the animals are insulated from all but the heaviest wave action.

More recently, growers have constructed longlines in intertidal waters by driving posts into the sediment and stretching cables between the posts. Injection-molded plastic baskets called Aquapurses (Tooltech PTY, Australia) clip on the cable and open like a clamshell to access the shellfish. The cables can be raised or lowered to adjust the time of air exposure needed to control fouling. Some species of oysters become adapted to a subtidal existence and when they are harvested they tend to gape; and once they become dehydrated and the gill dries out the animals soon die. By gradually raising the level where these oysters are grown, farmers are able to condition oysters to increasing periods of air exposure, which allows them to strengthen their adductor muscles so they maintain a tight seal while they are being shipped to consumers.

5.14   Fouling

Biofouling is a huge problem for shellfish growers because fouling organisms restrict the flow of water to the crop and often compete directly for food (Lesser *et al.* 1992). Severe fouling will reduce the condition index of the crop and can lead to mortalities. A recent national survey revealed that that biofouling control efforts can account for up to 20% of annual operating costs (gear-intensive floating or suspended-culture systems reported higher costs; Adams *et al.* 2011). In addition, encrusting organisms such as barnacles can detract from the visual appeal of the product, reducing the market value.

The principle fouling organisms include macroalgae, tunicates, sponges, barnacles, and other shellfish. Macroalgae can block the flow of water, but most are relatively easy to control and do not compete for food. In contrast, the many species of colonial and solitary tunicates (sea squirts) can pose significant challenges for growers. Some notable solitary tunicate species include the Asian stalked tunicate, *Styela clava*; the sea grape, *Molgula manhattensi*; the sea squirts, *Ascidiella aspersa* and *Ciona intestinalis*. Many of these are non-native, invasive species, thriving in new environments where there are few natural predators. They are all highly fecund, opportunistic filter-feeders with astounding growth rates, capable of coating the surfaces of any gear put into the water in a matter of weeks. Besides blocking flow and competing for food, heavy growths of tunicates can cause floating gear to sink and force growers to redesign handling gear to bear the added weight.

Barnacles (*Balanus* sp.) are another particularly nettlesome fouling organism and are difficult to control without killing the shellfish that they set on. In certain locales, oysters themselves can be fouling organisms, even when they are also the species in culture. Mid-Atlantic growers find that juvenile oyster overset on both their market-size adults and their culture gear interferes with growth and condition index, lowering the market value of the crop. Similarly, a heavy overset of mussels on culture gear can be devastating to the crop causing the producer to incur crippling removal costs.

Another fouling organism, the yellow boring sponge, *Cliona* spp., deserves special mention. In addition to competing for food and restricting the flow of water, this organism will encrust oysters and bore small holes into the shell by secreting acid. The labyrinth of holes weakens the shell, making the oysters more susceptible to predation. The oyster tries to secrete additional shell to protect itself, taking precious energy away from somatic growth. More importantly the holes reduce the market value of the oysters because the shell crumbles when shucked, leaving a nasty mess of shell fragments and yellow sponge in the meat.

5.15   Fouling control strategies

There are three basic approaches to controlling fouling: prevention, physical removal, and slaughter. Prevention can be done by physically avoiding the larvae of the fouling organisms, providing one can predict when and where sets of these organisms occur, but this is rarely practical. More commonly growers have tried various nontoxic antifouling coatings that discourage the settlement of invertebrate larvae. Removal techniques include pressure washing with high-pressure water, physically brushing, scraping, or scrubbing off the fouling, all of which are costly and labor intensive.

There are various approaches to killing fouling organisms on the crop and on the gear. Killing the fouling community without having to remove the shellfish from the gear obviously saves a lot of effort, but killing the fouling organisms without killing the crop can be a challenge. Air-drying is one of the most common approaches, especially for oyster growers who work in intertidal sites. By adjusting the height of the gear, growers can find the right exposure time that kills most of the fouling, but not the oysters. Growers who use intertidal rack-and-bag techniques usually flip the bags periodically, letting the sun do the work of killing the soft-bodied invertebrates. To kill tougher fouling organisms such as barnacles, oyster overset, or mussels it is usually necessary to restock the crop in clean gear and allow the fouled gear to dry out for several days.

Dipping oysters in their containers into a saturated brine solution for several minutes is another effective method of killing macroalgae and soft-bodied invertebrates, but to control boring sponge or mud blister worms (*Polydora websterii*) the dip needs to be followed by an extended period of air-drying to further concentrate the salt (MacKenzie & Shearer 1959). This method can also be effective for controlling very small oyster overset and barnacles, but once these fouling organisms reach a few millimeters in size they will tolerate the treatment (Debrosse & Allen 1993). Brine dipping and air-drying cannot be used to control fouling on scallops because their valves do not form a watertight seal.

Spraying with, or dipping in, acetic acid, hydrated lime, or chlorine can also be an effective control method for tunicates (MacKenzie 1977; Forrest *et al.* 2007), but the application of harsher chemicals may not be suitable for use on food or in the marine environment. Some growers have experimented with various means of biological control, reporting some success with placing a few killifish (*Fundulus heteroclitus*) or periwinkles (*Littorina littorea*) inside each tray or bag so they can graze on the fouling.

5.16 Predation

It seems as if almost everything likes to eat shellfish, and smaller seed are especially highly vulnerable. Planting unprotected, 1-mm seed is like ringing a dinner bell. Many species of crabs are quickly attracted to such plantings and won't leave until they reduce densities to background levels (Arnold 1984). As the seed get larger they will sometimes be able to resist attack by small crabs, or at the very least it will take the crabs longer to eat the crop. Some predators such as starfish (*Asterias* sp.) or the oyster drill (*Urosalpinx cinerea*) can attack adult shellfish several times their own size. Some growing areas are plagued by large predatory snails such as the whelks *Busycotypus* spp. (formerly *Busycon*), moon snails *Neverita duplicatus* and *Euspira heros*, or the cow nose ray (*Rhinoptera bonasus*), all of which can take market-size shellfish.

Even protecting the shellfish behind protective netting cannot guarantee that the predators won't take their share. Free-swimming larvae of some snails, crabs, and the oyster flatworm, *Stylochus ellipticus,* may settle inside the containers, sometimes growing fast enough to consume most of the contents. Starfish can evert their stomach, pushing it through 12-mm mesh Vexar bags to externally digest the shellfish in the bag while remaining on the outside. It may take several days to consume an oyster this way, but over a period of months the losses can be catastrophic. Growers need to be vigilant, learning about the various predators that afflict their particular growing area and developing appropriate strategies to combat each foe.

Mussel farmers can suffer severe losses from flocks of diving ducks such as the Scoter (*Mellanitta nigra*) or Eider duck (*Somateria mollissima*). These ducks, along with waders like the Oystercatcher (*Haematopus ostralegus*), can wreak havoc on unprotected clam and oyster seed.

Predator control strategies may include sweeping the ground clean with a hydraulic dredge before planting, placing baited traps (effective for starfish, whelks, and crabs), or mopping (dragging weighted cloth mops over the grounds to entangle starfish). Growers have also experimented with barrier fencing to exclude cow nose rays or whelks, and enclosing mussel rafts with netting to exclude diving ducks.

But the most pernicious and cunning predators of all are humans. Shellfish poachers are inventive and persistent, and they often cause tremendous damage as they "help" growers harvest their crops. Trapping and eliminating human predators can be expensive and time consuming.

5.17   Harvest

Shellfish farmers have several advantages over wild fishermen when it comes to harvesting. The farmers know where to find their crops and the densities are usually much greater than for wild stocks. Farmers are not typically constrained by harvest regulations designed to preserve wild stocks such as seasonal restrictions, bag limits, quotas, possession limits, harvest methods, or minimum size. While wild-harvest fishermen often resent this double standard, it makes little sense to enforce these restrictions on farmers who plant and own their crops. The harvest methods of a culturist rarely have a negative impact on wild populations. Sometimes enforcement authorities will insist on maintaining some restrictions, such as minimum size, because they cannot differentiate cultured product from wild and they believe relaxing the restriction for one group would invite unscrupulous wild harvesters to apply for aquaculture leases, simply so they can land sub-legal wild shellfish.

Shellfish farmers are also free from the "race to fish" tendency of quota-managed fisheries (FAO 1997). When fishermen are managed by a quota that they do not own, they are highly motivated to fish as fast as they can until the quota is filled and the fishery is closed. This results in short seasons and huge fluctuations in supply, which, in turn, causes huge fluctuations in price. In contrast, farmers are able to modulate their harvest to meet demand or to optimize price, quality, or yield.

Since shellfish are often sold live and fresh for raw consumption and since they are highly perishable, farmers have the added advantage of being able to harvest only what they need to meet immediate demand. They can effectively stockpile product in the water under optimal conditions, delivering product to the market that has been harvested that day instead of stockpiling inventory in coolers.

Shellfish farmers who use container culture are able to simply pull up their gear, select the market-sized individuals, and return the smaller ones to the water for additional growth. This means growers are able to provide a more uniform cull than wild harvesters. Growers who have opted for bottom planting typically use traditional harvest methods such as rakes, dredges, or hand-picking at low tide. This description makes it sound easy, but once farms develop to a significant scale, the work becomes an exercise in materials handling on an epic scale. Tonnes of shellfish must be handled on a weekly basis, taking care to keep the animals alive, undamaged and wholesome. Harvesting efforts can easily comprise 30% of operating costs. Mechanization can be critical to success, especially where labor costs are high. New Zealand mussel farmers have developed vessels and equipment capable of harvesting more than twenty tonnes in a single day with a crew of three. Clam farmers in the Pacific Northwest and Europe have modified small tractors designed to harvest tulip bulbs to cut their harvest costs by more than 80% (W. Dewey pers. comm.).

5.18   Food safety

Since shellfish are filter-feeders that can accumulate pathogenic viruses and bacteria from the water, and since they are often destined for raw consumption, there is a greater potential for food-borne illness outbreaks with shellfish than with many other foods. In the United States the Food and Drug Administration's National Shellfish Sanitation Program (NSSP) has established stringent regulations to minimize the risks associated with raw molluscan shellfish consumption (FDA 2009). The NSSP sets forth guidelines for the bacteriological monitoring and designation of harvest areas; depuration of product from uncertified waters; and the processing, handling, and transportation of molluscan shellfish. Most other countries have shellfish sanitation programs that are equivalent or similar to the NSSP.

Once the shellfish is harvested it is important to ensure that it is protected from temperature extremes and placed into refrigeration as soon as possible. At elevated temperatures bacteria can multiply rapidly, going from safe concentrations to unsafe levels in a matter of hours. However, freezing or even cooling shellfish too quickly can kill shellfish, and when they thaw they will gape, dry out, and begin to decay. In the United States all processing of shellfish must be done in accordance with a Hazard Analysis and Critical Control Point (HACCP) plan designed to identify and eliminate potential sources of contamination and protect the product from temperature abuse (FDA 2005).

5.19   Shellfish diseases

Historical records are replete with descriptions of mass mortalities and shellfish epizootics afflicting practically every species of shellfish, but none of these shellfish diseases are known to affect humans. Bivalves are susceptible to cancers (neoplasia), bacterial diseases (vibriosis), and viral diseases (herpes virus), as well as a wide variety of protozoan parasites that can devastate the crop and wild populations. To list all of the diseases, symptoms, and causative agents is beyond the scope of this chapter; however, Bower and McGladdery (1997) catalog over 150 diseases of commercially exploited mollusks.

It is important to differentiate between infection and disease. Many organisms may carry an infection without ever showing the clinical signs of disease. Some protozoan infections may not cause significant problems for years. Whether an infection becomes a disease is determined by the virulence of the pathogen, the susceptibility of the host, and other environmental factors.

While some diseases are found in the hatchery or may only affect young seed, the most devastating epizootics seem to be associated with protozoan parasites. Low-level infections may persist for years without ever causing significant mortalities, but under the right conditions they can proliferate and wipe out a crop in a matter of months. Pathological examination often reveals a number of parasites that are apparently living off the shellfish benignly, causing little harm to the host. There are, however a few notable parasites that have killed billions of shellfish, both wild and cultured.

There are several examples where a native oyster population has been driven to near extinction by disease. The native European flat oyster (*Ostrea edulis*) was all but eliminated by the protozoan *Bonamia ostraea*, forcing the European industry to import *Crassostrea gigas*, which is now the dominant oyster produced in Europe. In the mid-Atlantic region of the United States the parasites MSX (*Haplosporidium nelsoni*) and Dermo (*Perkinsus marinus*) have reduced the populations of the native Eeastern oyster to less than 1% of historical levels, causing regulators to consider listing the species as threatened.

5.20   Disease management options

Many epizootic outbreaks can be traced to the early practices involving the indiscriminate movement of shellfish between growing areas. Once this was recognized, growers and regulators backed the development of regulations mandating the pathological inspection of seed stock before it can be transported across state lines, and quarantine procedures to govern international transport of broodstock. Strict adherence to these regulations is the best way to minimize the risk of introducing diseases to new areas and is perhaps the best defense against new epizootic disease outbreaks.

Many diseases are already endemic in many growing areas around the world, and growers do what they can to manage around them. Diseases are often exacerbated by stressors such as overcrowding, pollution, poor nutrition, hypoxia, and temperature shock. In some cases mortalities can be reduced by thinning the crop, increasing flow rates, or simply by maintaining good hygiene. With progressive diseases growers can sometimes harvest before mortalities become too severe. Some parasites are intolerant of low or high salinities and mortalities can be avoided by moving the crop to appropriate waters.

Because shellfish have primitive immune systems it is unlikely that efforts to develop vaccinations to stimulate immunity will ever be successful. Also, once the animals are moved out of the hatchery it becomes impractical to deliver any type of drug or antibiotic. It appears that the best hope of improving resistance to disease lies in a long-term, coordinated, selective breeding program. The industry is working to develop lines that are not only disease resistant, but also fast growing and adapted to the very different environments across the species' entire range.

5.21 Genetics: selective breeding

The principal function of the hatchery is to allow growers to control the life cycle of an animal, from egg to adult, and back to egg again. This process is referred to as domestication. Many agricultural species have been domesticated for thousands of years, while others such as shellfish, for mere decades. Closing the life cycle in hatcheries invariably results in some measure of change in the genetic structure of the population because the process selects for animals that do well in hatchery conditions. Most hatcheries also select broodstock for traits such as fast growth, better meat yield, specialized coloration, disease resistance, and shell shape. Many culturists have worked for several generations to develop "strains" adapted to the conditions unique to their particular growout site.

By agricultural standards, professional shellfish breeding programs are in their infancy. The longest running professional breeding program is located at the Haskin Shellfish Research Lab, where attempts to select for resistance to MSX disease in *C. virginica* were begun by Dr. Hal Haskin in the late 1950s. There are now several MSX-resistant oyster strains being developed that are the living legacy of his work. In the last 20 years, breeding programs for the Pacific oyster, *C. gigas,* have also been established in the Pacific Northwest, Australia, and France, focusing on growth characteristics and resistance to summer mortality.

Sometimes selective breeding can have unintended consequences. In the process of selecting for a trait such as fast growth, hatcheries run the risk of narrowing the genetic diversity of the population and potentially eliminating some desirable trait such as resistance to disease. To avoid genetic bottlenecks some hatchery operators try to maintain genetic diversity by performing periodic out-crosses with wild populations and then performing back-crosses among the progeny to ensure that the desirable traits are not diluted.

5.22   Triploidy

In the 1970s researchers at the University of Maine demonstrated that shellfish could be induced to retain an extra set of chromosomes (Stanley *et al.* 1981). The vast majority of sexually reproducing organisms are referred to as "diploid" because they hold two sets of chromosomes, one inherited from the father's sperm and another from the mother's egg. For example, each human cell has two sets of 23 chromosomes for a diploid chromosome number of 46. Oysters have only 10 chromosomes and have a diploid chromosome number of 20. Triploid oyster cells have three sets, or 30 chromosomes. Only plants and lower animals, such as amphibians, fish, and mollusks, can tolerate the triploid condition.

Triploid animals are reproductively sterile, which prevents them from expending energy on forming eggs and sperm. Most bivalve species devote a considerable portion of their energy budget to the formation of gametes. Right before and right after spawning, shellfish can have an undesirable texture, being either fat and milky or thin and watery. Culturists find that triploids don't "waste" energy on reproduction so they grow faster, have improved marketability, and are generally more robust.

Today, only oysters are widely cultivated as triploids, since the process of making triploids has not yet been perfected for other shellfish taxa. Triploid oysters were first used commercially in the mid-1980s in the Pacific Northwest, but the process was not widely adopted because of difficulties scaling up the procedure for making triploids. This changed with the development of tetraploid oysters—those with four sets of chromosomes (Guo *et al.* 1996). For reasons not discussed here, tetraploids are fertile and produce sperm containing two sets of chromosomes, called di-haploid. When tetraploid sperm are used to fertilize normal (haploid) eggs, triploids result. Tetraploids have been used to make triploids in both *C. gigas* and *C. virginica,* and triploids now make up a substantial percentage of oyster production in the Pacific Northwest, France, and increasingly, the mid-Atlantic (Nell 2002).

Triploidy is also a valuable tool for culturists hoping to limit genetic interaction between cultured animals and wild populations. Sterile triploids were used in recent studies evaluating the ecological issues surrounding the proposed introduction of non-native *C. ariakensis* into Chesapeake Bay.

5.23   Harmful algal blooms

There are many species of phytoplankton that produce toxins and noxious substances. Often referred to generically as "red tides," a more appropriate term to encompass the wide range of species involved is "Harmful Algal Blooms" (HABs; reviewed by Shumway 1990). Shellfish will accumulate the algal toxins and when eaten by humans, the toxins can cause paralysis, amnesia, and a number of other maladies. HABs can also impact the shellfish themselves by interfering with filtration, or occasionally causing mass mortalities and set failures (Shumway *et al.* 2006; Hégaret *et al.* 2007).

Unfortunately, there is virtually nothing that can be done to prevent or disperse an algal bloom. At best, the toxin is detected and the harvest areas are closed before anyone gets ill. A temporary disruption in harvest and sales is easier to endure than a reputation sullied by illness (or death) associated with one's product. Blooms usually pass after a few weeks and the toxin in the tissue usually dissipates shortly thereafter, but in some species the toxins may persist for years (Bricelj & Shumway 1998; Hégaret *et al.* 2008). Unfortunately, the only management approaches available to growers are to select growing areas with infrequent HABs and to monitor for toxins in the tissues, since moving the crop to a different location is rarely a viable option.

5.24 Site selection

Perhaps the single most important decision facing a prospective grower is where to place the farm. As mentioned previously, the most important factors regulating growth are temperature and seston flux (the product of current speed and phytoplankton concentration). Within a region, temperatures are likely to be similar, but current speed and phytoplankton concentration can each vary tenfold over very short distances, resulting in huge differences in growth rates and in optimal stocking densities. Significant growth differences can sometimes be measured between sites that are 100 meters apart, usually because of physical features impacting the current profile. Rather than deploying expensive current meters and taking phytoplankton samples, perhaps the best way to evaluate and compare sites is to deploy small experimental cages for a growing season. Let the shellfish do the math of integrating the sinusoidal current speed function with the seasonal and diurnal phytoplankton data.

After seston flux, the next most important site characteristic is typically the predator assemblage. It makes little sense to broadcast shellfish seed in areas where they will await certain death. Conversely, why spend money on expensive gear if free-planted seed have good survival rates? Determining the predators specific to a site is a matter of trial and error with a few kilometers (or parts per thousand of salinity) potentially making huge differences in which predators are abundant, which diseases are a problem, and what survival rates can be expected.

Increasingly, hypoxia (low dissolved oxygen concentration) is becoming a concern in many coastal growing areas, especially those in shallow sheltered waters with excess nutrient inputs. The same phytoplankton that produces oxygen during the day and makes good shellfish food will consume oxygen in the dark, potentially depleting oxygen to dangerous levels. When algal cells die and sink to the bottom, bacterial decomposition will cause further oxygen depletion. Since oxygen solubility decreases in warmer waters, calm summer nights are most susceptible to hypoxic events. Most shellfish can tolerate a night of hypoxia, but some species are more susceptible than others, and repeated hypoxic events will lead to stress, impaired growth, and may cause mortalities.

Another consideration when choosing potential culture sites is the exposure to wind and waves. Sites with too much exposure will not be accessible for many days of the year and gear will need to be engineered to handle the forces associated with significant wave heights. Protected sites allow farmers to work with smaller boats, and can limit storm damage to gear and crops.

In an ideal world, growers would base their site selection decision on water quality and environmental parameters alone, but other factors, such as ease of access, distance from commercial dock space, and user conflicts, typically take precedence. Many growout sites have conflicts with other users of the marine resource, who may perceive a proposed shellfish farm to be a navigation hazard or a displacement of commercial or recreational fishing activities. Waterfront homeowners may object to the visual impact of the farm, or the potential for noise or lights while harvesting or processing. These types of user conflicts sharply constrain the choices of those wishing to farm shellfish. Federal law prohibits disturbing eelgrass beds because eelgrass has been designated as essential fish habitat. Similar laws protect endangered or threatened species that may use a desired site.

5.25   Carrying capacity

A significant concern relating to site selection is the cumulative impact of large populations of shellfish on primary production and other organisms in the local environment. Small-scale shellfish aquaculture has been shown to have many environmental benefits (Shumway *et al*. 2003), but excessive densities of shellfish can have adverse consequences (reviewed by McKindsey *et al*. 2006a). The intensity level of shellfish farming that starts to yield undesirable impacts is referred to as the "carrying capacity," and several carrying capacity terms have been defined depending on the impacts of concern (Inglis *et al*. 2000; Crawford *et al*. 2003; McKindsey *et al*. 2006b; Ferreira *et al*. 2007).

Farms that utilize large vertical arrays and suspended longlines can greatly increase the density of animals per m^2 compared with bottom-planted shellfish. Excessive food-competition can deplete local food supplies to a point where growth rates slow and shellfish on the farm have reduced meat yields. Further increases in planting density beyond this point no longer increase yield. This level is referred to as the "production carrying capacity" (Inglis *et al*. 2000). Once densities start to impact production, growers have a strong economic motivation to reduce densities, so examples of farms exceeding the production carrying capacity are rare and short-lived.

Dense populations of shellfish can cause the accumulation of fecal material on the bottom, potentially overwhelming the capacity of the environment to assimilate the organic matter, leading to local hypoxia and reduced diversity (Kaiser *et al*. 1998). This level is sometimes referred to as the "ecosystem carrying capacity" (Crawford *et al.* 2003). More recently, resource managers have turned their attention to the impacts of shellfish culture on other organisms, the structure of the food web, and nutrient cycles. The point at which these impacts start to occur is referred to as the "ecological carrying capacity" (McKindsey *et al.* 2006b).

To determine the production carrying capacity is an empirical exercise in monitoring the condition index and growth rates of the animals on the farm. Determining the assimilative carrying capacity involves extensive monitoring of characteristics such as benthic oxygen demand and diversity. Predicting the ecological carrying capacity involves complex mass balance modeling with data inputs that describe the biomass and production of every significant food web element as well as measurements of nutrient inputs and water current dynamics (Jiang & Gibbs 2005). A relatively simple first-order estimate can be made by comparing the total filtration capacity of the shellfish population with the residence time of water in a given embayment. If the cumulative filtration capacity of shellfish in an area exceeds the flow of seston to the area, then the system is more likely to be dominated by filter feeders, increasing the potential for carrying capacity issues (Dame & Prins 1998).

5.26 Permitting challenges

The number and types of permits required for shellfish culture vary greatly. Most applications will have elements of local, state, and federal authority that need to be addressed. In the United States, some states delegate lease permitting entirely to the towns while some have centralized permitting authorities. It is usually easier to navigate centralized permits because local authorities are more easily swayed by the often passionate and vocal pleas of "not in my backyard." Opponents will usually claim that they support shellfish aquaculture, but they believe that the proposed site is not appropriate. Since claims that the proposed farm may impair the view are rarely convincing, opponents will often suggest that the proposed farm has some sort of undesirable environmental impact.

Fortunately, there is a growing body of evidence that shellfish farms are environmentally benign or beneficial to the environment (Shumway *et al.* 2003). Shellfish filter all types of particles from the water, enhancing light penetration and reducing turbidity. The presence of shellfish, and the gear used to grow them, increases the productivity and the biodiversity of an area (DeAlteris *et al.* 2004). Shellfish aquaculture is usually supported by environmental groups because farming methods use no drugs, chemicals, or feeds. In the United States, shellfish farms are widely recognized as a sustainable method of food production with tangible environmental benefits (Seafood Watch 2008).

Permitting can take months or years depending on the municipality and their familiarity with the issues. Shellfish lease permits usually require consultation and approval from the Army Corps of Engineers (for navigation concerns), NOAA (for essential fish habitat designation and endangered or threatened species concerns), the Environmental Protection Agency (to address water quality issues), local or regional Fisheries Management Councils (to assess conflicts with commercial fishing), and local user groups (to evaluate recreational fishing and boating conflicts). This is not an exhaustive list, but it gives one an idea of what is involved. Some states have set up pre-permitted aquaculture zones that streamline the application process, but confine the applicant to predefined locations and methods of culture.

5.27   Non-native species

Over the past century there have been a great number of introductions of nonnative shellfish species for aquaculture production. Fisheries managers, anxious to increase production when local stocks became depleted, borrowed a page from terrestrial agriculture and brought in novel shellfish from other regions. Most of our major terrestrial food crops and domesticated animals are introduced species, so introducing new shellfish species seemed to be a natural choice (Carlton 1992).

More recently, resource managers have become more cautious and tried to slow the spread of non-native species because of the potential for disastrous unanticipated environmental consequences (Carlton & Ruiz 2005). While many intentional introductions of exotic shellfish were successful, the majority failed to produce established populations. The most prominent example is the Japanese oyster, *Crassostrea gigas,* which now reigns supreme as the single most cultured species on the planet (FAO 2007). *C. gigas* was introduced on the West Coast of North America in the 1800s after fisheries and pollution decimated their native oyster, the delicate Olympia, *Ostrea conchaphelia.* Seed of *C. cigas* were shipped from Japan packed in seaweed, and several other species were inadvertently introduced in the process. The Manila clam first showed up on the West Coast in the 1930s and now Manila clams and *C. gigas* make up the vast majority of the cultured shellfish harvested in the Pacific Northwest. Other hitchhikers in these seed shipments included two notable pests: the green crab, *Carcinus maenas,* and the stalked tunicate, *Styela clava* (Carlton 1992).

The catastrophic environmental and economic impacts of a number of unintentional introductions have caused resource managers around the world to become more cautious about intentional species introductions (Carlton & Ruiz 2005). The International Council for the Exploration of the Sea has established strict protocols for moving species intended for culture across international boundaries in an attempt to limit unintended consequences (ICES 2005). Most states have implemented strict regulations prohibiting the introduction of non-native species, and responsible growers in the United States have established Best Practices governing interstate transport of seed and shell stock. These measures are designed to limit the risk of spreading hitchhikers, non-native species, and diseases beyond their existing ranges.

Despite these reservations there is still a motivation to introduce novel shellfish species for aquaculture. In the mid-Atlantic region there has been a push by some states to introduce the Suminoe oyster *Crassostrea ariakensis* to replace the native *C. virginica,* whose stocks have been depleted by disease, overharvesting, and environmental degradation. This potential introduction was studied by a multidisciplinary team of National Academy of Sciences researchers who, after five years, decided that such an introduction "posed unacceptable ecological risks" (ACOE 2009).

5.28   References

Adams, C., Shumway, S.E., Whitlatch, R.B. & Getchis, T. (2011) Biofouling in marine molluscan shellfish aquaculture: a survey assessing the business and economic implications of mitigation. *Journal of the World Aquaculture Society* 42:242–52.

Army Corps of Engineers (2009) *Final programmatic oyster environmental impact statement.* http://www.nao.usace.army.mil/OysterEIS/Final_PEIS/homepage.asp.

Arnold, W.S. (1984) The effects of prey size, predator size, and sediment composition on the rate of predations of the blue crab, *Callinectes sapidus* Rathbun on the hard clams *Mercenaria mercenaria* (Linné). *Journal of Experimental Marine Biology and Ecology* 80(3):207–19.

Blinderman, C. (1998) *The Huxley File.* http://aleph0.clarku.edu/huxley/SM5/fish.html#cite3

Booth, J.D. & Cox, O. (2003) Marine fisheries enhancement in New Zealand: Our perspective. *New Zealand Journal of Marine and Freshwater Research* 37(4): 673–90.

Bower, S.M. & McGladdery, S.E. (1997) *Synopsis of Infectious Diseases and Parasites of Commercially Exploited Shellfish.* URL: http://www-sci.pac.dfo-mpo.gc.ca/shelldis/abstract_e.htm.

Bricelj, V.M. (1979) *Fecundity and related aspects of hard clam, (Mercenaria mercenaria) reproduction in Great South Bay, NY.* MSc Thesis, State University of New York at Stoney Brook.

Bricelj, V.M. & Shumway S.E. (1998) A review of paralytic shellfish poisoning toxins in bivalve molluscs: occurrence and transfer kinetics. *Reviews in Fisheries Science* 6:315–83.

Brock, V., Kennedy, V.S. & Brock, A. (1986) Temperature dependence of carbohydrase activity in the hepatopancreas of thirteen estuarine and coastal bivalve species from the North American east coast. *Journal of Experimental Marine Biology and Ecology* 103:87–101.

Carlton, J.T. (1992) Introduced marine and estuarine mollusks of North America: An end-of-the-20th-century perspective. *Journal of Shellfish Research* 11:489–505.

Carlton, J.T. & Ruiz, G.M. (2005) The magnitude and consequences of bioinvasions in marine ecosystems: Implications for conservation biology. In *Marine Conservation Biology: The Science of Maintaining the Sea's Biodiversity* (Ed. by E.A. Norse & L.B. Crowder), pp. 123–48. Island Press, Washington, DC.

Carriker, M.R. (2004) *Taming of the Oyster: A History of Evolving Shellfisheries and the National Shellfisheries Association.* Sheridan Press, Hanover.

Crawford, C.M., MacLeod, C.K. & Mitchell, I.M. (2003) Effects of shellfish farming on the benthic environment. *Aquaculture* 224:117–40.

Dame, R.F. & Prins, T.C. (1998) Bivalve carrying capacity in coastal ecosystems. *Aquatic Ecology* 31(4):409–21.

DeAlteris, J.T., Kilpatrick, B.D. & Rheault, R.B. (2004) A comparative evaluation of the habitat value of shellfish aquaculture gear, submerged aquatic vegetation, and a non-vegetated seabed. *Journal of Shellfish Research* 23(3):867–74.

Debrosse, G.A. & Allen, S.K. (1993) Control of overset on cultured oysters using brine solutions. *Journal of Shellfish Research* 12:29–33.

Epifanio, C.E. (1981) Phytoplankton and yeasts as foods for juvenile bivalves: A review of research at the University of Delaware. In *Proceedings of the Second International Conference on Aquaculture Nutrition: Biochemical and Physiological Approaches to Shellfish Nutrition* (Ed. by G.D. Pruder, C.J. Langdon & D.E. Conklin), pp. 292–301. Louisiana State University, Baton Rouge.

FAO (2010) The State of the World Fisheries and Aquaculture 2010. *FAO Fisheries and Aquaculture Department.* FAO, Rome. http://www.fao.org/docrep/013/i1820e/i1820e00.htm.

FAO (2009) *Fisheries and Aquaculture Information and Statistics Service (FIES).* http://www.fao.org/fishery/statistics/global-aquaculture-production.

FDA (2009) *National Shellfish Sanitation Program Guide for the Control of Molluscan Shellfish 2009 Revision.* http://www.fda.gov/Food/FoodSafety/Product-Specific Information/Seafood/FederalStatePrograms/NationalShellfishSanitationProgram/ucm 046353.htm.

FAO (2007) The State of the World Fisheries and Aquaculture 2006. *FAO Fisheries and Aquaculture Department ISSN 1020-5489.* FAO, Rome.

FAO (2006) *Cultured Aquatic Species Information Programme, Mytilus galloprovincialis (Lamark, 1819).* http://www.fao.org/fishery/culturedspecies/Mytilus_galloprovincialis/en.

FAO (1997)Individual quota management in fisheries—methodologies for determining catch quotas and initial quota allocation. *FAO Fisheries Technical Paper 371.* FOA, Rome. http://www.fao.org/docrep/003/W7292E/w7292e00.HTM.

Ferreira, J.G., Hawkins, A.J.S., Montereiro, P., Service, M., Moore, H., Edwards, A., Gowen, R., Lourenco, P., Mellor, A., Nunes, J.P., Pascode, P.L., Ramos, L., Sequeira, A., Simas, T. & Strong, J. (2007) *Sustainable Mariculture in Northern Irish Lough ecosystems.* Institute of Marine Research, Portugal.

Forrest, B.M., Hopkins, G.A., Dodgshun, T.J. & Gardner, J.P.A. (2007) Efficacy of acetic acid treatments in the management of marine biofouling. *Aquaculture* 262(2–4):319–32.

Grizzle, R.E. & Lutz, R.A. (1989) A statistical model relating horizontal seston fluxes and bottom sediment characteristics to growth of *Mercenaria mercenaria. Marine Biology* 102(1):95–105.

Guillard, R.R.L. (1975) Culture of phytoplankton for feeding marine invertebrates. In *Culture of Marine Invertebrate Animals* (Ed. by W.L. Smith & M.H. Chanley), pp. 29–60. Plenum Book Publishing Corp., New York.

Guo, X., DeBrosse, G.A. & Allen, S.A. (1996) All-triploid Pacific oysters (*Crassostrea gigas* Thunberg) produced by mating tetraploids and diploids. *Aquaculture* 142(3–4):149–61.

Hégaret, H., Wikfors, G.H. & Shumway, S.E. (2007) Diverse feeding responses of five species of bivalve molluscs when exposed to three species of harmful algae. *Journal of Shellfish Research* 26(2):549–59.

Hégaret, H., Wikfors, G.H. & Shumway, S.E. (2008) Biotoxin contamination and shellfish safety. In *Shellfish Safety and Quality* (Ed. by S.E. Shumway & G.E. Rodrick), pp.43–80. Woodhead Publishing, UK.

ICES (2005) Code of Practice on the Introductions and Transfers of Marine Organisms 2005.

Imai, T. (1978) *Aquaculture in Shallow Seas: Progress in Shallow Sea Culture.* A.A. Balkema, Rotterdam.

Inglis, G.J., Hayden, B.J. & Ross, A.H. (2000) *An Overview of Factors Affecting the Carrying Capacity of Coastal Embayments for Mussel Culture.* NIWA, Christchurch. Client Report CHC00/69: vi.

Jiang, W.M. & Gibbs, M.T. (2005) Predicting the carrying capacity of bivalve shellfish culture using a steady, linear food web model. *Aquaculture* **244**:171–85.

Kaiser, M.J., Burnell, G. & Costello, M. (1998) The environmental impact of bivalve mariculture: A review. *Journal of Shellfish Research* **17**(1):59–66.

Kurlansky, M. (2006) *The Big Oyster—History on the Half Shell.* Ballantine Books, New York.

Lesser, M.P., Shumway, S.E., Cucci, T. & Smith, J. (1992) Impact of fouling organisms on mussel rope culture: Interspecific competition for food among suspension-feeding invertebrates.*Journal of Experimental Marine Biology and Ecology* **65**(1): 91–102.

Lou, Y. (1991) China. In *Scallops: Biology, Ecology and Aquaculture* (Ed. by S.E. Shumway), pp. 809–22. Elsevier, Amsterdam.

MacKenzie Jr., C.L. (1977) Use of quicklime to increase oyster seed production. *Aquaculture* **10**(1):45–51.

MacKenzie, C.L. & Shearer, L.W. (1959) Chemical control of *Polydora websteri* and other annelids inhabiting oyster shells. *Proceedings of the National Shellfish Association* **50**:105–11.

Mackie, G.L. (1984) Bivalves. In *The Mollusca Vol. 7 Reproduction* (Ed. by A.S.M Saleuddin & K.H. Wilbur), pp. 351–418. Academic Press, New York.

Matthiessen, G.C. (2001) *Oyster Culture.* Fishing News Books, Blackwell Science, London.

McKindsey, C.W., Anderson, M.R., Barnes, P., Courtenay, S., Landry, T. & Skinner, M. (2006a) *Effects of shellfish aquaculture on fish habitat.* DFO Canadian Science Advisory Secretariat Research Document 2006/011:viii.

McKindsey, C.W., Thetmeyer, H., Landry, T. & Silvert, W. (2006b) Review of recent carrying capacity models for bivalve culture and recommendations for research and management. *Aquaculture* **261**:451–62.

Nell, J.A. (2002) Farming triploid oysters. *Aquaculture* **210**(1–4):69–88.

Nie, Z. (1982) Artificial culture of bivalves in China. In *Bivalve Culture in Asia and the Pacific* (Ed. by F.B. Davy & M. Graham), pp. 21–7. DRC, Ottawa, Canada.

Persoone, G. & Claus, C. (1980) Mass culture of algae: A bottleneck in the nursery culturing of molluscs. In *Algae Biomass* (Ed. by G. Schelef & C.J. Soeder), pp. 265–85. Elsevier, New York.

Peterson, C.H., Summerson, H.C. & Duncan, P.B. (1984) The influence of seagrass cover on population structure and individual growth rate of a suspension-feeding bivalve, *Mercenaria mercenaria. Journal of Marine Research* **42**:123–38.

Rheault, R.B. & Rice, M.A. (1996) Food-limited growth and condition index in the Eastern oyster, *Crassostrea virginica* (Gmelin 1791) and the bay scallop, *Argopecten irradians irradians* (Lamarck 1819). *Journal of Shellfish Research* **15**(2): 271–83.

Rheault, R.B. & Rice, M.A. (1995) Transient gear shellfish aquaculture. *World Aquaculture* **26**(1):26–31.

Seafood Watch (2008) *Seafood recommendations—What to eat and why.* Monterey Bay Aquarium Foundation, 886 Cannery Row, Monterey. http://www.monterey bayaquarium.org/cr/seafoodwatch.aspx.

Shumway, S.E. (1990) A review of the effects of algal blooms on shellfish and aquaculture. *Journal of the World Aquaculture Society* 21:65–104.

Shumway, S.E., Burkholder, J.M. & Springer, J. (2006) Effects of the estuarine dinoflagellates *Pfiesteria shumwayae* (Dinophyceae) on survival and grazing activity of several shellfish species. *Harmful Algae* 5(4):442–58.

Shumway, S.E., Davis, C., Downey, R., Karney, R., Kraeuter, J., Parsons, J., Rheault, R. & Wikfors, G. (2003) Shellfish aquaculture—in praise of sustainable economies and environments. *World Aquaculture* 34(4):15–8.

Smith, B.C. (1992) The relationship between gross biochemical composition of cultured algal foods and growth of the hard clam, *Mercenaria mercenaria* (L.). *Aquaculture* 108(1-2):135–154.

Stanley, J.G., Allen, Jr., S.K. & Hidu, H. (1981) Polyploidy induced in the American oyster *Crassostrea virginica* with cytochalasin B. *Aquaculture* 23:1–10.

Webb, K.L. & Chu, F.E. (1983) Phytoplankton as a food source for bivalve larvae. In *Proceedings of the Second International Conference on Aquaculture Nutrition: Biochemical and Physiological Approaches to Shellfish Nutrition* (Ed. by G.D. Pruder, C.J. Langdon & D.E. Conklin), pp. 272–91. Louisiana State University, Baton Rouge.

Wikfors, G.H., Ferris, G.E. & Smith, B.C. (1992) The relationship between gross biochemical composition of cultured algal foods and growth of the hard clam, *Mercenaria mercenaria* (L.). *Aquaculture* 108(1–2):135–54.

Yuhas, C. (2002) The status of shellfish beds in the NY-NJ Harbor Estuary. NJ Sea Grant College Extension Program/ NY-NJ Harbor Estuary Program. http://www.seagrant.sunysb.edu/HEP/pdf/hepshellfish.pdf.

Chapter 6

Cage Culture in Freshwater and Protected Marine Areas

Michael P. Masser

Cages are systems that retain the species cultured in a confined area, usually at high density, while excluding unwanted animals from the larger body of water. Cages depend on water movement through them (from the water body) to supply the culture organisms with water of sufficient quality for sustained growth and to remove problematic metabolic wastes. Historically, many of the failures of cage operations and environmentally founded criticisms of cage systems stem from improperly placing them at sites with poor water circulation (Stickney 2002). Sites with poor water circulation have led to low dissolved oxygen conditions and to the buildup of metabolic wastes causing poor water quality conditions, reduced growth, dense algal blooms, sediment contamination, devastation of benthic biota, disease outbreaks, and often significant mortalities.

The water bodies the cage or cages are placed in provide most basic functions needed to sustain culture. The temperature regime is that of the water body and may be fairly stable depending on location and volume of the water body. The concentrations and variability of water quality parameters depend on the water body, currents, and any external inputs. In particular, the delivery of dissolved oxygen and the removal and decomposition of wastes from cages depend on the water body and its physical, chemical, and environmental characteristics. In some cases (extensive culture), the water body provides food for the cultured species. Therefore, the success of cage culture is very dependent on the water body in which the cages are placed and environmental impacts on that water body from the culture practices, the climate, water quality, and possibly human activities.

Aquaculture Production Systems, First Edition. Edited by James Tidwell.
© 2012 John Wiley & Sons, Inc. Published 2012 by John Wiley & Sons, Inc.

(a)

(b)

Figure 6.1 Large marine cage complex off South Korea (a) and small freshwater catfish cages in Texas, United States (b).

Strictly defined, cages are rigid structures or have a frame on which the netting or mesh is held in place while net pens utilize flexible netting or mesh with no complete or encasing frame. However, the terms cage and net pen are often used interchangeably with no regard to these distinctions (fig. 6.1 a, b)

Considering that over 70% of the Earth's surface is covered with water, the opportunity for cage culture should be obvious. This foreseen opportunity has led to many advances in cage design, species cultured, and production volume in recent years. This chapter will discuss the species currently being cultured, the advantages and disadvantages of cage culture, and the management issues of cage aquaculture.

6.1 Current status of cage culture

In 2007, the United Nations Food and Agricultural Organization (FAO) Fisheries and Aquaculture Department held a Special Session on Cage Aquaculture and published a global overview on the status of cage aquaculture (Halwart *et al.* 2007). This FAO document divided the information by geographic regions (e.g., northern Europe, Latin America, etc.) with the exception of China, for which a country review was developed and included due to the recent expansion and magnitude of cage aquaculture in that country.

The FAO report was based on data available from 2005 with a total of sixty-two countries providing information (Tacon & Halwart 2007). While these data can in some cases be considered incomplete (not all countries reported) they are considered the best estimate of the scope and importance of global cage aquaculture to date.

Total global cage production was estimated to be approximately 3.4 million metric tonnes in 2005. Based on this report the countries that led in cage aquaculture were China, Norway, Chile, Japan, United Kingdom, Vietnam, Canada, Turkey, Greece, Indonesia, and the Philippines (Tacon & Halwart 2007). While China led in total cage production, it was only 2.3% of total Chinese aquaculture by volume (Chen *et al.* 2007). By contrast, Canada reported almost 70% of its production by volume was from cages (Masser & Bridger 2007).

Finfish species cultured in cages are predominately high-valued species. The vast majority of finfish cultured in cages are carnivorous species requiring high-protein, high fish-meal diets. These include salmon (Atlantic, *Salmo salar*, Coho, *Oncorhynchus kisutch*, and Chinook, *O. tshawytscha*), seabass (*Dicentrarchus labrax* or *Lates calcarifer*), seabream (*Sparus aurata*), amberjack (*Seriola* spp.), snapper (*Cromileptes* and *Epinephelus* spp.), cobia (*Rachycentron canadum*), rainbow trout (*O. mykiss*), mandarin fish (*Siniperca* spp.), and snakehead (*Channa* spp.). However, the culture of omnivorous species in cages is expanding and includes species of carps (*Cyprinus carpio, Hypophthalmichthys molitrix, Aristichthys nobilis,* and *Ctenopharyngodon idella*), tilapia (*Oreochromis* spp.), *Colossoma*, and catfish (*Pangasius* spp.; Tacon & Halwart 2007). Globally, five families of finfish (i.e., *Salmonidae, Sparidae, Carangidae, Pangasidae, and Cichlidae*) make up 90% of all cage production. Salmonids are the dominant cage cultured species by both volume (66%) and value (>US$4.7 billion; Tacon & Halwart 2007).

Cage aquaculture is practiced at all socio-economic levels. The diversity of cage producers ranges from artisanal farmers producing for their families, to family enterprises, to large industrial scale farms utilizing both private and public waters (De Silva & Phillips 2007; Masser & Bridger 2007). Commercial cage culture has expanded rapidly since the Norwegians developed modern salmon farming methods in the 1970s. Salmonids cage farming grew from a production of only 294 tonnes in 1970 to 1,235,972 tonnes in 2005 (Tacon & Halwart 2007). Counties leading this dramatic expansion were Norway, Chile, and the United Kingdom. The development of new cage designs and emerging species, especially marine, has led to expanded interest in cage aquaculture. This expansion is particularly notable in developing countries with tropical climates, and continued growth is anticipated (Tacon & Halwart 2007).

6.2   History and evolution of cage culture

Cages were undoubtedly first used as holding and transporting pens for wild caught fish and shellfish awaiting sale (Beveridge & Little 2002). The actual culturing of fish in cages supposedly was first practiced in China during the Han Dynasty (100 to 200 BC; Li 1994). Most of the early cages were constructed of wood or bamboo and some used cloth to hold and culture fry (Hu 1994). Feed consisted of human food scraps & trash fish (Beveridge 2004). This type of cage culture has been practiced for centuries and is still practiced in parts of Asia (Halwart *et al.* 2007). For a more complete review of the origins of cage culture see *Cage Aquaculture* by Malcolm Beveridge (2004).

Cages have been utilized for all stages of aquaculture production: holding broodstock, spawning, rearing fry and fingerlings, and production of foodfish. Cage mesh size is chosen based on the size of fish at stocking and often fish are moved to cages of larger mesh size as they grow. In general, fish should be stocked in the largest mesh size that will retain them. Tilapias are often spawned and the fry reared in fine-meshed cages called "hapas" (fig. 6.2; Smith *et al.* 1985).

Figure 6.2 Hapas for rearing of fish fry.

While cages are used to produce fingerlings of many species, their primary use is for growout to market sizes (Beveridge 2004).

Cages have also been utilized in research and environmental monitoring. Cages have been adopted for nutrition research in order to eliminate or minimize the fish's consumption of other foods (Stickney 1979; Webster *et al.* 1994) or to monitor water quality (Chamberlain 1978; Grizzle *et al.* 1988).

Cages can be categorized as extensive, semi-intensive or intensive based on their stocking density and source of food (Beveridge 2004). Extensive cages receive no external feed and rely on only natural foods such as plankton (phyto and zoo), seston, and detritus. Extensive cage systems have usually been placed in highly eutrophic freshwater lakes, reservoirs, and some sewage effluent lagoons in order to take advantage of the natural productivity (Costa-Pierce & Effendi 1988; Edwards 1992). The species cultured extensively have been primarily carps and tilapias (Li 1994; Costa-Pierce 2002). Extensive cage production can yield up to 1.9 kg/m^3 per month (Beveridge 1984).

Semi-intensive cage culture is very common in freshwater systems in tropical climates (Beveridge 2004). Like extensive culture the species commonly cultured include the tilapias and the carps (i.e., silver, bighead, grass, and common). In addition to available natural foods, the fish are fed other (usually nutritionally incomplete) foodstuffs. These have included food scraps, rice bran, brewery byproducts, wheat middlings, and other agricultural byproducts (Pantulu 1979; De Silva 1993; McAndrew *et al.* 2002). Semi-intensive production in cages can yield over 100 kg /m^3.

Intensive cage culture requires the use of nutritionally complete feeds. Intensive culture usually is limited to high-valued species because of the cost associated with feeding. In freshwater these include catfish (Ictalurids and Pangasids), salmonids, snakeheads, carps, and tilapias. In marine waters, species cultured include salmonids, seabass, seabream, yellowtail, amberjack, cobia, and croaker. Increasing interest in cage culture of other marine species such as tuna, grouper, and snapper has begun. Yields in intensive cages are greatest and can be in the range of 250 to 600 kg/m^3 (Schmittou *et al.* 2004).

6.3   Advantages and disadvantages of cages

All culture systems have advantages and disadvantages. Advantages of cage aquaculture include the exploitation of almost any existing water body (e.g., ocean, bay, estuary, river, lake, or pond); relatively low construction costs compared to building ponds, raceways, or recirculation systems; observation, feeding, and harvest are relatively straightforward; and disease treatment (if necessary) can be effected more accurately and economically (Masser 2008).

Most disadvantages associated with cages can be attributed to the relatively high densities that culture species are stocked. These can include crowding related abrasions, rapid disease spread, localized water quality issues, attractiveness to predators and poachers, and communal interactions of the cultured species that may cause reduced growth (Masser 2008). Fouling of the cage netting is also a common problem. Fouling organisms can diminish the volume of the cage and severely decrease water flow circulation through the cage. Fouling organisms can include algae or sessile organisms such as bryozoans and sponges. Often cages have to be periodically scrubbed or the netting replaced to cope with this problem (Milne 1970). In addition, cage culture has little or no control over water quality parameters like temperature, pH, alkalinity, and hardness. Dissolved oxygen can be supplemented with aeration or circulation devices and location (i.e., currents and depth) can influence the concentrations of particulates and nitrogen waste products in and around the cage (Masser 1997a, b; Beveridge 2004; Masser & Woods 2008).

Another possible disadvantage to cage culture, if public waters are to be used, can be permitting. Most countries require permits to place and operate cages in public water. The requirements to obtain permits vary greatly from country to country and can include the preparation of environmental impact assessment and public hearings. Currently in the United States, for example, it is extremely difficult to obtain permits for cage culture in public waters and legislation to streamline the process in federal waters has not been passed by the United States Congress (Masser & Bridger 2007).

6.4   Site selection

Cage production is a very attractive culture system where existing water bodies can be utilized with little or no economic costs. However, the proper siting of cages is essential for their functionality, particularly in relation to proper water quality within the cage and reduced environmental impacts around the cage. Research on actual site characteristics and modeling of water currents and waste dispersion may be used to help predict impacts from cages in public waters (Landless & Edwards 1977; Ervik *et al.* 1997; Turner 2000; Stickney 2002; Olsen *et al.* 2005; Belle & Nash 2008).

Cage culture confines fish or shellfish at relatively high densities but relies on the surrounding water body to maintain suitable water quality within the cage and to perform functions of waste removal (Stickney 2002). When sited in unsuitable locations, cages have caused environmental degradation to the surrounding waters and benthic zone (Axler *et al.* 1992, 1994; Kelly 1995; Tsutsumi 1995). This reliance on the water body to provide ecosystem functions to private aquaculture ventures in public waters has, at times, led to conflicts with local communities and environmental groups, often resulting in negative publicity (Lumb 1989; Stickney 1990). Most criticisms of cages located in public waters are related to possible water pollution, escapement, benthic community alteration, spread of diseases to native populations, navigation issues, or simply the aesthetics of seeing the cages from the shore (referred to as "sight" or "visual" pollution; Stickney 1990, 2002).

6.5   Stocking cages

The densities of cultured species stocked in cages are highly variable and little research has been done to establish optimum stocking densities for many species (Beveridge 2004). Fish can either be stocked at the density needed to reach final production goals (i.e., desired number and harvest size minus anticipated mortalities) or in a two-stage process wherein small fish can be stocked at high densities and then restocked into additional cages at lower densities as their biomass increases. Which strategy is used often depends on the length of time it takes to get fish to the final desired market size.

Often cages are discussed in terms of fish density versus cage volume. In general terms, cages can be low-density/high-volume or high-density/low-volume. Marine cage culture operations are usually low-density/high-volume. In these cases, densities of 5 to 20 fish/m^3 are common, realizing that often the desired harvest size is several kilograms (e.g., Atlantic salmon are usually harvested at 4 kilograms or larger).

High-density/low-volume cages are more common in freshwater cage culture. In these systems cages are often stocked at densities of 150 to 450 fish per cubic meter with the target harvest weight of one kilogram or less (fig. 6.3). Species commonly cultured at these densities include catfish, tilapia, carps, and other freshwater species (Schmittou 1993; Duarte *et al.* 1993; Masser 1997c). These higher densities are often necessary for some species (like channel catfish) to minimize aggressive interactions, which may occur at lower densities (Masser 2004).

Figure 6.3 Channel catfish shown at a common density of 50 kilograms per cubic meter to illustrate fish density in low volume cages.

6.6   Feeding caged fish

In semi-intensive and intensive cage culture, feed costs represent 25 to 50% of total production costs (Beveridge 2004). Feeds must meet nutritional requirements, be palatable and of appropriate size, and not pollute the environment to the point of causing deleterious consequences.

As with most culture systems, caged fish are fed based on fish size, water temperature, and feeding response. In semi-intensive cage culture numerous kinds of feedstuffs have been utilized in the diets (Lovell 1989; New *et al.* 1993; Beveridge 2004). Development of modern intensive culture diets did not begin until the 1950s (Lovell 1989) and have yet to be formulated for many desired culture species. "Trash fish" (i.e., fish of little commercial value, rarely consumed by people, and would otherwise be discarded) and invertebrates are often fed to certain species in areas where formulated diets are are too expensive or too difficult to obtain. Common problems associated with feeding trash species include availability, nutritional suitability, storage, seasonal quality, and possible introduction of diseases (Beveridge 1984; Tacon 1992; Wu *et al.* 1994). The utilization of trash species and high fish meal and fish oil content in diets may be a concern and is often a criticism of aquaculture (Tidwell & Allan 2001; Pike & Barlow 2003; Belle & Nash 2008).

Over or underfeeding can be problematic in any type of culture system and can be particularly challenging in cage culture. Feed can quickly wash out of cages due to currents or, in the case of sinking feeds, can fall through the cage. For these reasons caged fish are often hand fed, fed by automatic fixed-ration systems, fed utilizing demand feeders, or the cages can be equipped with interactive systems that monitor feeding activity and adjust feeding rate to accommodate appetite (Beveridge 2004; fig. 6.4). In the salmonid industry improvements in feed formulation, manufacture, and automated feeding systems have greatly improved feeding efficiencies so that feed conversion ratios have decreased from around 2.5:1 in the 1970s to below 1:1 by 2005 (Grøttum & Beveridge 2007).

Fish behavior must also be considered and many species of fish have preferred feeding times. Channel catfish, like many other species, are crepuscular (i.e., dawn and dusk) feeders and will consume more feed and grow faster if feed in low-light conditions (Woods 1994).

6.7   Polyculture and integrated systems

"Polyculture" involves culturing more than one species in the same cage while "integrated culture" involves using more than one aquaculture system in association to culture multiple species. With polyculture the species may or may not directly benefit from the other and may actually compete. With integrated culture the species benefit in some way from the presence of the other. Cage polyculture has been limited in application mostly because of species' competition for

Figure 6.4 Hand feeding of sea bass in cages off Corsica.

whatever natural food or artificial feed is applied. In China, polyculture of silver carp and bighead carp (both plankton feeders) in cages has been practiced since the 1970s (Chen *et al.* 2007). Often, common carp or tilapia are also stocked into silver/bighead cages to reduce fouling (Chen *et al.* 2007).

Channel catfish have been successfully polycultured with rainbow trout and tilapia. In both cases the polyculture improved catfish growth and it was suggested that the other species stimulated increased feed consumption by the catfish (Williams *et al.* 1987; Beem *et al.* 1988; Woods 1994). Sea urchins have been placed in net enclosures hung inside Atlantic salmon cages and grew well feeding on waste feed (Kelly *et al.* 1998). Seaweeds have been cultured in shrimp cages in Brazil and provided shade and hiding places for the shrimp while benefiting from the shrimp wastes (Lombardi *et al.* 2001). Integrated culture of salmonids with seaweed and mussels has been practiced in Canada (Chopin & Bastarache 2002). The seaweed and mussels benefit from the fish wastes and significantly reduce the impact on water quality of the fish wastes, thus improving the overall environment.

6.8 Problems with cage culture

As noted previously, many problems associated with cage culture are related to the high stocking densities. Skin abrasions can occur as fish rub against cage materials or through fish/fish bumping or biting. Abrasions combined with stress and high nutrient loads often result in the opportunity for diseases development and diseases spread rapidly at high densities (Masser 2004; Masser & Woods 2008). Feeding hierarchies, like pecking orders in poultry, are usually observed when densities are too low and lead to reduced feed consumption and slowed growth in subordinate animals (Schwedler *et al.* 1989; Lazur 1996). High densities and the associated high feeding rates can exacerbate water quality problems: especially, low dissolved oxygen, high ammonia, and increased turbidity. The potential effects of densities that are too low or too high illustrate the need for research to identify optimal densities for different species and cage types.

Other common problems associated with cage culture include net fouling. The nutrients from the feed and fish wastes stimulate the growth of algae and many other sessile or benthic type organisms, which attach to and colonize around the cage. Fouling of the cage reduces volume and can severely reduce water movement through the cage, which further exacerbates water quality, stress, and disease issues.

Currents and drifting debris pose additional problems in cage culture. While good water exchange is critical for maintaining a quality cage environment, excessive currents cause problems with moorings and deformation of cage structures and integrity (i.e., collars and netting), loss of feed, and stress on the fish. Currently, the maximum velocity of water is about 60 cm/s (Beveridge 2004). Drifting debris (e.g., logs, floating aquatic plants, etc.) can impede water movement, and deform, damage or destroy cages. Vigilant removal of drifting objects and deflection barriers are often employed to mediate this problem.

Genetics can play a significant role in selection or development of strains of fish well suited for cage culture. Selection of species strains that tolerate and thrive with confinement in cages and its associated stressors is critical for success of cage culture (Woods 1994). The salmon industry leads other cage farming industries in selecting fish with traits that prosper in the cage environment (Jobling 1995; Gjøen & Bentsen 1997). This is an area that needs much more research for many species before their successful cage culture can be achieved.

Cages can be attractive to other species in the water body as hiding places, attachment structures, and as sources of food (Oakes & Pondella 2009). Competitors, scavengers, and predators also are attracted to cages. Often wild fish will be attracted to the cage at feeding times and push against the cage netting trying to get to the feed. This often intimidates the culture fish and reduces their feeding activity (Woods 1994). Predators and scavengers can cut cage nets, releasing the culture fish into the wild (Rueggeberg & Booth 1989). The cage culture industry has had problems with seals, piscivorous birds, and many other species of fish, invertebrates (e.g., squid), reptiles, and mammals (Chua & Teng 1980; Quick *et al.* 2004). Cage damage, disease issues, and fish losses can have had significant economic impacts on cage operations (Nash *et al.* 2000; Würsig & Gailey 2002).

Finally, poaching and vandalism can be devastating problems in cage aquaculture in both public (Gooley *et al.* 2000) and private waters (Masser 1999;

Masser & Woods 2008). Security is an element that all cage operations must consider and plan for all contingencies.

6.9   Economics of cage culture

The costs associated with cage construction and mooring vary greatly upon depending materials and sizes used. As with many types of construction, the cost per unit volume generally decreases as size increases. Cage netting or mesh can be relatively inexpensive if plastic or fiber netting is used, but it increases considerably if coated welded wire, stainless steel, or copper-nickel mesh is utilized. High-energy environments also increase construction and mooring costs (Lisac & Muir 2000; Belle & Nash 2008; see chapter 7 for cages and net pens in offshore and high-energy environments).

Variable costs for cage operations can vary widely depending on size of operation, site, species, culture intensity, and management costs. As with most intensive culture systems, feed costs are usually the highest variable cost averaging around 50 to 60% of total costs. The nutritional requirements of the cultured species and transport costs can greatly impact feed costs. The second highest variable cost is usually seed or fingerling costs and can range from 10 to 40% of variable cost, dependent on the species cultured (Escover & Claveria 1985; Heen 1993; Marte *et al.* 2000; Woods & Masser 2009). Severe fouling, more problematic with marine cages, increases labor costs while wooden cages do not have as long a lifespan as cages constructed from plastics and metals.

In general, cage farms are less expensive to build and operate compared to other systems (Beveridge 2004). This has been documented in salmon (Shaw & Muir 1986; Heen 1993), sea bass and sea bream (Lisac & Muir 2000), and catfish (Collins & Delmendo 1979; Bernardez 1995). These analyses do not take into consideration any environmental costs associated with the ecological services provided by the surrounding cage environment (i.e., water quality and benthic alterations). These costs may be increasingly evaluated in the future and possibly recovered through environmental taxes.

6.10   Sustainability issues

Sustainability issues surrounding cage culture are similar in many ways to those of other aquaculture production systems and include: sustainable sources of seed, diet composition (e.g., high fish meal and oil), and environmental impacts (Tucker & Hargreaves 2008). However, cage systems can be under more scrutiny as they often involve public waters and wild populations. Still, considering all the challenges cage aquaculture faces, it remains an excellent culture system for many species. Especially since the development of improved seed production techniques, genetically improved culture strains, improved feed quality, and better knowledge of proper siting considerations.

6.11   References

Axler, R., Larson, C., Tikkanen, C. & McDonald, M. (1992) Water quality changes in mine pit lakes used for salmonid culture. In *Abstracts of Aquaculture 92*, Annual Meeting of the Worlds Aquaculture Society, pp. 32–3. Orlando, Florida.

Axler, R., Larson, C., Tikanen, C., McDonald, M., Yokom, S. & Aas, P. (1994) Water quality issues associated with aquaculture: A case study in Minnesota mine pit lakes. *Water Environment Research* 68:995–1011.

Beem, M.D., Gebhart, G.E. & Maughan, O.E. (1988) Winter polyculture of channel catfish and rainbow trout in cages. *The Progressive Fish-Culturist* 50:49–51.

Belle, S.M. & Nash, C.E. (2008) Better management practices for marine shrimp aquaculture. In *Environmental Best Management Practices for Aquaculture* (Ed. by C. S. Tucker), pp. 261–330. Blackwell Publishing, Oxford.

Bernardez, R.G. (1995) Evaluation of an in-pond raceway system and its economic feasibility for fish production. Master's thesis. Auburn University, Auburn.

Beveridge, M. (2004) *Cage Aquaculture*, 3rd edition. Blackwell Publishing, Oxford.

Beveridge, M.C.M. (1984) Cage and pen fish farming: Carrying models and environmental impact. In *FAO Fisheries Technical Paper No. 255*. FAO, Rome.

Beveridge, M.C.M. & Little, D.C. (2002) History of aquaculture in traditional societies. In *Ecological Aquaculture* (Ed. by B. A. Costa-Pierce), pp. 3–29. Blackwell Publishing, Oxford.

Chamberlain, G.W. (1978) The use of caged fish for mariculture and environmental monitoring in a power plant cooling water system. Master's thesis. Texas A&M University, College Station.

Chen, J., Guang, C., Xu, H., Chen, Z., Xu, P., Yan, X., Wang, Y. & Liu, J. (2007) A review of cage and pen aquaculture: China. In *Cage Aquaculture—Regional Reviews and Global Overview* (Ed. by M. Halwart, D. Soto & J. R. Arthur), pp. 50–68. FAO Fisheries Technical Paper No. 498. FAO, Rome.

Chopin, T. & Bastarache, S. (2002) Finfish, shellfish and seaweed mariculture in Canada. *Bulletin of the Aquaculture Association of Canada* 102:119–24.

Chua, T.E. & Teng, S.K. (1980) Economic production of estuary grouper, *Epinephelus salmoides* Maxwell, reared in floating cages. *Aquaculture* 20(3):187–228.

Collins, R.A. & Delmendo, M.N. (1979) Comparative economics of aquaculture in cages, raceways and enclosures. In *Advances in Aquaculture* (Ed. by T.V.R. Pillay & W.A. Dill), pp. 472–7. Fishing News Books, Oxford.

Costa-Pierce, B.A. (2002) Sustainability of cage aquaculture ecosystems for large-scale resettlement from hydropower dams: An Indonesian case study. In *Ecological Aquaculture* (Ed. by B.A. Costa-Pierce), pp. 286–313. Blackwell Publishing, Oxford.

Costa-Pierce, B.A. & Effendi, P. (1988) Sewage fish cages of Kota Cianjur, Indonesia. *NAGA: The ICLARM Quarterly* 11(2):7–9.

De Silva, S.S. (1993) Supplementary feeding in semi-intensive aquaculture systems. In *Proceedings of the FAO/AADCP Regional Expert Consultation on Farm-Made Aquafeeds*. Bangkok, Thailand: FAO-RAPA/AADCP.

De Silva, S.S. & Phillips, M.J. (2007) A review of cage aquaculture: Asia (excluding China). In *Cage Aquaculture—Regional Reviews and Global Overview* (Ed. by M. Halwart, D. Soto & J.R. Arthur), pp.18–48. FAO Fisheries Technical Paper No. 498. FAO, Rome.

Duarte, S.A., Massel, M.P. & Plumb, J.A. (1993) Seasonal occurrence of diseases in cage-reared channel catfish, 1989–1991. *Journal of Aquatic Animal Health* 5(3):223–9.

Edwards, P. (1992) Reuse of human waste in aquaculture: A technical review. In *Water and Sanitation Report 2*, UNDP-World Bank Program. Washington, DC.

Ervik, A., Hansen, P.K., Aure, J., Stifebrandt, A., Johannessen, P. & Jahnsen, T. (1997) Regulating the local environmental impact of intensive marine fish farming. I. The concept of the MOM system (Modelling-Ongrowing fish farms-Monitoring). *Aquaculture* 158(1–2):85–94.

Escover, E.M. & Claveria, R.L. (1985) Economics of cage culture in Bicol freshwater lakes. In *Proceedings of the PCARRD/ICLARM Tilapia Economics Workshop*, U.P. Los Banos, Laguna, Philippines, 10-13 August 1983. (Ed. by I.R. Smith, E.B. Torres & E.O. Tan), pp. 50-64. ICLARM, Manila.

Gjøen, H.M. & Bentsen, H.B. (1997) Past, present, and future of genetic improvement in salmon aquaculture. *ICES Journal of Marine Sciences, Journal du Conseil* 54(6):1009–14.

Gooley, G.J., De Silva, S.S., Hone, P.W., McKinnon, L.J. & Ingram, B.A. (2000) Cage aquaculture in Australia: A developed country perspective with reference to integrated aquaculture development within waters. In *Cage Aquaculture in Asia: Proceedings* Liao, I.C.Lin, C.K. (Eds.), *Asian Fisheries Society, Manila and World Aquaculture Society— Asia Branch*, Bangkok. Tungkang, Taiwan.

Grizzle, J.M., Horowitz, S.A. & Strength, D.R. (1988) Caged fish as monitors of pollution: Effects of chlorinated effluent from a wastewater treatment plant. *Journal of the American Water Resource Association* 24(5):951–9.

Grøttum, J.A. & Beveridge, M. (2007) A review of cage aquaculture: Northern Europe. In *Cage Aquaculture—Regional Reviews and Global Overview* (Ed. by M. Halwart, D. Soto & J.R. Arthur), pp. 126–154. FAO Fisheries Technical Paper No. 498. FAO, Rome.

Halwart, M., Soto, D. & Arthur, J.R. (2007) *Cage Aquaculture—Regional Reviews and Global Overview*. FAO, Rome.

Heen, K. (1993) Comparative analysis of the cost structure and profitability in the salmon aquaculture industry. In *Salmon Aquaculture* (Ed. by K. Heen, R.L. Monaghan & F. Utter), pp. 220–38. Fishing News Books, Oxford.

Hu, B.T. (1994) Cage culture development and its role in aquaculture in China. *Aquaculture and Fisheries Management* 25:305–10.

Jobling, M. (1995) Simple indices for the assessment of the influences of social environment on growth performance, exemplified by studies on Arctic char. *Aquaculture International* 3(1):60–5.

Kelly, L.A. (1995) Predicting the effects of cages on nutrient status of Scottish freshwater lochs using mass-balance models. *Aquaculture Research* 26(7):469–77.

Kelly, M.S., Brodie, C.C. & McKenzie, J.D. (1998) Somatic and gonadal growth of the sea urchin *Psammechinus miliaris* (Gmelin) maintained in polyculture with Atlantic salmon. *Journal of Shellfish Research* 17(5):1557–62.

Landless, P.J. & Edwards, A. (1977) Some simple methods for surveying a marine farm site. *Fish Farming International* 4(1):32–4.

Lazur, A.M. (1996) The effects of periodic grading on production of channel catfish cultured in cages. *Journal of Applied Aquaculture* 6(4):17–24.

Li, S.F. (1994) Fish culture in cages and pens. In *Freshwater Fish Cultures in China: Principals and Practices* (Ed. by S. Li & J. Mathias), pp. 305–46. Elsevier, Amsterdam.

Lisac, D. & Muir, J. (2000) Comparative economics of offshore and mariculture facilities. Mediterranean Offshore Mariculture. *Options Méditerranéenes, Series B* 30: 203–11.

Lombardi, J.V., Marques, H.L.A. & Barreto, O.J.S. (2001) Floating cages on open sea water: An alternative for promoting integrated aquaculture in Brazil. *World Aquaculture* 32:47, 49–50.

Lovell, T. (1989) *Nutrition and feeding of fish*. Van Nostrand Reinhold, New York.

Lumb, C.M. (1989) Self-pollution by Scottish salmon farms? *Marine Pollution Bulletin* 20(8):375–9.

Marte, C.L., Cruz, P. & Flores, E.E.E. (2000) Recent developments in freshwater and marine cage aquaculture in the Philippines. In *Cage Aquaculture in Asia, Proceedings of The First International Symposium in Cage Aquaculture in Asia*. Asian Fisheries Society, World Aquaculture Society—Asia Branch, Bangkok. Tungkang, Taiwan.

Masser, M.P. (1997a) Cage culture: Site selection and water quality. *Southern Regional Aquaculture Center No. 161*.

Masser, M.P. (1997b) Cage culture: Cage construction, placement, and aeration. *Southern Regional Aquaculture Center No. 162*.

Masser, M.P. (1997c) Cage culture: Species suitable for cage culture. *Southern Regional Aquaculture Center No.163*.

Masser, M.P. (1999) Cages and in-pond raceways as sustainable aquaculture in watershed ponds. In *Proceedings of the Ninth National Extension Wildlife, Fisheries, and Aquaculture Conferences*, Portland, Maine (Ed by R.M. Timm & S.L. Dann), pp. 171–80.

Masser, M.P. (2004) Cages and in-pond raceways. In *Biology and Culture of Channel Catfish* (Ed. by C.S. Tucker & J.A. Hargreaves), pp. 530–43. Elsevier Science, Amsterdam.

Masser, M.P. (2008) What is cage culture? *Southern Regional Aquaculture Center No. 160*.

Masser, M.P. & Bridger, C.J. (2007) A review of cage aquaculture: North America. In *Cage Aquaculture—Regional Reviews and Global Overview* (Ed by M. Halwart, D. Soto & R.J. Arthur), pp. 105–25. FOA Fisheries Technical Paper 498. FAO, Rome.

Masser, M.P. & Woods, P. (2008) Cage culture problems. *Southern Regional Aquaculture Center No. 165*.

McAndrew, K.I., Brugere, C., Beveridge, M.C.M., Ireland, M.J., Ray, T.K. & Yasmin, K. (2002) Improved management of small-scale tropical cage systems in Bangladesh: Potential benefits of an alliance between an NGO and a western research institute. In *Rural Aquaculture* (Ed. by P. Edwards, D.C. Little & H. Demaine), pp. 129–41. CABI Publishing, Cambridge.

Milne, P.H. (1970) *Fish Farming: A Guide to the Design and Construction of Net Enclosures*. HMSO, Edinburgh.

Nash, C.E., Iwamoto, R.N. & Mahnken, C.V.W. (2000) Aquaculture risk management and marine mammal interactions on the Pacific Northwest. *Aquaculture* 183(3–4):307–23.

New, M.B., Tacon, A.G.J. & Csavas, I. (Eds) (1993) Farm-Made Aquafeeds. *Proceedings of the FAO/AADCP Regional Expert consultation on Fish-Made Aquafeeds* December 14–18, 1992, *Bangkok*. FAO Regional Office for Asia and the Pacific/ASEAN-EEC Aquaculture Development and Coordination Programme (AADCP), Bangkok. Reprinted in 1995 as *FAO Fisheries Technical Paper* 343. FAO, Rome.

Oakes, C.T. & Pondella II, D.J. (2009) The value of a net-cage as a fish aggregating device in southern California. *Journal of the World Aquaculture Society* 40(1):1–21.

Olsen, Y., Slagstad, D. & Vadstein, O. (2005) Assimilative carrying capacity: Contributions and impacts on the pelagics system. In *Lessons from the Past to Optimize the Future*. European Aquaculture Society, special publication No. 35. pp. 50–52. Oostende, Belgium.

Pantulu, V.R. (1979) Floating cage cultures of fish in the lower Mekong River Basin. In *Advances in Aquaculture* (Ed. by T.V.R. Pillay & W.A. Dill), pp. 423–7. Fishing News Books Ltd., Farnham, UK.

Pike, I. H. & Barlow, S. M. (2003) Impact of fish farming on fish stocks. *Fish Farmer* **Jan–Feb**: 14–6.

Quick, N.J., Middlemas, S.J. & Armstrong, J.D. (2004) A survey of anti-predator controls at marine salmon farms in Scotland. *Aquaculture* **230**:169–80.

Rueggeberg, H. & Booth, J. (1989) Interactions between wildlife and salmon farms in British Columbia: Results of a survey. *Technical Report Series 67* (Pacific and Yukon Region, Canadian Wildlife Service).

Schmittou, H.R. (1993) *High Density Fish Cultures in Low Volume Cages*. American Soybean Association, Singapore.

Schmittou, H.R., Cremer, M.C. & Zhang, J. (2004) *Principals and practices of high density fish culture in low volume cages*, pp. 1–4. American Soybean Association, Singapore.

Schwedler, T.E., Tomasso, J.R. & Collier, J.A. (1989) Production characteristics and size variability of channel catfish reared in cages and open ponds. *Journal of the World Aquaculture Society* **20**:158–61.

Shaw, S. & Muir, J.F. (1986) *Salmon Economics and Marketing*. Croom Helm, London.

Smith, I.R., Torres, E.B. & Tan, E.O. (1985) *Phillippine Tilapia Economics, Proceedings of the ICLARM, Conference Proceedings 12*.

Stickney, R.R. (1979) *Principles of Warm Water Aquaculture*. John Wiley & Sons, Inc., New York.

Stickney, R.R. (1990) Controversies in salmon aquaculture and projections for the future of the aquaculture industries. In *Proceedings of the Fourth Pacific Congress on Marine Science and Technology*, July 16–20, pp. 455–61. PACON, Tokyo.

Stickney, R.R. (2002) Impacts of cage and net-pen culture on water quality and benthic communities. In *Aquaculture and the Environment in the United States* (Ed. by J.R. Tomasso), pp. 105–18. US Aquaculture Society, Louisianna.

Tacon, A.G.J. (1992) Nutritional fish pathology. In *Morphological Signs of Nutrient Deficiency and Toxicity in Farmed Fish*. FAO fisheries Technical Paper No 330. FAO, Rome.

Tacon, A.G.J. & Halwart, M. (2007) Cage aquaculture: A global overview. In *Cage Aquaculture—Regional Reviews and Global Overview* (Ed. by M. Halwart, D. Soto & J.R. Arthur), pp. 1–16. FAO Fisheries Technical Paper No. 498. FAO, Rome.

Tacon, A.G.J., Hasan, M.R. & Subasinghe, R.P. (2006) Use of fishery resources as feed inputs to aquaculture development: Trends and policy's implications. *FAO Fisheries Circular* **1018**:99.

Tidwell, J.H. & Allan, G L. (2001) Fish as food: Aquaculture's contribution. EMBO Reports. **2(11)**:958–63.

Tsutsumi, H. (1995) Impact of fish net-pen culture on the benthic environment of a cove on South Japan. *Estuaries* **18**:108–15.

Tucker, C.S. & Hargreaves, J.A. (2008) *Environmental Best Management Practices for Aquaculture*. Wiley-Blackwell, Oxford.

Turner, R. (2000) Offshore mariculture: Mooring system design. *Mediterranean Offshore Mariculture, Options Méditerranéennes, Series B* **30**:141–57.

Webster, C.D., Goodgame-Tiu, L.S., Tidwell, J.H. & Reed Jr., E.B. (1994) Effects of dietary protein level on growth and body composition in channel catfish reared in cages. *Applied Aquaculture* **4**(2):73–86.

Williams, K., Gebhart, G.E. & Maughan, O.E. (1987) Enhanced growth of cage cultured channel catfish through polyculture with blue tilapia. *Aquaculture* **62**:207–14.

Woods, P. (1994) *Comparison of cage cultured commercial strains of channel catfish for survival, growth, feed conversion, and percent marketability*. Master's thesis, Auburn University, Auburn.

Woods, P. & Masser, M.P. (2009) Cage culture: Harvesting and economics. *Southern Regional Aquaculture Center No. 166*.

Wu, R.S.S., Lam, K.S., Mackay, D.W., Lau, T.C. & Yam, V. (1994) Impact of marine fish farming on water quality and bottom sediment: A case study in the sub-tropical environment. *Marine Environmental Research* **38**:115–45.

Würsig, B. & Gailey, G.A. (2002) Marine mammals and aquaculture: Conflicts and potential resolutions. In *Responsible Marine Aquaculture* (Ed. by R.R. Stickney & J.P. McVey), pp. 45–59. CABI Publishing, New York.

Chapter 7

Ocean Cage Culture

Richard Langan

It is widely acknowledged that future increases in seafood production will likely come from farming, not fishing. The growth of land-based and nearshore marine aquaculture in many developed countries is constrained by space, economics, and environmental concerns. Open ocean or offshore waters offer a tremendous potential for expansion of the marine farming sector, and developments to date indicate that it is indeed feasible to install, maintain, and operate cage culture systems in high-energy offshore waters. Despite evidence that open ocean farming is possible, production has been limited thus far and a number of technical, operational, economic and political challenges must be addressed before large-scale production in true open ocean conditions can be realized.

7.1 The context for open ocean farming

Population growth and consumer preference have resulted in a growing demand for seafood, a trend that is projected to continue into the future (FAO 2006). Production from capture fisheries has leveled off and by most projections will remain stagnant or decline depending on management and regulatory measures implemented by fishing nations (NOAA 2005a; Worm *et al.* 2006). In contrast, aquaculture production has increased by nearly 10% each year since 1980, and has played an important role in filling the gap between seafood supply and demand. There are signs, however, that the rate of growth may have peaked for

Aquaculture Production Systems, First Edition. Edited by James Tidwell.
© 2012 John Wiley & Sons, Inc. Published 2012 by John Wiley & Sons, Inc.

land-based and nearshore marine culture due to political, environmental, economic, and resource constraints (FAO 2006). Expansion of land-based culture is limited primarily by economics, particularly in developed countries where costs associated with land, capital equipment, and energy required to pump and filter water are prohibitive. For nearshore marine cage culture, available space is the primary limiting factor. Suitable sites for marine faming in protected coastal waters are, for most countries, quite limited to begin with and those that do exist are used for a multitude of recreational and commercial activities with which aquaculture must compete for space. Expansion of large-scale finfish farming in coastal waters is also constrained by environmental concerns, engendered primarily by the unintentional and undesirable environmental effects of salmon farming that occurred during a period of rapid industry growth in the 1980s and 1990s. Incidents of seafloor pollution from uneaten feed and fish wastes (Hargrave *et al.* 1993), outbreaks of deadly diseases (Hovland *et al.* 1994), interaction with marine mammals and other predators (Nash *et al.* 2000), overuse of antibiotics and biocides, and escapement of fish from sea cages were documented. While the worst conditions were associated with inappropriate sites or poorly managed farms (NOAA 2001), the growing opposition to cage culture was nonselective and all salmon farming—and by way of extension, cage culture for many species—has been deemed environmentally "unsustainable" by opponents of marine fish farming.

Responding to criticism from environmental groups and pressure from regulatory agencies, the industry began to improve management practices to address environmental concerns. Better feed formulations and careful monitoring of the feeding response of fish resulted in improved feed conversion ratios and less waste (DFO 2005). Vaccines drastically reduced the use of antibiotics (Knapp *et al.* 2007). More informed site selection led to a reduction in benthic impacts and fallowing allowed impacted sites to recover. Voluntary codes of conduct that embraced environmental protection were developed in the United States, Canada, Europe, and Australia. While environmental performance has greatly improved, public perception of coastal fish farming has not, and opposition persists—if anything, it has increased in recent years. In the current climate, new permits to farm salmon or other species in coastal waters are difficult, if not impossible, to obtain.

In developed countries, conflict with coastal residents and tourist-related businesses over aesthetic values, primarily over water views from shorefront property, have also affected the permitting of new cage culture sites. As the demographic of coastal communities continues to change and new residents place more value on views and recreation than food production, these conflicts will only increase. Given the constraints on expansion of current methods of production, it is clear that alternative approaches are needed in order for the marine aquaculture sector to make a meaningful contribution to the world's seafood supply.

Farming in open ocean marine waters (as used in this chapter, synonymous with "offshore farming") has been identified as one potential option for increasing production and has been a focus of international attention for more than a decade. Despite this global interest, industry development in open ocean waters has been measured, primarily due to the significant technical and operational challenges posed by wind and wave conditions in most of the world's oceans (Ryan 2004). Open ocean farming requires a completely new engineering approach since equipment and methods currently used for fish production in protected nearshore waters are largely unsuitable for the open ocean. In addition, the scale of investment required to develop and demonstrate new technologies and methods for offshore farming is yet to be determined, though most engaged in this endeavor would agree that it will likely be substantial.

Despite these challenges, there is sufficient rationale for pursuing the development of open ocean cage culture. Favorable features include ample space for expansion, tremendous carrying capacity, reduced conflict with many user groups, lower exposure to human sources of pollution, the potential to reduce some of the negative environmental impacts of coastal fish farming (Ryan 2004; Helsley & Kim 2005; Ward *et al.* 2006; Langan 2007) and optimal environmental conditions for a wide variety of marine species (Ostrowski & Helsley 2003; Ryan 2004; Benetti *et al.* 2006; Howell *et al.* 2006). For many countries where cost, environmental concerns, limited space and competing uses have restricted growth of land-based and nearshore marine farming, few other options for significant expansion exist.

7.2 Characterization and selection of open ocean sites

7.2.1 Definition

Before discussing approaches to open ocean aquaculture development, it is important to establish a clear definition of the term. For most engaged in this sector, the terms *open ocean* and *offshore* are interchangeable and are generally accepted to mean farming in locations that are subjected to ocean waves and currents and are removed from any significant influence of land masses, rather than a set distance from shore. Clearly, a wide range of sea conditions falls under this broad definition. Ryan (2004) reported on a site classification system for marine waters developed in Norway that is based on significant wave height exposure (table 7.1). While this classification method is instructive, knowledge

Table 7.1 Norwegian classification of offshore waters based on significant wave heights (Ryan 2004).

Site Class	Significant Wave Height (Meters)	Degree of Exposure
1	<0.5	Small
2	0.5–1.0	Moderate
3	1.0–2.0	Medium
4	2.0–3.0	High
5	>3.0	Extreme

of the full range of conditions that occur at a particular site is needed to design and sufficiently test robust engineered systems and to develop safe and efficient operating procedures.

7.2.2   Site selection criteria and methods

The suitability of sites for open ocean farming is dependent on a number of criteria, many of which are also considerations for nearshore sites. These include proximity to infrastructure such as ports, processing, and distribution centers; physical and biological criteria such as bathymetry, seabed characteristics and contour, current velocities, temperature profiles, dissolved oxygen, turbidity, and the frequency of occurrence of harmful algal blooms. The most important additional feature of offshore sites is wave climate. Significant wave heights, wave periods, the frequency and duration of high energy storm conditions and combined forcing of waves and currents must be known in order to determine whether a site is suitable, and if so, what type of technology is required for farming. For example, some sites may be relatively calm most of the time and infrequently experience occurrences of severe weather such as tropical cyclones. Other sites may never have waves greater than three meters but they may experience short period waves in this range most of the time—conditions that would cause excessive wear and tear on equipment and make surface operations such as feeding and harvesting difficult. For the former scenario, technology operated at the surface with the option to submerge for short periods would be appropriate, while for the latter it is likely that submersible cages supported by automated technologies would be needed.

It is imperative that a thorough evaluation that includes the parameters described above be conducted before proceeding with development of a site for farming. The requirements for data and subsequent analysis can be substantial; however, the use of advanced oceanographic technologies can greatly facilitate this task. Multibeam sonar and three-dimensional visualization can generate a wealth of data on seafloor contours and texture to inform mooring system design and placement. Collection of time-intensive data on temperature, salinity, dissolved oxygen, turbidity, and fluorescence can be greatly facilitated by strategic deployment of in-situ instrumentation at appropriate depth intervals in the water column. Additional instrumentation should include Acoustic Doppler Current Profiler (ADCP) current meters that can profile current velocity and direction throughout the water column, wave sensors that can give precise data on wave height, steepness, direction and period, and meteorological sensors to measure air temperature and wind speed and direction. Many countries have buoy arrays in coastal and shelf waters that can provide long-term data on regional climatology to aid site evaluation; however, collection of site-specific data is critical. Assessment of the potential for the effects of global climate change on critical site features such as water temperature and storm frequency and intensity should also be considered.

The data collection period required for site evaluation will vary, depending on local and regional environmental and meteorological conditions. Good baselines for some parameters can be established in a relatively short time frame (one year), others such as the frequency, duration, and severity of storms or blooms of toxic algae are less predictable and it may take longer to determine the suitability of a particular site.

In addition to physical, chemical, and biological characteristics of a site, other human uses in the area such as shipping, fishing, and mining must be identified in order to avoid conflicts. Other factors including use of the area by marine mammals, the likelihood of encounters with large predators, the location of important spawning grounds for indigenous fish, and proximity of sensitive biological communities must also be considered. Many countries require characterization of the benthic community and sediment quality to establish of preoperational baselines of environmental quality.

7.3 Technologies for open ocean farming

Initial attempts at offshore farming relied to a large extent on trial and error. Cages and mooring systems used in sheltered sites were simply moved to ever more exposed locations, and, as might be expected, many failures occurred. In addition to catastrophic failures, excessive wear and tear on cages, nets, and mooring components meant that crews had to spend more time on maintenance, adding to the production cost for the farm owner. It became clear that new technologies were needed to farm in open ocean waters.

7.3.1 Engineering design and assessment

Beginning in the early 1990s, several groups began to apply a more sophisticated engineering approach to cage (Loverich & Goudey 1996; Lisac 1996) and mooring design (Fredricksson *et al.* 2004), assessment of the structural integrity of cage materials (DeCew *et al.* 2005), and modeling the effects of hydrodynamic forcing on cages and netting (Lader & Fredheim 2003; Swift *et al.* 2006). An approach that includes numerical modeling (e.g., finite element modeling), scale model testing and in-field measurement of line tensions and physical forcing on cage and mooring components, has been shown to effectively inform the design, materials selection, and integrity of offshore systems and to reduce the possibility of system failure (Fredricksson *et al.* 2003).

7.3.2 Mooring and cage systems

Mooring systems for ocean cages include modifications of multi-cage grid systems commonly used for nearshore cages (fig. 7.1), single point moorings for individual cages (fig. 7.2), or cage arrays (fig. 7.3) and novel constructs such

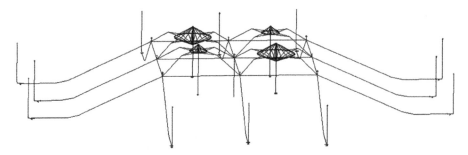

Figure 7.1 A schematic of the submerged grid mooring system designed by the University of New Hampshire, United States. This system can accommodate up to four submersible cages.

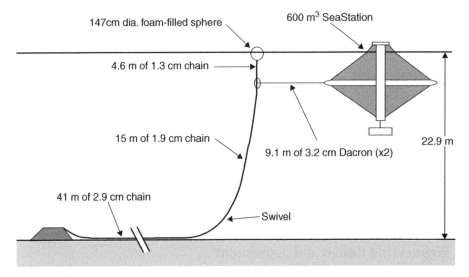

Figure 7.2 A schematic of the single-point mooring system designed by MIT to anchor an individual SeaStation cage.

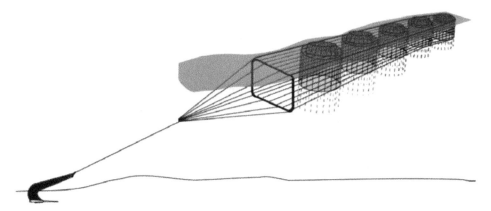

Figure 7.3 The single point mooring system designed by the Israeli company SUBflex to anchor an in-line array of submersible cages.

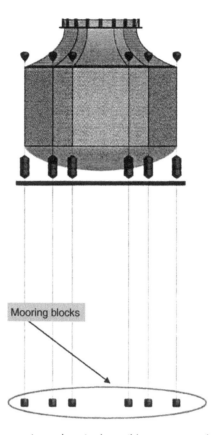

Mooring blocks

Figure 7.4 The tension leg mooring and semi-submersible cage system developed RefaMed, Italy.

as tension leg moorings (fig. 7.4). Mooring components typically include dead-weight anchors or embedment or plow-type anchors, heavy-chain ground tackle, synthetic braided or twisted rope, and specialized rings and plates to connect mooring component lines. Submerged grid systems also utilize buoyancy at critical junctions to provide tension and rigidity for the mooring system (fig. 7.5).

A wide array of cage technologies has been applied or proposed for use in open ocean sites. These technologies have been described in detail by Scott and Muir (2000), and more recently by Ryan (2004), though a number of new technologies have emerged in the concept or prototype phase since those documents were published. The previously mentioned authors parsed cage technologies into several categories based on structural and operational properties; however, for the purpose of simplicity, most cages can be divided into one of two main categories: (1) surface referenced or gravity cages that use steel, high density polyethylene (HDPE), or rubber collars to float the cage and net; and (2) submersible cages. Further subdivision can be made based on how they are moored to the seafloor and whether they are of flexible or rigid design and construction. In addition, some amount of hybridization has blurred the distinction between

Figure 7.5 A linear grid mooring system developed by the US company Ocean Spar to anchor multiple SeaStation cages. The buoyancy attached to the grid corners provides the tension to maintain the desired grid geometry.

the two categories and some older technologies and recent innovations defy categorization.

Given the wide range of sea conditions that fall under the definition of open ocean aquaculture, no single cage technology can be considered ideal or even appropriate for use under all circumstances. Currently, the greatest production in exposed locations is achieved with gravity cages. Large rubber (e.g., Bridgestone, Dunlop) and HDPE (e.g., PolarCirkel, Fusion, AquaLine) collar cages (fig. 7.6)

Figure 7.6 A photo of a large diameter HDPE collar gravity cage used for tuna penning at a semi-exposed site in the Mediterranean Sea.

are in use for salmon production in high-energy sites in Ireland, Scotland, the Faeroe Islands, and New Brunswick, Canada. Similar technologies are used in the Canary Islands and the Mediterranean Sea for bass and bream culture. Tuna fattening operations in Australia, the Mediterranean, and Mexico also use large HDPE collar cages in exposed and semi-exposed sites. The trend in recent years has been toward increasingly larger diameter (e.g., 50 meters) HDPE collar cages. The increased size results in greater flexibility in response to waves, as well as enormous production volumes that now exceed 60,000 cubic meters.

There are a number of advantages to the use of HDPE collar gravity cages: a relatively long history of use and operational familiarity at sheltered sites; their ability in some circumstances to use existing automated infrastructure such as air-piped centralized feeding systems; and their low cost relative to containment volume. However, there are also limitations to their use (Fredricksson *et al.* 2007). These include structural failures, operational difficulties related to feeding, harvesting, fish monitoring in rough weather, and increased maintenance to repair and replace system components due to excessive wear and tear. All of these limitations can affect production schedules and increase operational costs. Surface conditions such as waves and high currents during storms can also compress cage volume and can have detrimental effects on the fish.

Manufacturers of rubber and HDPE collar gravity cages continue to make structural improvements and several companies have developed submersible versions of their cages. These adaptations are likely to result in more robust systems and the option to submerge during storms will expand the range of sites in which these systems can be operated. It is likely that in the near term, these technologies will continue to be used for open-ocean farming at suitable sites.

Variations on the gravity cage include the Farmocean (fig. 7.7), which uses a rigid steel umbrella-like frame for floatation. The SADCO-SHELF (fig. 7.8) uses a similar steel superstructure and can also be operated in a submerged position. Both cages incorporate automated feeders in the structural framework. Due in part to high cost per volume as well as some structural and operational issues, neither of these cages has achieved wide-scale adoption.

Fully submersible cages include the semi-rigid SeaStation from Ocean Spar and the relatively new, rigid construction AquaPod from Ocean Farm Technologies (fig. 7.9). Moored individually or in a submerged grid system (fig. 7.5), SeaStations have been used successfully for nearly a decade in very rough conditions, including a site off the coast of New Hampshire in the northwest Atlantic, where significant wave heights can exceed nine meters (Chambers *et al.* 2007). There are currently more than fifty SeaStations deployed in sixteen countries for growout of a wide variety of species.

Despite their demonstrated ability to withstand extreme sea conditions and provide a stable environment for fish, fully submersible cages have not achieved widespread use for a number of reasons: The cages are small by commercial salmon farming standards (3,000 to 6,000 cubic meters), are relatively expensive, and are considered by some in the industry to be difficult to feed, harvest, and clean. Many routine operations require scuba diving, which is time consuming, expensive, and dangerous. Recent improvements to both the SeaStation

Figure 7.7 A photo of the modified gravity cage system developed by Farmocean International of Sweden. The top center component of the structure is the built-in feeding system.

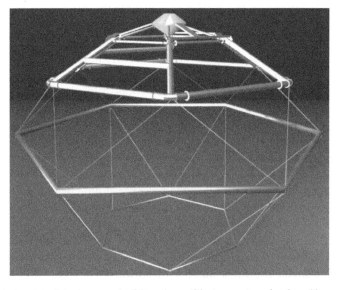

Figure 7.8 A drawing of the framework of the submersible cage system developed by SADCO-SHELF of Russia.

Figure 7.9 An artist's rendition of the submersible AquaPod cage from Ocean Farm Technologies (United States) anchored in a two-point mooring.

and AquaPod cages have made some operations such as net cleaning, removal of mortalities, and harvesting safer and easier, and the companies continue to address the operational aspects of their cages. Both companies have developed larger versions of their cages, though even the largest of these designs is substantially smaller than gravity cages now in use. Farms that are currently using fully submersible systems tend to be small (two to eight cages) with relatively low production volumes of high-value niche species.

A number of new designs for submersible cages have emerged in recent years, such as the 40,000 cubic meter Ocean Globe from Byks of Norway, but few of them have been built at full scale and virtually none have been tested in field situations. Innovation in submersible cage technology continues and several new designs are due for unveiling in the near future; however, reluctance by the industry to embrace submerged technologies has hampered their development.

There are several cage technologies—some implemented and others in the design phase—that do not fall neatly into either the gravity cage or submersible categories. Ocean Spar developed a 20,000-cubic-meter anchor tension cage in the 1990s called the AquaSpar that uses spar buoys rather than a floatation collar to provide buoyancy and create the cage volume. RefaMed in Italy has developed the tension-leg mooring mentioned previously to secure what is essentially an inverted gravity cage (Lisac 1996). A number of these cages are in use

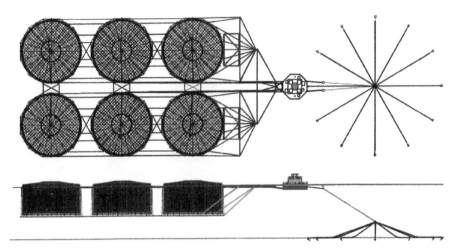

Figure 7.10 Drawings of top and side views of a mooring and cage system developed by Aquaculture Engineering Group from New Brunswick, Canada.

in the Mediterranean for bass and bream culture. The New Brunswick, Canada, company Aquaculture Engineering Group (AEG) has developed a unique farming concept that integrates an array of modified gravity cages with an automated feeding system and current velocity deflector designed for use in the high-current open waters of the Bay of Fundy (fig. 7.10). The entire system is anchored to the seafloor by a single-point mooring and orients itself in the direction of the current (fig. 7.10). The Israeli company SUBflex has developed a similar design that utilizes a single-point mooring and an in-line array of cages, though the SUBflex system is fully submersible (fig. 7.3). A concept for an untethered, ocean drifter cage was developed by Ocean Spar and the Massachusetts Institute of Technology in the 1990s (Goudey 1999), but it has not advanced beyond the concept and design phase for more than a decade.

7.3.3   Supporting technologies

In addition to the challenge of developing sufficiently robust containment and mooring systems, offshore farming presents challenges for nearly all aspects of day-to-day farm operation. Methods and equipment developed for routine operations such as feeding, harvesting, and monitoring at sheltered sites have been designed for calm sea conditions, and, for the most part, cannot be directly transferred to the open ocean environment. Development of alternative operational systems has not kept pace with ocean cage development and farmers have struggled to integrate existing, as well as new and unproven, supporting technologies into offshore installations.

Of all operations, feeding is probably the most important. Inshore approaches, which include dispensing feed by cannons from a service vessel or automated

Figure 7.11 A photo of the multi-cage automated feeding system developed by the University of New Hampshire and Ocean Spar at the university's experimental open-ocean farm.

feeding with blowers mounted on centralized feed barges, are severely hampered by rough seas. An ideal feeding system for offshore aquaculture would be robust, remotely controlled, fully automated, able to accommodate the volume of food needed for a two- to three-week period and have a hydraulic rather than pneumatic feed delivery system. It would also ideally be capable of wireless transmission of in-cage video, environmental monitoring data or other information critical to farm operation. Though no system as described currently exists, some progress has been made. The Scottish company Gael Force has developed the Sea Cap, a concrete feed barge that has operated successfully in semi-exposed locations for several years. The University of New Hampshire (UNH) has developed two small (single cage), remotely operated feeders that have been in use since 2001 (Rice *et al.* 2003) and designed and built a larger multi-cage feeder in 2007 (fig. 7.11). Developed in conjunction with Ocean Spar, the feeder has four separate feed silos and can dispense feed hydraulically to four submerged cages (Turmelle *et al.* 2006). It also incorporates a two-way remotely operated communications system that operates in-cage video cameras to monitor fish behavior and response to feed introduction. The system also houses a unique acoustic tracking system that can continuously monitor behavioral and physiological responses of tagged fish to changes in environmental conditions (Howell *et al.* 2006). Farmocean and SADCO also have integrated automated feeding systems into their cage designs, and the Canadian company Aquaculture Engineering Group has developed an automated hydraulic feeder that is integral to their single-point mooring cage array.

Other routine offshore operations such as grading, harvesting, biofouling control, and removal of mortalities are complicated by sea conditions and additional strategies must be developed to make these practices safer and more efficient.

7.4 Finfish species cultivated in open ocean cages

Fish cultivated in open ocean cages include warm and cool temperate as well as tropical and subtropical species, though at present, cool temperate species predominate. In terms of production volume, the leading species is Atlantic salmon (*Salmo salar*), which are produced in gravity cages in Ireland, Scotland, Canada, Norway, and the Faeroe Islands. Sea bass (*Dicentrarchus labrax*) and sea bream (*Sparus auratus*) are grown in gravity, tension leg, and submersible cages throughout the Mediterranean (Ryan 2004). Fattening of Northern Bluefin tuna (*Thunnus thynnus*) in large high-density polyethylene (HDPE) collar gravity cages takes place in exposed locations throughout the Mediterranean and has expanded dramatically in recent years. Other tuna fattening operations that use HDPE cages in exposed locations include Southern Bluefin tuna (*Thunnus maccoyii*) in South Australia, Pacific Bluefin (*Thunnus thynnus orientalis*), yellowfin (*Thunnus albacares*) and bigeye tuna (*Thunnus obesus*) in Mexico. In the northeastern United States, the University of New Hampshire operates an experimental open-ocean farm and has produced small quantities of Atlantic cod (*Gadus morhua*), haddock (*Melanogrammus aeglefinis*), and Atlantic halibut (*Hippoglossus hippoglossus*) in submerged SeaStation cages (Howell *et al.* 2006).

Warm temperate and tropical species produced in ocean cages include milkfish (*Chanos chanos*) in the Philippines; Florida pompano (*Trachinotus carolinus*) in Belize; summer flounder (*Paralichthys dentatus*) in Mexico; parrotfish (*Oplegnathus faciatus*) and olive flounder (*Paralichthys olivaceus*) in Korea; and golden pompano (*Trachinotus blochii*) in China. In Hawaii, Pacific Threadfin (*Polydactylus sexfilis*) and amberjack (*Seriola rivoliana*) are being grown commercially in submerged cages at open-ocean sites off Oahu and the Island of Hawaii. Another warm-water species of interest for open-ocean farming is cobia (*Rachycentron canadum*), which is currently produced in Puerto Rico and the Bahamas.

A number of additional marine species have been identified as having potential for open ocean production. In the United States, Sablefish (*Anoplopoma fimbria*) is a leading candidate for ocean farming in the Pacific northwest and California yellowtail (*Seriola dorsalis lalandi*), striped bass (*Morone saxatilis*), California halibut (*Paralichthys californicus*), and tuna (*Thunnus* sp.) have been proposed for open ocean culture in Southern California. Red drum (*Sciaenops ocellatus*), Florida pompano (*Trachinotus carolinus*), and tunas are being considered for the Gulf of Mexico in the United States. Mulloway (*Argyrosomus hololepidotus*) and yellowtail kingfish (*Seriola lalandi*) are under consideration for open ocean culture in Western Australia.

It is likely that many more species will come into production as offshore farming technologies are further developed; however, it should be noted that aside from a few species such as Atlantic salmon, sea bass, and sea bream (which are produced from established broodstock lines) many of the marine species under cultivation are the offspring of wild fish. Additional broodstock and hatchery work is needed to achieve desirable traits such as better and more uniform growth rates, improved food conversion ratios, delayed maturation, and disease resistance.

7.5 Environmental considerations

Like all forms of food production, the culture of marine species, whether practiced in land-based, nearshore, or offshore locations will have some effect on the environment. The effects can be both negative and positive and can vary depending upon the species, location, and farming practices. Concerns over the potential impacts of open ocean finfish farming essentially mirror those of nearshore cage culture (table 7.2; however, the degree of impact will likely be different, as will society's perceptions).

7.5.1 Cage effluents

Due to the limited number and relatively small size of the farms operating in open ocean sites, scientific knowledge of possible effects on offshore environments is not, as yet, well known. Data gathered to date would indicate that water column impacts from inorganic nutrients are unlikely (Helsley & Kim 2005; Langan 2007) although effects will vary depending upon production volume and the trophic status of receiving waters. Also, inputs from individual farms must be examined in the context of inputs from other sources in the region, including other farms. Benthic impacts from deposition of organic materials from waste feed and feces may also be reduced due to greater dispersion by ocean currents. Ryan (2004) cited a 2001 *Review of Benthic Conditions at Irish Fish Farms* conducted by Aquafact International Services Ltd., which reported benthic impacts were greatly reduced, if not more or less absent, beneath farms at exposed sites. At an experimental US offshore farm off the New Hampshire coast, no significant differences have been observed in sediment organic content or benthic community diversity between projected impact and reference sites after seven years of multi-species farming (Ward *et al.* 2006; Langan 2007). Alston *et al.* (2005) reported only minor changes to the benthos directly beneath cobia cages at an offshore farm in Puerto Rico, and Rapp (2006) found no changes in the organic content of the sediments beneath the cages at the same farm. Due to the small size of the farms in these studies, the results should not be viewed as convincing evidence that the potential for benthic impacts from offshore farms can be dismissed. A case in point is a benthic study conducted at an offshore

Table 7.2 Potential environmental impacts form marine net pen farming (NOAA 2005b).

Effects	Sources
Increased organic loading	• Particulate organic loading Fish fecal material Uneaten fish feed Debris from biofouling organisms Decomposed fish mortalities on the farm • Soluble organic loading Dissolved components of uneaten feed Harvest wastes (blood)
Increased inorganic loading	• Nitrogen and phosphorus from fish excretory products • Trace elements and micronutrients (e.g., vitamins) in fish fecal matter and uneaten feed
Residual heavy metals	• Zinc compounds in fish fecal material • Zinc compounds in uneaten feed • Copper compounds in antifouling treatments
The transmission of disease organisms	• Indigenous parasites and pathogens • Exotic parasites and pathogens
Residual therapeutants	• Treatment by inoculation • Treatment in feed • Treatment in baths
Biological interaction of escapes with wild populations	• Unplanned release of farmed fish • Unplanned release of gametes and fertile eggs • Cross infection of parasites and pathogens • Planned release of cultured fish for enhancement or ranching
Physical interaction with marine wildlife	• Entanglement with lost nets and other jetsam • Entanglement with nets in place, structures, and moorings, etc. • Attraction of wildlife species (fish, birds, marine mammals, reptiles) • Predator control
Physical impact on marine habitat	• Buoyant fish containment structures and mooring lines • Anchors and moorings
Using wild juveniles for growout	• Harvest of target and nontarget species as larvae, juveniles, and subadults
Harvesting industrial fisheries for fish feed	• Increased fishing pressure on the shoaling small pelagic fish populations

farm in Hawaii. Lee *et al.* (2006) reported an increase in opportunistic polychaete species and some loss of diversity in the benthic community, though these changes were spatially limited to areas immediately beneath the cages. As this study demonstrates, benthic impacts may be reduced in offshore environments; however, an expectation of "no impact" is unrealistic.

The degree of environmental change that will be tolerated must be decided by governing bodies in their respective jurisdictions, although it would be advantageous to establish international agreement on environmental performance standards with the caveat that some flexibility in the measured parameters is needed to account for site differences. There is no need to create standards from scratch. Based on existing knowledge of benthic impacts from nearshore farms, standards that have been developed for ecosystem protection in countries including Australia, Ireland, Norway, Scotland, and the United States have many common features and provide a good starting point for addressing environmental concerns.

Models, particularly those that integrate hydrodynamic, physical, and biological processes, can play a major role in predicting potential impacts of offshore farming. They enable scenarios to be run to forecast possible effects at specific farm sites as well as broader ecosystem effects. Farm-focused models such as AquaModel (Rensel *et al.* 2006) and DEPOMOD (Cromey *et al.* 2002) that use site-specific hydrodynamic, sediment, and biological data to simulate water column and benthic effects can be used to evaluate different management and production strategies. Site-specific models are also useful for identifying depositional areas and locating monitoring stations. Cumulative effects must also be considered as the offshore sector scales up so ecosystem models will be needed for broad area management.

It is reasonable to assume that monitoring will be required to ensure environmental compliance and to satisfy stakeholders that individual farms and the collective industry are operating in an environmentally responsible manner. Monitoring methods to determine the degree of organic enrichment vary in cost, clarity of interpretation, and the ability to standardize methods and establish meaningful performance standards. Detailed enumeration of benthic species obtained from sediment grab samples may be required during the initial site assessment to establish baselines and during the first few stocking cycles; however, this type of monitoring is very expensive and time consuming. Ideally, video or photographic analyses coupled with some type of chemical measurement such as Total Organic Carbon (TOC) or Loss on Ignition (LOI) that have been calibrated against this information would replace benthic monitoring and reduce costs for farm owners.

7.5.2 Diseases and parasites

Open ocean culture may offer some fish health benefits, which in turn may reduce the risk of transmission of parasites and diseases to wild populations as well as reduce the need for treatments and therapeutics. Ryan (2004) reported lower incidence of sea lice at open ocean salmon farms in Ireland, which he attributed to dispersion of the planktonic stage of the ecto-parasites. One species of sea lice, *Lepeophtheirus salmonis,* is known to aggregate at the mouths of estuaries; therefore, placing salmon farms further offshore would greatly reduce exposure to this parasite (Costello *et al.* 2004). In addition, lower stress levels and better fish health observed at offshore farms have been attributed to the more stable temperature and salinity regimes, as well as the higher oxygen concentration and reduced ammonium levels that result from the greater water exchange through cages (Ryan 2004; Benetti *et al.* 2006; Howell *et al.* 2006). Bricknell (2006) concluded, "Offshore aquaculture offers many opportunities to reduce disease interactions between wild and farmed fish."

7.5.3   Escapement

The high-energy conditions of offshore sites increase the risk of fish escapement due to catastrophic equipment failure; therefore, the robustness of engineered systems (e.g., cages and moorings) must be carefully matched to site conditions. Some fish species under development for offshore farming are particularly prone to escapement. Cod, for example, will bite holes in netting material and escape from cages (Moe *et al.* 2007), so more durable alternatives to woven netting are needed. Also, since visual inspections of equipment by divers are more difficult and dangerous at offshore sites, automated systems such as video surveillance or acoustic abundance estimators may be required to monitor escapement. The major concern over salmon escapement from nearshore net pens is biological interactions with wild salmon. Moving farms offshore and away from salmon spawning rivers would greatly reduce this risk.

7.5.4   Marine mammal and predator interactions

Farming offshore may reduce exposure to coastal predators such as pinnipeds; however, exposure to other large predators such as sharks may increase. This risk can be managed through careful site selection to avoid known aggregation areas of local predators, prompt removal of dead or moribund fish, and the use of more robust containment barriers. The chance of encounters with protected species (e.g., whales, sea turtles) may also increase; however, entanglement with the mooring lines typically used for ocean cages is unlikely. Endangered whales are frequently sighted in close proximity to an open-ocean farm off the coast of New Hampshire and no adverse interactions have occurred in nine years of farming at this site (Ward *et al.* 2006).

7.5.5 Seed and feed issues

Issues such as the use of wild juveniles, which has been standard operating procedure for tuna penning and the use of industrial fish as feed ingredients, are essentially the same for offshore as for nearshore farming. There have been some advances in hatchery production of juvenile tuna species; however, it may be years before large quantities of hatchery produced juveniles are available for ongrowing. Significant progress has been made in developing alternatives to fish meal and fish oil, and continued research and development is likely to produce economically and nutritionally viable substitutes.

7.6 Future prospects and challenges

Developments over the past two decades indicate that fish farming open ocean environments is feasible and that there is potential for large-scale production of a wide variety of species. Conflicts with other uses can be significantly reduced, though they are not totally eliminated. Potential conflicts with capture fishing, navigation, and offshore energy production must be considered when selecting sites for offshore farming. There is also some evidence to support the premise that environmental impacts can be reduced by farming in offshore environments. Additional information on the potential effects of large-scale production is needed is needed to inform the development of rational policies and counter the negative perceptions of marine farming. Ideally, development of offshore farming should take place within the context of overall ocean management in order assure compatibility with other uses and consistency with broader goals to restore and sustain the health, productivity, and biological diversity of the oceans.

Significant progress has been made in the development of new marine species and engineered systems for offshore farming; however, a number of technical challenges remain. Hatchery technology has been developed for a number of marine fish species, though additional time and investment is needed to establish broodstock lines and the infrastructure to produce and transport large quantities of robust fingerlings. Open ocean farms may need to be coupled with nearshore farms to provide appropriate sites for intermediate stages of stocks, as larger juveniles may be needed for offshore cages. Nearshore farms can also be used as holding facilities for market-ready stock when weather conditions prohibit safe harvesting at offshore sites. There are a number of proven cage and mooring systems available, although they need to scale up in order to achieve the production volumes required for economic viability. Further development of supporting technologies is needed, including ocean-going service vessels, automated feeders, and remote-control observation and communication systems. Therefore, substantial and sustained investment in research and development from public and private sectors must be secured. In particular, research should focus on the development of highly mechanized and fully integrated offshore farming systems, to achieve greater efficiency and ensure worker safety in the conduct of routine operations. Until "turnkey" systems that are essentially autonomous are available and economic viability of offshore farming can be demonstrated, expansion of this sector in the near future will be limited. A logical next step to hasten the development of offshore farming would be to establish commercial-scale offshore demonstration farms supported by a combination of public and private funds where technologies can be tested at reduced financial risk for the private sector and a greater understanding of the environmental effects of offshore farming can be gained to inform a rational regulatory framework. Demonstration farms would also be useful for training vessel operators and farm personnel to work safely and efficiently in offshore environments.

Though the complexity and scale of the challenge for offshore aquaculture development is formidable, the potential economic and health benefits of a sustainable ocean farming industry are enormous. Realizing the vision for open-ocean farming will require creativity and innovation supported by substantial investment, as well as close collaboration between nations and the business, government, and research communities.

7.7 References

Alston, D.E., Cabarcas, A., Capella, J., Benetti, D., Keene-Maltzof, S., Bonilla, J. & Cortes, R. (2005) *Environmental and Social Impact of Sustainable Offshore Cage Culture Production in Puerto Rican Waters*. Final Report for NOAA Grant No. NA16RG1611, April 4.

Benetti, D., O'Hanlon, B., Brand, L., Orhun, R., Zink, I., Doulliet, P., Collins, J., Maxey, C., Danylchuk, A., Alston, D. & Cabarcas, A. (2006) Hatchery, ongrowing technology and environmental monitoring of open ocean aquaculture of cobia (*Rachycentron canadum*) in the Caribbean. Abstract No. 10424, World Aquaculture 2006, Florence Italy. World Aquaculture Society, Baton Rouge.

Bricknell, I. (2006) Disease Implications of Offshore Aquaculture. Breaking the Disease Cycle? In *Offshore Mariculture 2006 Conference Programme*. October 11–13, 2006, Malta.

Chambers, M.D., Langan, R., Howell, W., Celikkol, B., Watson, K.W., Barnaby, R., DeCew J. & Rillihan, C. (2007) Recent developments at the University of New Hampshire open ocean aquaculture site. *Bulletin of the Aquaculture Association of Canada* 105(3):233–245.

Costello, M.J., Burridge, L., Chang, B. & Robichaud, L. (2004) Sea Lice 2003—Proceedings of the sixth international conference on sea lice biology and control. *Aquaculture Research* 35(8):711–2.

Cromey, C.J., Nickell, T.D. & Black, K.D. (2002) DEPOMOD—modeling the deposition and biological effects of waste solids from marine cage farms. *Aquaculture* 214(1–4):211–39.

DeCew, J., Fredriksson, D.W., Bugrov, L., Swift, M.R., Eroshkin, O. & Celikkol, B. (2005) A case study of a modified gravity type cage and mooring system using numerical and physical models. *IEEE Journal of Oceanic Engineering* **30**(1): 47–58.

DFO (2005) Myths and realities of salmon farming-updated. *Fisheries and Oceans Canada*. http://www.dfo-mpo.gc.ca/media/backgrou/2005/salmon_e.htm.

FAO (2006) State of World Aquaculture: 2006. *FAO Fisheries Technical Paper No. 500*. FAO, Rome.

Fredricksson, D.W., Swift, M.R., Irish, J.D., Tsukrov, I. & Celikkol, B. (2003) Fish cage and mooring system dynamics using physical and numerical models with field measurements. *Aquacultural Engineering* **27**(2):117–46.

Fredricksson, D.W., DeCew, J. Swift, M.R., Tsukrov, I., Chambers, M.D. & Celikkol, B. (2004) The design and analysis of a four-cage, grid mooring for open ocean aquaculture. *Aquacultural Engineering* **32**(1):77–94.

Fredricksson, D.W., DeCew, J.C., Tsukrov, I., Swift, M.R. & Irish, J.D. (2007) Development of large fish farm numerical modeling techniques with in-situ mooring tension comparisons. *Aquacultural Engineering* **36**(2):137–48.

Goudey, C.A. (1999) Design and analysis of a self-propelled open-ocean fish farm. In *Proceedings of the Third International Conference on Open Ocean Aquaculture* (Ed. by R.R. Stickney), pp. 65–77. TAMU-SG-99-103.

Hargrave, B.T., Duplisea, D.E. & Pfeiffer, E. (1993) Seasonal changes in benthic fluxes of dissolved oxygen and ammonium associated with marine cultured Atlantic salmon. *Marine Ecology Progress Series* **96**:249–57.

Helsley, C.E. & Kim, J.W. (2005) Mixing downstream of a submerged fish cage: A numerical study. *IEEE Journal Of Oceanic Engineering* **30**(1):12–9.

Hovland, T., Nylund, A., Watanabe, K. & Endresen, C. (1994) Observations of infectious salmon anemia virus in Atlantic salmon, *Salmo salar* L. *Journal of Fish Diseases* **17**(3):291–6.

Howell, W.H, Watson, W.H. & Chambers, M.D. (2006) Offshore production of cod, haddock, and halibut. CINEMar/Open Ocean Aquaculture Annual Progress Report for the Period from Janurary 1, 2005, to December 31, 2005. Final Report for NOAA Grant No. NA16RP1718, interim Progress Report for NOAA Grant No. NA04OAR4600155. http://ooa.unh.edu.

Knapp, G., Roheim, C. & Anderson, J. (2007) *The Great Salmon Run: Competition Between Wild and Farmed Salmon*. TRAFFIC North America. World Wildlife Fund, Washington, DC.

Lader, P. & Fredheim, A. (2003) Modeling of net structures exposed to three-dimensional waves and current. In *Open Ocean Aquaculture: From Research to Commercial Reality* (Ed. by C.J. Bridger & B.A. Costa-Pierce), pp. 177–189. World Aquaculture Society, Baton Rouge.

Langan, R. (2007) Results of environmental monitoring at an experimental offshore farm in the gulf of Maine: Environmental conditions after seven years of multi-species farming. In *Open Ocean Aquaculture—Moving Forward* (Ed. by C.S. Lee & P.J. O'Bryen), pp. 57–60. Oceanic Institute, Waimanalo, HI.

Lee, H.W., Bailey-Brock, J.H. & McGurr, M.M. (2006) Temporal changes in the polychaete infaunal community surrounding a Hawaiian mariculture operation. *Marine Ecology Progress Series* **307**:175–85.

Lisac, D. (1996) Recent developments in open-sea cages: Practical experience with the tension leg cage. In *Open Ocean Aquaculture: Proceedings of an International Conference* (Ed. by M. Polk), pp. 513–522. May 8–10, Portland, Maine. New Hampshire/Maine Sea Grant College Program Rpt. UNHMP-CP-SG-96-9.

Loverich, G.F. & Goudey, C. (1996) Design and operation of an offshore sea farming system. Open ocean aquaculture. In *Open Ocean Aquaculture: Proceedings of an International Conference* (Ed. by M. Polk), pp. 495–512. May 8–10, Portland, Maine. New Hampshire/Maine Sea Grant College Rpt. UNHMP-CP-SG-96-9.

Moe, H., Dempster, T., Sunde, L.M., Winther, U. & Fredheim, A. (2007) Technological solutions and operational measures to prevent escapes of Atlantic cod (*Gadus morhua*) from sea cages. *Aquaculture Research* 38(1):91–9.

Nash, C.E., Iwamoto, R.N. & Mahnken, C.V.W. (2000) Aquaculture risk management and marine mammal interactions in the Pacific Northwest. *Aquaculture* 183(3–4):307–23.

NOAA (2001) *The Net-pen Salmon Farming Industry in the Pacific Northwest* (Ed. by C. Nash). NOAA Technical Memorandum NMFS-NWFSC-49.

NOAA (2005a) *Fisheries of the United States—2003*. NOAA, Washington, DC. http://www.st.nmfs.gov/st1/fus/fus03/index.html.

NOAA (2005b) *Guidelines for Ecological Risk Assessment of Marine Fish Aquaculture* (Ed. by C.E. Nash, P.R. Burbridge & J.K. Volkman). US Dept. of Commerce, NOAA Technical Memorandum, NMFS-NWFSC-71.

Ostrowski, A.C. & Helsley, C.E. (2003) The Hawaii offshore aquaculture research project: Critical research and development issues for commercialization. In *Open Ocean Aquaculture: From Research to Commercial Reality* (Ed. by C.J. Bridger & B.A. Costa-Pierce), pp. 285–291. The World Aquaculture Society, Baton Rouge.

Rapp, P. (2006) Measurement of the benthic loading and the benthic impact from an open-ocean fish farm in tropical waters. Preliminary Report for NOAA Grant No: NA040AR4170.

Rensel, J.E., Buschmann, A.H., Chopin, T., Chung, I.K., Grant, J., Helsley, C.E., Kiefer, D.A., Langan, R., Newell, R.I.E., Rawson, M., Sowles, J.W., McVey, J.P. & Yarish, C. (2006) Ecosystem-based management: Models and mariculture. In *Aquaculture and Ecosystems: An Integrated Coastal and Ocean Management Approach* (Ed. by J.P. McVey, C-S. Lee & P.J. O'Bryen), pp. 207–20. The World Aquaculture Society, Baton Rouge.

Rice, G., Stommel, M., Chambers, M.D. & Eroshkin, O. (2003) The design, construction, and testing of the University of New Hampshire feed buoy. In *Open Ocean Aquaculture: From Research to Commercial Reality* (Ed. by C.J Bridger & B.A. Costa-Pierce), pp. 197–203. The World Aquaculture Society, Baton Rouge.

Ryan, J. (2004) *Farming the Deep Blue*. Bord Iascaigh Mhara—Irish Sea Fisheries Board Technical Report.

Scott, D.C.B. & Muir, J.F. (2000) Offshore cage systems-a practical overview. In *Mediterranean Offshore Mariculture* (Ed. by J. Muir & B. Basurco), pp 79–89. Zaragoza: CIHEAM (Centre International de Hautes Etudes Agronomiques Mediterraneennes), Serie B:Etudes et Recherches, No. 30 Options Mediterraneennes.

Swift, M.R., Fredriksson, D.W., Unrein, A., Fullterton, B., Patursson, O. and Baldwin, K. (2006) Drag force acting on biofouled net panels. Aquacultural Engineering 35(3):292–9.

Turmelle, C., Swift, M.R., Celikkol, B., Chambers, M., DeCew, J., Fredriksson, D., Rice, G. & Swanson, K. (2006) Design of a 20-ton Capacity Finfish Aquaculture Feeding Buoy. *Proceedings of the Oceans 2006*, MTS/IEEE Conference. Boston, MA.

Ward, L.G., Grizzle, R.E. & Irish, J.D. (2006) UNH OOA Environmental Monitoring Program, 2005. CINEMar/Open Ocean Aquaculture Annual Progress Report for the Period from January 1, 2005, to December 31, 2005. Final Report for NOAA Grant No. NA16RP1718, Interim Progress Report for NOAA Grant No. NA04OAR4600155, Submitted January 23, 2006. http://ooa.unh.edu.

Worm, B., Barbier, E.B., Beaumont, N., Duffy, J.E., Folke, C., Halpern, B.S., Jackson, J.B.C., Lotze, H.K., Micheli, F., Palumbi, S.R., Sala, E., Selkoe, E.K., Stachowicz, J.J. & Watson, R. (2006) Impacts of biodiversity loss on ocean ecosystem services. *Science* **314**:787–90.

Chapter 8

Reservoir Ranching

Steven D. Mims and Richard J. Onders

Global aquaculture continues to grow at a rapid rate, increasing 8.7% annually since 1970, whereas production from capture fisheries peaked in the mid-1980s (FAO 2008). Seventy six percent of global freshwater fish production is by some form of aquaculture. Reservoir ranching is a relatively new culture system contributing to inland freshwater fish production, and it is a newly defined term in this chapter to denote an extensive aquaculture practice that can provide an alternative supply of freshwater food fish under eco-friendly, minimum input, and sustainable conditions.

8.1 Reservoir ranching vs. culture-based fisheries

Reservoir ranching is an extensive culture system in which young hatchery-produced fish are stocked in existing freshwater impoundments, feed on naturally available foods, and are harvested after a period of time (Semmens & Shelton 1986; Onders *et al.* 2001). The fish may be owned by a sole proprietorship, a cooperative group, or a corporation. Private groups may gain access to reservoirs as permittees or lessees for profit or for benefit of a rural community. Reservoir ranching as a specific culture system for food fish production is part of a larger group of practices known as culture-based fisheries.

Culture-based fisheries are defined as "enhancement practices which are maintained by stocking one or more aquatic species for supplementing or sustaining

Aquaculture Production Systems, First Edition. Edited by James Tidwell.
© 2012 John Wiley & Sons, Inc. Published 2012 by John Wiley & Sons, Inc.

their recruitment and raising the total production or production of selected elements of a fishery beyond a level that is sustainable through natural processes" (FAO 1997). Culture-based fisheries include enhancement measures that take the form of introduction of new and sometimes exotic species, stocking natural and man-made water bodies, supplementing with fertilizer, engineering the environment for habitat modifications and improvements, altering species composition by eliminating undesirable species and stocking only select species, and introducing genetically manipulated or modified species.

Culture-based fisheries encompass a broad spectrum of enhancement practices by stocking to increase food fish production, increase depleted stocks, provide conservation and restoration of special concern species, and supplement recreational fisheries. However, the application of culture-based fisheries depends on both the desired outcomes of the stocking entity and the attitudes of the public and government agencies. In developing countries, culture-based fisheries are primarily used to provide a low-cost protein source to rural communities and provide job opportunities. However, in developed countries, culture-based fisheries are primarily directed toward recreational fishing, or on a limited basis, enhancement of capture fisheries.

"Ranching" is a term coined in the American West to describe an extensive method of livestock production on natural forage with little input of grain-based feed. In recent years, a similarly extensive type of culture-based fishery has been demonstrated in small bodies of water (generally less than 100 ha) as a cost- and resource-effective way of increasing supplies of food fish in rural areas (De Silva *et al.* 2006). Therefore, the term "reservoir ranching" will be used in the remainder of this chapter to describe an aquaculture (farming) system that is based on continued ownership of the stocked fish, limited management input, and generation of revenue through sales of the resulting fishery products.

8.2 Reservoir

A reservoir is defined as a structure, usually man made, where water is collected and stored for future use. Reservoirs were primarily built for flood control; water supply for potable, agricultural, and industrial uses; navigation; and in some cases, generation of hydroelectric power. They have been credited with significant advancements in human development through socio-economic viability and environmental sustainability (World Commission on Dams 2000).

Reservoirs are formed from the regulated containment of streams or rivers behind structures (dams) of reinforced concrete, earth and rock fill, or a combination of materials. Reservoir dams are classified as major, large, or small. A major dam is defined as having at least one of these criteria: height of >150 m, volume $>15,000,000$ m^3, reservoir storage of >25 km^3, and/or electric power generation capacity >3.6 million MJ. A large dam is defined as >15 m high from foundation to crest and small dams are defined as <15 m (ICOLD 1998; Rosenberg *et al.* 2000). Since the end of World War II, global construction of large dam reservoirs has soared to over 40,000 in 140 countries. This represents more than 10,000 km^3 of combined storage, which is about five times the volume of water in all the world's rivers (Chao 1995). China has been the leader with 24,671 dams, followed by the United States (6,375 dams) and India (4,010 dams). Various estimates exist for the total surface area of reservoirs in the world ranging from 400,000 km^2 to 1,500,000 km^2 (Shiklomanov 1993; St. Louis *et al.* 2000). Further, water impounded in reservoirs with small dams is substantial. McCully (1996), using data from USCOLD (1995) and ICOLD (1998), calculated that for every large dam there are seventeen small dams in the world. The global number of reservoirs with small dams has been estimated to be 800,000. In the United States, the US Army Corps of Engineers National Inventory of Dams, which lists large and small dams, indicated a total of 79,000 dams having about 274,000 km^2 of reservoir surface, with about 75% of the total in the small dam category.

8.3   Natural processes of reservoirs

As a culture system, a reservoir provides oxygen, regulates temperature, supplies food for fish, and processes waste entirely by natural processes. These processes are highly complex and interrelated and are beyond the scope of this chapter; however, a brief summary is offered.

Although oxygen is supplied primarily as a product of photosynthesis by algae, and to a lesser extent, vascular plants, diffusion from the atmosphere is also an important source of oxygen in reservoirs. Local weather patterns affect diffusive inputs because wind produces wave action and waves increase surface area, thereby increasing the water/air interface. Further, solubility of oxygen in water increases non-linearly with decreasing temperature. The altitude of a reservoir can affect the amount of oxygen present in the water as well. In general, solubility of oxygen will decrease with increasing altitude due to the decrease in atmospheric pressure. The seasonal patterns of water circulation or "turnover" within reservoirs are important factors in maintaining oxygen levels adequate for fish production. Reservoirs in temperate latitudes used for reservoir ranching often experience dimictic (spring and fall) turnover patterns that disrupt thermal stratification and redistribute oxygen throughout the water column. In reservoirs having high organic load with high bacterial oxidative rates, the circulation events can have negative effects on fish if mixing of the water column results in low oxygen levels.

Regulation of temperature in reservoirs is largely a function of solar radiation, and, therefore; temperature is a function of climate and weather. However, wind energy and convection are important in the distribution of heat through the water column, accounting for as much as 90% of the distribution (Wetzel 1983). In temperate region reservoirs, seasonal temperature changes gradually overcome the low thermal conductivity of water, resulting in spring and fall turnover events interspaced with the formation of thermally stratified zones through the water column.

Spring turnover in the smaller reservoirs most suited to reservoir ranching is actually a number of circulation events that take place over a short time period, perhaps only a few days, depending on weather. The lack of a thermal gradient and presence of uniform density in the cold waters of early spring present little resistance to circulation. Surface waters are warmed slightly on sunny days or similarly cooled during chilly nights, and wind provides the necessary energy for mixing. However, as warming continues and surface waters become less dense, a thermal gradient becomes established that resists circulation. Eventually, two major stratified zones are formed: the deeper hypolimnion, with temperature largely established by the water temperature at the end of spring turnover; and the warmer epilimnion, which floats on top of the higher density hypolimnion and continues to circulate. Separating these layers is a third transitional zone known as the metalimnion. A common term applied to this zone is "thermocline"; however, this term more correctly refers to the plane of greatest temperature change with increasing depth.

The thermal gradient remains in place until cooler fall air temperatures or cold rain increases surface layer density and mixing (fall turnover) occurs once more. In the interim period, oxygen typically becomes depleted in the hypolimnion, especially if light is restricted by turbidity or oxidation levels are high. As a result, fish become isolated to the epilimnion and a portion of the metalimnion. This factor must be considered when stocking rates are considered for reservoir ranching. Further, as mentioned above, low oxygen levels in the hypolimnion can result in insufficient oxygen for fish throughout the water column after the fall turnover.

The processes by which food is provided for fish in reservoir ranching and waste is recycled are so closely interrelated that they require being part of one discussion: that of food chains, which interlock to become a food web. Two food chains are easily identified in a reservoir: the grazing food chain and the detritus food chain. Both of these food chains find their origin with the primary producers, mostly phytoplankton, that synthesize new organic matter from inorganic carbon and the energy of the sun (fig. 8.1). When living, this new organic matter enters the grazing food chain, and if nonliving, enters the detritus food chain. In the grazing food chain, zooplankton consume living phytoplankton, along with the bacteria that colonize particulate matter, organic and inorganic, suspended in the water. The zooplankton is then consumed by invertebrates, such as predacious insects, and by zooplanktivorous fish. In the detritus food chain, nonliving organic matter is consumed by bacteria, which are then consumed by detritivores, such as insect larvae and mollusks, as they also consume the nonliving organic matter. In reservoir ranching, these detritivores can be an important food for benthic-feeding fish. The food chains interlock in multidirectional fashion to form the reservoir ranching food web (fig. 8.1).

The waste products of metabolism and respiration produced by ranched fish are not a consideration in the management strategy for maximum yield because the stocking densities are low. Further, water quality is not affected by reservoir ranching because the fish are not fed. However, it is worth mentioning that these waste products join normal biogeochemical cycles, including the nutrient

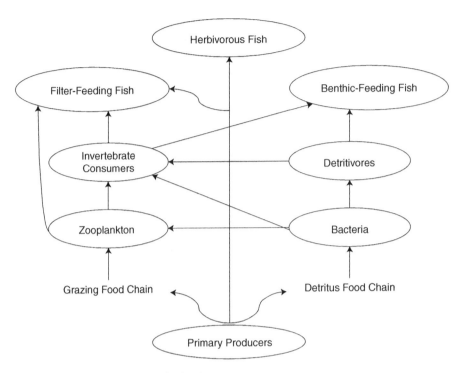

Figure 8.1 The reservoir ranching food web.

cycles that move elements and organic compounds essential to life, from the environment, to living organisms, and back to the environment. These cycles include, among others, the nitrogen, phosphorus, and carbon cycles commonly discussed in basic ecology texts.

8.4   Selection of reservoirs for reservoir ranching

There are millions of hectares of impounded water in many countries that have the potential to be used for reservoir ranching. However, not every impoundment may be suitable. Physical, physio-chemical, biological, and socio-economic factors must all be considered to determine the suitability of a reservoir for ranching of fish.

8.4.1   Physical factors

Reservoirs should store enough water to safely raise fish to market size each year. Because reservoirs often have multiple users, water consumption by the infrastructure (i.e., drinking water, irrigation, etc.) of the associated community should be estimated to ensure adequate water for the stocked fish. Also, any historical information (i.e., age of dam and its structural integrity) should be gathered for future decision-making.

In general, most ranching has been practiced in small (<100 ha) and medium-sized (100 to 670 ha) reservoirs (Li & Xu 1995). The depth should be sufficient to prevent aquatic plant growth but not so deep as to prevent efficient use of harvest gear. Fish barriers (i.e., netting) should be in place at any outflow area of the reservoir to prevent escapement of fish or entry of unwanted fish. Reservoirs with tree stumps and other submerged obstacles should be avoided unless harvest gear and its efficiency are not compromised (i.e., seining vs. gill nets). Presence of aquatic plants can negatively affect fish production (except herbivorous species). Floating plants reduce light penetration, suppressing phytoplankton production, which is essential to primary production in the system. Aquatic plants also bind up nutrients essential for plankton growth, can provide roosts for predatory birds, and potentially hinder or prevent harvest (De Silva *et al.* 2006). Further, the reservoir should be sampled for numbers and types of predatory fishes that could prey on stocked fish or reduce their growth as a result of predatory stress. Practices such as increasing the size of stockers, increasing the number of fish stocked to compensate for losses, or reducing the number of predatory fishes should be followed, or the use of reservoirs where predators are present should be avoided. Lastly, accessibility and transportation are important considerations in reservoir selection. Obtaining and bringing in seed stock and shipment and marketing the harvested products require ready access to facilities (i.e., hatchery, end-user markets) with minimum transportation requirements (De Silva *et al.* 2006).

8.4.2 Physio-chemical and biological factors

Reservoirs have many physio-chemical and biological factors, which can cause rapid, frequent, and irregular changes in productivity so that only average rates can be determined over time (Wetzel 1983). Natural foods such as phytoplankton, zooplankton, detritus, bacteria, periphyton, benthos, and aquatic macrophytes are constantly changing in the reservoir based on environmental changes such as, but not limited to, temperature, light, nutrient inputs and recycling, generation of dissolved oxygen and carbon dioxide, as well as losses to grazing by herbivores and planktivores or foraging by omnivores and carnivores. Because fish depend completely on the natural food supply in a ranching operation, individuals compete for a finite food supply and therefore, their growth is largely dependent on population density (Lorenzen 1995).

Measurement of primary production has become the most popular way to gauge overall productivity in reservoirs and can assist in determining if the body of water is favorable for ranching. Phytoplankton abundance often has been found to have a direct correlation with fish productivity in culture-based fisheries (Lorenzen 1995; Li & Xu 1995; De Silva *et al.* 2006).

Reservoirs can be classified into four groups based on their productivity or trophic status. In increasing order of productivity, the four groups are oligotropic, mesotrophic, eutrophic, and hypereutrophic. Green, low turbidity waters (i.e., hypereutrophic and eutrophic) normally have greater fish yields and survivals than clear or high turbidity (muddy) waters (De Silva *et al.* 2006). Though visual observation of a water body can give an initial prediction of productivity levels, a Secchi disk is a simple tool that can be used to estimate phytoplankton densities and primary productivity and has been effective in estimating fish yields. Hasan *et al.* (1999) reported that fish yield was inversely linear to Secchi disc depth in Bangladesh. Further, he demonstrated that 1.5 to 2.0 times greater fish production was achieved when Secchi disk readings were less than 100 cm, compared to readings that were more than 100 cm.

8.4.3   Socio-economic factors

Reservoir ranching requires minimal resources and less technical skills than traditional aquaculture. Therefore, lower inputs of capital are needed for start-up, and available labor can be utilized with a minimum of training. In areas with available reservoir resources, this could present an attractive management strategy for communities with minimal available capital and a local labor force in need of additional employment. Reservoir ranching is centered on water bodies that generally are communal properties or common pool property (De Silva *et al.* 2006). These water bodies often have multiple uses. Ownership of fish must be clearly identified at the planning stage so the local community is aware of this farming strategy and conflicts are avoided. Success of reservoir ranching depends on a number of important steps and factors. Initially, the producer should conduct sufficient and effective community forums to educate leadership and other multiple users of the reservoir. It is important to have suitable time for preparation of the reservoir such as installation of escape barriers and removal of unwanted fishes. The producer must identify or develop a dependable supply of seed stock at the appropriate time. Further, to utilize the local workforce the producer must train employees in the management strategy of reservoir ranching. Finally, it is essential that the producer have adequate and appropriate laws, regulations, and contracts in place to protect the enterprise and its fish and market strategies in place for the fish products.

8.5   Fish species selection

Once a water body has been selected, the type(s) of fish species can be determined. Selected species should be native, feed low on the food pyramid, accumulate biomass rapidly, and reach market size in a relatively short period of time. The species should be easily propagated and lend itself to efficient harvesting using conventional fishing gear. It should be desirable in the market place, lack the ability to reproduce in the system (i.e., monosex, hybridization, insufficient physiological-environmental cues), and be able to coexist with other selected species (i.e., polyculture) to maximize the available space and food resources (De Silva *et al.* 2006). Internationally, the species that best fit these criteria for reservoir ranching are the major Chinese carps (grass carp, *Ctenopharyngodon idella*;

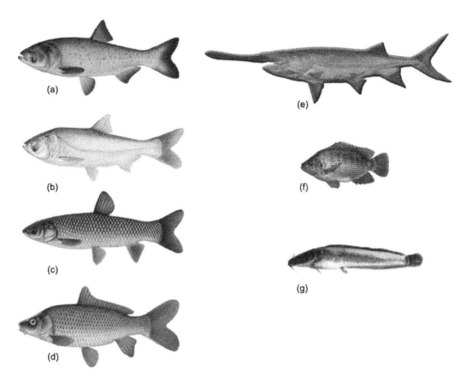

Figure 8.2 Some fish species used for reservoir ranching: (a) bighead carp (*Aristichthys nobilis*); (b) silver carp (*Hypophthalamichthys molitrix*); (c) grass carp (*Ctenopharyngodon idella*); (d) common carp, (*Cyprinus carpio*); (e) paddlefish (*Polyodon spathula*); (f) tilapia (*Oreochromis niloticus*); (g) African catfish (*Clarias gariepinus*).

bighead carp, *Aristichthys nobilis*; silver carp, *Hypophthalamichthys molitrix*, fig. 8.2; and the Indian carps including catla, *Catla catla*; mrigal, *Cirrhinus mrigal*; and rohu, *Labeo rohita*); the common carp, *Cyprinus carpio* (fig. 8.2); and the crucian carp, *Carassius carassius*. Other species are the African catfish, *Clarias gariepinus*; tilapia *Oreochromis niloticus*; and paddlefish, *Polyodon spathula* (fig. 8.2).

8.6   Stocking density and size

Stocking density should be based on the trophic status of the reservoir, desired market sizes of fish, and projected yields. Li and Xu (1995) suggested an empirical method for determining stocking density. Fast growing, fat laden fish would indicate underutilization of a food resource, suggesting a higher stocking density of individuals or all species; whereas, slow growth or presence of emaciated fish would indicate insufficient food, requiring a reduced stocking density. Stocking densities and species ratios are manipulated until optimum production of market-size fishes are reached. With over forty years of harvest data in China, Li and Xu (1995) demonstrated the use of a theoretical relationship between stocking density (D) and fish yield (F), average harvest size (W) and return (i.e., recapture) rate (S): D = F/WS. For example, if F = 500 kg/ha, W = 0.5 kg, and S = 50%, then 2,000 fish/ha would be stocked. In addition, Lorenzen (1995) developed models for populations of carp to determine the effects of stocking densities and sizes of seed stock, which have direct application for optimizing ranching operations.

Selecting the size of seed stock is a crucial aspect of reservoir ranching. The best recapture or harvest rate is usually obtained by stocking large (>14 cm) healthy fishes that ensure faster growth, higher survival, greater yield, and better economic returns than the stocking of smaller fish (<14 cm). Chen (1982) conducted a series of experiments on stocking efficiency of various sizes of bighead carp fingerlings. The experiments demonstrated that the return rates of stocking 14.7 cm average length bighead carp were two to four times higher than rates of stocking fingerlings of average length 11.6 and 9.1 cm. Cao *et al.* (1976) found that 90% of the carp preyed on by carnivorous fish were less than 13 cm in length. Only in newly constructed reservoirs, when predatory fishes are less abundant and plankton populations are high, should smaller (3 cm) fingerlings be stocked. China currently recommends 15 to 17 cm and 25 to 30 g, respectively, as best management practices in reservoirs. Though larger fish would probably provide even higher return rates, limited fingerling production facilities and associated higher fingerling costs are constraints to this practice.

In temperate climates, fish are usually stocked in late winter or early spring. There are several advantages (Li & Xu 1995) to stocking fingerlings when water temperatures are low: Fish are less active and have less physiological stress due to handling and transport; fish predation is minimized at lower temperatures resulting in less adverse effects on newly stocked fish; and plankton populations are highest in early to late spring providing food supplies for best growth and survival of fingerlings.

8.7   Status of reservoir ranching around the world

In this section, representative countries from Asia, Africa, and North America will be used to compare and contrast reservoir ranching in different regions of the world. China has the most complete published information specific to this culture system with other Asian countries following similar methodology. Africa has little published information but it is included to emphasize the need for more impoverished countries to evaluate and invest in the use of freshwater reservoirs for food fish production. Lastly, the United States historically permitted only culture-based fisheries for recreational fishing in reservoirs but now has limited testing in small reservoirs for paddlefish as food fish.

8.7.1   China

China produces more freshwater fish than any other country in the world. Though best known for their polyculture of fish in ponds, reservoir ranching

Table 8.1 Comparison of different reservoir sizes and their trophic levels to species stock proportions, species densities, and yields using the reservoir ranching system. Table adapted from Li and Lu (1995).

Size of Reservoirs (ha)	Trophic Level	BH & SC	GC, WF, & CC	MC	Total Stocking Density (fish/ha)	Fish Yield (kg/ha)
Small (<70)	Eutrophic	45	40	15	7,500	3,000
	Mesotrophic	35	30	35	5,000	1,500
	Oligotrophic	10–15	10–15	70–85	3,000	750
Medium (70–670)	Eutrophic	45	40	15	3,000	750
	Mesotrophic	50	30	20	2,300	600
	Oligotrophic	40	20	40	1,500	450
Large (670–6,670)	Eutrophic	50	35	15	1,500	450
	Mesotrophic	50	30	20	1,100	350
	Oligotrophic	40	20	40	750	225
Extra Large (>6,670)	Eutrophic	55	30	15	750	225
	Mesotrophic	55	25	20	600	185
	Oligotrophic	40	20	40	450	150

BH = bighead carp; SC = silver carp; GC = grass carp; CC = common carp; WF = bluntnose bream; and MC = mudcarp.

has provided a relatively new system that has increased freshwater fish production in the region. Though other Asian countries practice reservoir ranching (De Silva 2003), the most comprehensive information for best management practices of this system is available for China (Li & Xu 1995).

China has over 86,000 reservoirs, encompassing more than 2 million hectares of freshwater and representing about 40% of the total inland water surface area (De Silva 2003). Most of them are small (<70 ha) or medium (70 to 670 ha) in size totaling about 1.5×10^6 ha. Over 1 million tonnes of fish are annually produced in these reservoirs (De Silva 2003). Small- and medium-size reservoirs have been shown to have higher fish yields (>450 kg/ha/year) than larger size reservoirs (>670 ha; table 8.1). Li and Xu (1995) suggested that larger reservoirs have proportionally less shoreline and often less inflow of nutrients, which would result in lower fish production.

Most reservoirs in China are stocked primarily with bighead and silver carps, which account for 60 to 80% of the harvest (Li & Xu 1995). Other secondary species such as grass carp, black carp (*Mylopharyngodon piceus*), common carp, and mud carp (*Cirrhina molitorella*) are often mixed in to feed on aquatic plants and benthic organisms. Recently, mandarin fish (*Siniperca chuatsi*), a predator, have been stocked because of a greater demand by consumers and interest as a sport fish, but currently production is insignificant. Most reservoirs have their own hatchery and seed stock facilities to ensure a regular supply of fingerlings.

The fish species ratios and densities are often adjusted annually according to water fertility and the composition of food organisms (Li & Xu 1995). Bighead carp have been found to have greater individual growth and yields than silver carp, indicating a larger food supply of zooplankton than phytoplankton. Table 8.1 provides guidelines for stocking ratios and densities to obtain a wide variety of yields in reservoirs of different trophic status.

Yields in Chinese reservoirs have improved with the increase in stocking sizes of the fish and installation of escapement barriers. Li and Xu (1995) reported that stocking a larger size fish (>14 cm) not only resulted in higher yields (i.e., up to fourfold increase in return rate) than with smaller stockers, but individual sizes were also larger at harvest. Further, escapement barriers have provided an improvement in fish yields by limiting and ideally preventing the loss of fish from the reservoir during high water events. A variety of barriers is used, such as wooden or wire fence, polyethylene/polyvinyl net, and electrified screens (Li & Xu 1995).

A variety of gear and technologies has been developed for harvesting pelagic (e.g., bighead carp) and demersal (e.g., common carp and mud carp) species from Chinese reservoirs (Li & Xu 1995; De Silva 2003). Combinations of block net, driving gear (such as air and electrical curtains), gillnets, and seines are best for pelagic species, whereas, trawling and trammel nets are best for demersal species. At least two trawling boats and four to six john-boats with thirty to forty workers are needed to harvest a medium- to large-size reservoir. Fish harvest is most efficient during the winter months when water temperature is low and the fish are lethargic and often schooled together. After harvest, fish are held in net pens in the reservoir for subsequent distribution into the marketplace.

8.7.2 African countries

Africa, with nearly 1 billion people in fifty-four different countries, is considered the world's poorest inhabited continent. Of the 175 countries reviewed in the Human Development Report (UNDP 2008), twenty-five African nations ranked lowest among the nations of the world. About 60% of African workers are employed by the agricultural sector, with about three-fifths of African farmers considered to be subsistence level. Reservoir ranching has the potential to advance in several African countries and contribute to commerce, as well as provide much needed low-cost protein to local communities.

Anderson (1989) reported that there were over 20,000 small reservoirs (<100 ha) in Africa. More than half were concentrated in Botswana, Malawi, Tanzania, Zambia, and Zimbabwe, with over 80% of these in Zimbabwe. These small reservoirs were considered to have the greatest potential for supplementing the fish supply, especially in rural areas that were long distances from large lakes where capture fisheries were well established (FAO ALCOM 1995). Small reservoirs usually have high productivity due to nutrient runoff (i.e., agriculture and sewage effluent discharge). Ownership of these reservoirs varies from private ownership for farming purposes, to community ownership for livestock water supply and irrigation, to government ownership for domestic water supply for municipalities and schools.

The fish species most often found in the small reservoirs were common carp, tilapias of the genera *Oreochromis* spp. and *Tilapia* spp., and African catfish, *Clarias gariepinus* (Marshall & Maes 1995). The Food and Agriculture Organization ALCOM (1995) developed a work plan to evaluate and implement management strategies for these reservoirs in order to better utilize these existing water resources and increase availability of food fish as well as revenue for communities. However, no information is available to determine if such plans were implemented.

Government support and/or international aid have been limited in the development of management strategies and infrastructure (i.e., private hatcheries and distribution locations) to expedite the growth of reservoir ranching in Africa. Several countries have existing small reservoirs suitable for increasing fish production but a major bottleneck is availability of seed stock. Some government-owned hatcheries can provide a limited number of seed stock, but not on the regular basis needed for proper production.

Small reservoirs were reported to have an annual mean fish yield of 329 kg/ha based on limited information collected in the 1980s and 1990s (Marshall & Maes 1995). This average yield was over three times higher than the average yield (112 kg/ha) from capture fisheries in large (1,000 to 8,000 ha) reservoirs. If these small reservoirs could be stocked and properly managed, Bellemans (1989) estimated yields of one to three million tonnes per year. Therefore, reservoir ranching should be encouraged as a production method to at least improve the nutritional status of the rural population and therefore have a positive economic impact in some countries of Africa.

8.7.3   United States

The United States has millions of hectares of surface water impounded by dams. US reservoirs were constructed for much the same reasons as elsewhere in the world: flood control, water storage, and hydropower. However, state and federal regulations and resistance from some user groups have prevented the use of reservoirs for aquaculture in any form in the United States, and recreational fishing, boating, and scenic views have taken precedence over commercial production of fish. In the United States, there is a history of commercial fishing in large reservoirs, primarily for such species as catfish, *Ictalurus* spp.; buffalo, *Ictiobus* spp.; common carp; and paddlefish. However, state agencies have gradually restricted these fisheries over several decades primarily because of the perception that they interfere with recreational fishing. This has resulted in opposition to reservoir ranching in the United States, as stocking fish for commercial harvest would be perceived as a reversal of the decline in inland capture fisheries, and this would be viewed as undesirable by recreational fishing groups and state fishery managers

In the mid-1990s, the United States Department of Agriculture (USDA) funded a pilot research project in which paddlefish were stocked in small dam reservoirs (14 to 40 ha) on private land in Kentucky and harvested after eighteen months

(Onders *et al.* 2001). Paddlefish are members of the sturgeon family and are zooplanktivores. They are valued as a source of black caviar and the meat is white and boneless (Mims *et al.* 2009). This project showed that paddlefish would survive and grow when stocked in reservoirs as Phase II (>150 g) juveniles and could be harvested with conventional gear. Subsequently, a more advanced project was proposed for large public reservoirs; however, the project was not funded at least partially due to concerns voiced by the regulatory agency responsible for management of the reservoirs.

In 2006, new regulations were issued allowing private individuals to contract with small municipalities on a profit sharing basis for ranching of paddlefish in potable water supply reservoirs (Mims *et al.* 2006). These reservoirs are owned by the municipalities; however, the management agency retains control over the fisheries. There are between 1,600 and 2,000 ha of this type of reservoir in Kentucky; however, only about two-thirds are suitable for reservoir ranching. The regulations specify that only paddlefish can be stocked and only once every ten years. Harvest gear is restricted to 127 mm bar mesh gill nets, preventing harvest of smaller fish for meat. In addition, the management agency provides no enforcement protection for the paddlefish and allows taking by archery fishing.

Despite these constraints, private individuals have contracted with municipalities and stocked over 800 ha throughout Kentucky. The reservoirs range in size from 20 to 270 ha and are stocked at up to 50 paddlefish/ha. A minimum stocking size of 150 g was selected to minimize mortality from predation. After three years, sampling has produced paddlefish up to 6 kg. Researchers at Kentucky State University are sampling two of the largest stocked reservoirs to monitor for any changes that may occur in the reservoir or sport fishery that would indicate negative effects from the paddlefish stocking. To date, no negative effects have been reported. Based on anecdotal information, one municipality has observed a reduction in blue green algae in 2008 and 2009 and has significantly reduced their cost of adding an algaecide. The paddlefish themselves are also being monitored for survival, growth, and progression to sexual maturity (≥8 years), when the females can be harvested for roe.

The results of a state-wide survey mandated by the Kentucky Legislature showed strong public support for the concept of paddlefish ranching in the state's large reservoirs (Dasgupta *et al.* 2006). Reservoir ranching of paddlefish could become an economically viable alternative to current river fisheries for wild paddlefish that are being increasingly restricted to the point of closure. Current estimates show potential revenues of US$5,000/ha, assuming 50% survival to maturity for females (harvesting 12.5 females/ha), 1.5 kg roe/harvested female, and a wholesale price of US$275/kg for caviar.

8.8 Summary

- Reservoir ranching is an extensive culture system in which young fish that feed on naturally available foods are stocked in existing freshwater impoundments and harvested after a period of time.

- Ranching provides an alternative supply of freshwater food fish for rural communities as well as in some cases for international trade under eco-friendly sustainable conditions.
- Small (<100 ha) and medium (100 to 670 ha) size reservoirs are usually best suited for ranching purposes because they generally have higher primary productivity and fish biomass per ha and are less complicated to harvest than larger reservoirs.
- Selected species should feed low on the food pyramid (i.e., plankton), grow rapidly, and be native to the region, easily propagated, harvested efficiently using conventional fishing gear, and desirable in the marketplace.
- Effective stocking densities can be determined using qualitative indicators (e.g. fast growing, fat laden fish would indicate underutilization of a food resource suggesting a higher stocking density; whereas, slow growth or presence of emaciated fish would indicate insufficient food and require stocking density reduction).
- China has the longest history of reservoir ranching using >2,000,000 ha of freshwater and producing over 1 million tonnes per year from over 86,000 small and medium-size reservoirs.
- The United States could benefit from paddlefish ranching in reservoirs by producing a regular supply of consumer-safe caviar and meat for domestic and international trade as well as relieving pressure on wild stocks, which are under strict regulations (i.e., CITES Appendix II).

8.9 References

Anderson, A. (1989) The development and management of fisheries in small water bodies in Africa. In *Proceedings of the Symposium on the Development and Management of Fisheries in Small Water Bodies* (Ed. by M. Giasson & J.L. Gaudet), pp. 15–9. FAO Fisheries Report No. 425. FAO, Rome.

Bellemans, M. (1989) Problems associated with the gathering of information on small water bodies in Africa. In *Proceedings of the Symposium on the Development and Management of Fisheries in Small Water Bodies* (Ed. by M. Giasson & J.L. Gaudet), pp. 33–7. FAO Fisheries Report No. 425. FAO, Rome.

Cao, F., Huang, K. & Zhu, Z. (1976) Fisheries and its utilization in Qingshan reservoir. *Chinese Journal of Zoology Sinica* 3:31–5 (In Chinese).

Chao, B.F. (1995) Anthropogenic impact on global geodynamics due to reservoir water impoundment. *Geophysical Research Letters* 22:3529–32.

Chen, D. (1982) Review of the techniques to increase fish production in Dongfeng Reservoir. *Freshwater Fisheries* 3:44–8 (In Chinese).

Dasgupta, S., Mims, S.D. & Onders, R.J. (2006) Reservoir ranching of paddlefish, *Polyodon spathula*: Results of a public opinion survey in Kentucky. *Journal of Applied Aquaculture* 18:81–9.

De Silva, S.S., Amarasinghe, U.S. & Nguyen, T.T.T. (Eds.) (2006) *Better-practice approaches for culture based fisheries development in Asia*. ACIAR Monograph No. 120. Australian Centre for International Agricultural Research, Canberra.

De Silva, S.S. (2003) Culture-based fisheries: An underutilized opportunity in aquaculture development. *Aquaculture* 221:221–43.

FAO ALCOM (1995) Appendix 5 Work Plan 1995. In *Report of the Eighth Steering Committee Meeting*. ALCOM Report No. 20. Dar-es-Salaam, Tanzania, February 13–16, 1995.

FAO (1997) *Aquaculture Development*. FAO Technical Guidelines for Responsible Fisheries No. 5. FAO, Rome.

FAO (2008) The state of the world fisheries and aquaculture. Food and Agriculture Organization of the United Nations, electronic publishing policy and support branch. FAO, Rome.

Hasan, M.R., Bala, N. & De Silva, S.S. (1999) Stocking strategy for culture-based fisheries: A case study from the oxbow lakes fisheries project. In *Sustainable Inland Fisheries Management in Bangladesh* (Ed. by H.A.J. Middendorp, P.M. Thompson & R.S. Pomeroy), pp. 157–62. ICLARM Conference Proceedings 58. Manila, Philippines.

Li, S. & Xu, S. (1995) *Culture and Capture of Fish in Chinese Reservoirs*. Southbound, Penang, Malaysia/International Development Research Centre. Ottawa, Canada.

Lorenzen, K. (1995) Population dynamics and management of culture-based fisheries. *Fisheries Management and Ecology* 2:61–73.

Marshall, B. & Maes, M. (1995) *Small Bodies and Their Fisheries in Southern Africa*. CIFA Technical paper No. 29. FAO, Rome.

McCully, P. (1996) *Silenced Rivers: The Ecology and Politics of Large Dams*. Zed Books, London.

Mims, S.D., Onders, R.J. & Shelton, W.L. (2009) Propagation and culture of paddlefish (Part 4). In *Paddlefish Management, Propagation, and Conservation in the 21st Century: Building from 20 Years of Research and Management* (Ed. by C.P. Purkert & G.D. Scholten), pp. 357–83. American Fisheries Society, Symposium 66, Maryland.

Mims, S.D., Onders, R.J., Parrott, B.T. & Stickney, J. (2006) Caviar from paddlefish grown in water supply lakes. *Waterproof* 8(4):12–13.

Onders, R.J., Mims, S.D., Wang, C. & Pearson, W.D. (2001) Reservoir ranching of paddlefish. *North American Journal of Aquaculture* 63:179–90.

Rosenberg, D.M., McCully, P. & Pringle, C.M. (2000) Global-scale environmental effects of hydrological alterations: Introduction. *BioScience* 50(9):746–51.

Semmens, K.J. & Shelton, W.L. (1986) Opportunities in paddlefish aquaculture. In *Paddlefish Status, Management, and Propagation* (Ed. by J.G. Dillard, K. Graham & T.R. Russell), pp. 106–13. American Fisheries Society, Special Publication No. 7, Modern Litho-Print Company, Missouri.

Shiklomanov, I.A. (1993) World fresh water resources. In *Water in Crisis: A Guide to the World's Fresh Water Resources* (Ed. by P.H. Gleick), pp. 13–24. Oxford University Press, New York, NY.

St. Louis, V.L., Kelly, C.A., Duchemin, E., Rudd, J.W.M. & Rosenberg, D.M. (2000) Reservoir surfaces as sources of greenhouse gases to the atmosphere: A global estimate. *BioScience* 50(9):766–75.

US Commission on Large Dams (USCOLD) (1995) *US and World Dams, Hydropower and Reservoir Statistics*. USCOLD, Colorado.

Wetzel, R.G. (Ed.) (1983) *Limnology*, 2nd edition. Saunders College Publishing, Harcourt Brace College Publishers, Pennsylvania.

World Register of Dams (ICOLD) (1998) *International Commission on Large Dams* (ICOLD). Paris, France.

World Commission on Dams. (2000) *Dams and Development: New Framework for Decision-making*. Earthscan Publication, Ltd., London and Sterling, VA.

Chapter 9

Flow-through Raceways

Gary Fornshell, Jeff Hinshaw, and James H. Tidwell

A raceway in its simplest form is just a flume for carrying water. Raceways for fish culture are tanks that are relatively shallow and rely on a high water flow in proportion to their volume in order to sustain aquatic life. The focus of this chapter is on fish culture systems that pass water through the systems once, provide waste treatment as required, and then discharge the water rather than treat and recirculate it. For successful aquaculture, the inflowing water must be within the temperature tolerance of the species being cultured and should match the optimum temperature for the target species as closely as possible. Oxygen is also provided by the incoming water and is removed by the fish as the water progresses down the raceway. In most raceway systems dissolved oxygen is replenished by allowing the water to fall into subsequent tanks within the raceway. Dissolved metabolites from animals in the system are carried out in the effluent, while settleable particulate wastes can be captured by settling, or less frequently, by other means of filtration. Depending on the water chemistry, the accumulation of ammonia, carbon dioxide, or fine particulates can eventually become limiting to animal production within the system. No natural foods are generated in these systems and nutritionally complete diets are an essential requirement for successful raceway aquaculture.

Aquaculture Production Systems, First Edition. Edited by James Tidwell.
© 2012 John Wiley & Sons, Inc. Published 2012 by John Wiley & Sons, Inc.

9.1 Types of raceways

9.1.1 Earthen flow-through ponds

In their earliest forms, flow-through systems were created by placing screens or bars above and below a section of stream to confine fish within that area. True raceway production systems in the aquaculture industry evolved from earthen ponds, the most popular shape of which was long and narrow, with sufficient slope to allow for aeration by gravity between ponds if the water was used in more than one pond. Simple earthen flow-through ponds are still in use in many regions today (fig. 9.1), particularly for smaller farms. Earthen ponds have the advantage of being relatively inexpensive to build, but water flow is maintained at low velocities to help minimize erosion. Without supplemental oxygenation, the lower water velocity limits the carrying capacity per unit of water volume (fish density), which in turn reduces overall production. However, because of lower-stocking densities, fish from these systems are reported by producers to be more healthy and colorful in appearance (Hinshaw *et al.* 2004). Earthen raceways are subject to problems with erosion, which is sometimes mitigated using rocks, wood, or concrete structures. Solids management and removal is also difficult due to the irregularity of shapes and surfaces, and a tendency for settling to occur throughout the raceway due to low water velocity. In most instances, solid wastes are left on the bottom to decompose naturally, and then are dredged from the system after several years. Earthen systems also present

Figure 9.1 An earthen flow-through raceway.

Figure 9.2 A series raceway arrangement.

challenges for carrying out other fish management practices such as grading, harvesting, or application of therapeutants.

9.1.2 Concrete raceways and tank arrangements

Raceways can be constructed of wood and prefabricated modular units built of fiberglass are also available. Most modern raceways are constructed of concrete. Compared to earthen systems, concrete raceways can increase production 25 to 40% using the same quantity of water (Fornshell 2002). A typical raceway production system consists of a series of tanks, usually rectangular with water flow along the long axis. On farms, raceways may be divided into two or more tanks at each step in the series, but on smaller farms the tanks are usually in pairs for ease of access (fig. 9.2). Larger farms may construct access on the tanks and have multiple raceways in parallel series (fig. 9.3). The water in raceway production systems is rarely recirculated, but is nearly always "reused" serially with aeration or oxygenation between tanks.

9.1.3 Advantages of raceways

Compared to ponds, which are also semi-closed production systems, raceways have several advantages. Per unit of space, production in raceway systems is much higher. Raceways also offer a much greater ability to observe the fish,

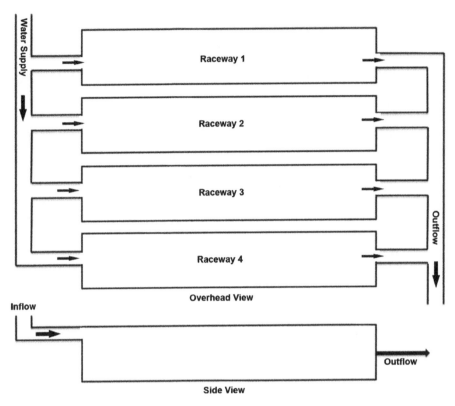

Figure 9.3 A schematic of a parallel raceways system.

making feeding potentially more efficient and disease problems easier to detect and at earlier stages. If disease signs are observed, disease treatments in raceways are easier to apply and require a lesser amount of chemicals than a similar number of fish in a pond (due to the higher fish density in the raceway). Raceways also allow closer monitoring of growth and mortality and better inventory estimates than ponds. Management inputs such as size grading are much more practicable in raceways than they are in ponds, and harvesting is also easier.

9.1.4   Limitations of raceways

The disadvantages of raceways are primarily related to their need for a large, consistent flow of high-quality water. Since such resources are not common, locating and securing a proper water supply is a major consideration. Also, commercial viability often requires that the water flows by gravity through multiple tanks or raceways before it is released. This adds a requirement for an elevation of the water source and suitable topography for the gravity flow between tanks. Another limitation compared to ponds is the release of a high volume of effluent containing low concentrations of metabolites. While ponds largely process fish wastes within the culture systems, raceways, with their low retention times, do not.

9.2   Physical requirements

9.2.1   Site characteristics

Most raceway production systems are found where three criteria can be satisfied: (1) an inexpensive and reliable supply of flowing water is available; (2) sufficient slope is available or can be created to allow gravity flow of water through and away from the system; and (3) a culture species of sufficient value and tolerance to high density to be grown. These requirements can be satisfied by careful site selection, by capturing an artesian water source such as a spring or well, or by diverting water from a natural water body such as a river or the outflow from a lake. Due to the cost of pumping and the need for redundancy in pumps and power supplies, water sources that can provide flow to the raceways using gravity flow are preferred and often dictate the specific farm location. Locations with geologic features conducive to formation of springs such as the Thousand Springs area of the Snake River Canyon in Idaho are well suited to raceway aquaculture development. Regions such as the north slope of the Central Anatolian plateau of Turkey or the Appalachian Mountains of eastern United States have numerous rivers and streams with protected water sources that also are suitable for raceway aquaculture. Farms based on diverted surface waters tend to be smaller in scale than the spring-fed systems in the Snake River Canyon area, due to limitations on water withdrawal, availability of sites with suitable slope and proximity to the water supply, and variations in seasonal water flow. Raceway systems using saltwater are nearly non-existent as saltwater springs or streams are also nearly nonexistent.

9.2.2   Raceway characteristics and arrangements

Individual raceways are long, narrow, shallow troughs, typically divided into multiple rearing units. There are several basic designs but a common length to width ratio is 10:1 (Soderberg 1995). To include depth, a ratio of 30:3:1 is recommended (lenth:width:depth; Hinshaw *et al.* 2004). Common sizes of individual units are 7- to 9-m long by 3- to 5-m wide by 1.1-m deep (Mazik & Parker 2001). These individual units are grouped together in different arrangements based on water flows, elevation of water source, and topography.

9.2.3   Parallel raceway systems

In a parallel arrangement (fig. 9.3) multiple raceways are grouped in such a way that they all receive water from the same source. It makes one pass through each raceway then the flows are combined into a common outflow. The advantage to this arrangement is that each unit operates independently. The potential for disease transmission from unit to unit is greatly reduced. Also, the water quality in each unit is not affected by the tank location in the farm, as the units do not receive water from, or release water into, other production units (Landau 1992).

9.2.4   Serial reuse raceway systems

To maximize production per unit of water flow, another design is to arrange the individual units in a step arrangement that allows water exiting the distal end of a raceway to drop down and enter the proximal end of the next (fig. 9.2). Since oxygen is the first limiting factor in production intensification in flow-through raceways (Colt *et al.* 2009), the drop from one raceway into the next is used to re-aerate the water. To accomplish this, the system is designed with a 0.6 m drop between sequential raceways (Mazik & Parker 2001). Several devices have been developed and tested for increasing oxygen transfer over a gravity fall, including splash boards, stair steps, and cascades (Soderberg 1995). The drop from one raceway to the next replaces approximately 50% of the oxygen consumed by fish in the previous section (Mazik & Parker 2001), depending upon the height of the fall and other factors. The oxygen content of the incoming water combined with this gravity aeration between raceways accounts for nearly all the oxygen budget of a serial reuse raceway system.

To increase production, some producers also add mechanical aeration or oxygenation, especially during peak feeding periods. However, aeration or oxygenation can add significantly to the cost of production. Once oxygen needs are satisfied, the accumulation of waste products now becomes the next limiting factor in system intensification. Since flow-through raceways are mechanical rather than biological systems, there is no in situ removal or conversion of nitrogenous waste products. If the oxygen removed by the animals' respiration is replaced, the water can continue to be used until the un-ionized ammonia accumulates to harmful levels, approximately 0.03 mg/L as un-ionized ammonia-N.

In the primary trout-producing region of the Hagerman Valley of Idaho in the United States, the water can be sequentially used in four to five raceways before un-ionized ammonia concentrations become too high (Fornshell 2002). In contrast, the waters of trout producing areas of the southern Appalachian Mountains such as North Carolina are soft (poorly buffered) and slightly acidic in pH. This ionized form of ammonia is less toxic to the fish. Therefore, higher levels of total ammonia can be tolerated, which equates to more water reuses before ammonia becomes limiting. Up to twenty reuses are possible in these regions if oxygen demands are met (Fornshell 2002) and provisions are made to remove carbon dioxide. Usually stocking rates are reduced in downstream raceways due to decreasing water quality and accumulated metabolites. If there is sufficient flow, multiple serial reuse raceways can be built side by side. The

Figure 9.4 A series raceways arrangement illustrating the tiers of raceways at each level.

raceways at each elevation or step are known as a tier (Mazik & Parker 2001). Raceways at each level share walls to reduce construction cost (fig. 9.4).

9.3   Water requirements

9.3.1   Required flows

Since flow-through raceways rely on inflowing water to supply oxygen and water leaving the system to remove ammonia, the amount of water available and its physical/chemical characteristics are the primary factors determining the production capacity of each system. As stated previously, water resources of a commercial scale are hard to locate and dictate where flow-through raceway aquaculture systems can be located. Minimum water flows of 5,678 to 7,570 liter per minute (Lpm) are needed to efficiently operate aquaculture raceways with the typical dimensions described previously (unit volumes of 21 to 45 m^3; Mazik & Parker 2001). Smaller production systems can operate efficiently using smaller water supplies but production will also be proportionally lower. These flow rates should represent at least four water exchanges ("turnovers") for each unit per hour (Fornshell 2002) and up to nine turnovers per hour (Soderberg 1995). If there are multiple production units in series, each tier will require this much flow. If there is a seasonal variation in flow, projected production should be based on the lower flow season.

9.3.2 Water sources

Water quality and quantity are the primary determinates of the carrying capacity of raceway systems. Important water quality variables include not only oxygen content, alkalinity, hardness, and pH, but also suspended solids and temperature. Surface waters are used in many regions, especially in locations that receive regular heavy rainfall (Soderberg 1995). Water temperatures can vary seasonally, as can flow rates. For trout culture in the United States, requiring water temperatures of 10 to 15°C, surface waters in the southern Appalachian region are considered suitable at high elevations (>900 m). Below this elevation water temperatures can become high enough in summer to reduce survival and low enough in winter to slow growth (Soderberg 1995). When available, ground water sources are preferable to surface waters, as they tend to have more stable temperatures and reduced risk of contamination from wild fish or pathogens (Hinshaw *et al.* 2004). Ground water temperatures are almost constant throughout the year. The temperature of a region's ground water can be estimated from the annual mean air temperature of the region (Soderberg 1995), unless influenced by geothermal activity. The negative aspects of groundwater sources are that they can be low in some gases (such as oxygen) and high in others (such as carbon dioxide, nitrogen, or hydrogen sulfide; Hinshaw 2000). This is especially true if the water has been confined within impermeable geologic strata. Another negative is that where ground waters emerge as springs, they tend to be at low elevations, which do not allow the gravity flow required for efficient serial reuse.

Another source of ground water has been under evaluation in recent years. In the coalfields of the Appalachian region of the eastern United States, there are hundreds of deep mines that have been abandoned. After mining ceases, these mines usually fill up with ground water and even overflow. This represents a significant source of readily accessible ground water, which can often be used without pumping, either through gravity flow or siphon. Initially, there were concerns about low pH (acid waters) or the potential of heavy metals. However, in sites that have not been actively mined for decades, these do not seem to be significant issues. It appears that these groundwater resources represent a largely untapped economic resource for a region in dire need of economic opportunities (D'Souza *et al.* 2004).

9.4 Carrying capacity

9.4.1 Indices of carrying capacity

A number of methods have been developed to estimate carrying capacity of flow-through raceway systems. To date, most of these models have been based on metabolic characteristics of rainbow trout (Haskell 1955; Willoughby 1968; Piper *et al.* 1982), the species most commonly associated with raceway systems. The carrying capacity of each tank in a raceway system is dependent on several factors including water flow, tank volume, exchange rate, water temperature, oxygen content, pH, fish species and size, production targets (feeding rates), and accumulation of waste products. Piper *et al.* (1982) recommend using flow and density indices to calculate carrying capacity in flow-through systems. Both indices are important to know as spatial requirements and water requirements of fish have been found to be independent (Soderberg 1995).

9.4.2 Flow index

The flow index (an index of weight of fish per unit fish size and water flow) is calculated as follows: $F = W/(L*I)$, where F = flow index, W = known permissible weight of fish, L = length of fish, and I = water flow in units of volume per minute. Standard flow index tables are available that have calculated the flow index at various temperatures and elevations with the assumption the incoming water is at or near 100% saturation with atmospheric oxygen levels. By rearranging the formula to: $W = F*L*I$, the permissible weight can be calculated. The total fish biomass per unit water flow in a tank is referred to as the tank "loading."

9.4.3 Density index

The density index calculates the maximum allowable weight of fish per unit of tank volume. The level of density that aquatic organisms will tolerate is affected by behavioral aspects of the species, as well as their physical and/or physiological traits (Soderberg 1995). Too much crowding can cause behavioral problems, physical problems, or increased disease and disease transmission, even if water quality conditions are adequately maintained. The density index is calculated as: $W = D*V*L$, where W = permissible weight of fish, D = density index, V = tank volume, and L = fish length. Annual production or yield is generally two to three times the total farm carrying capacity.

9.4.4 Estimating carrying capacity

The flow and density indices were developed at state and federal hatcheries where production goals differ from commercial facilities and are conservative relative to maximum potential carrying capacity. Many commercial aquaculturists base their carrying capacity on empirical observations relative to available oxygen, where a predetermined limit of dissolved oxygen flowing out of one unit in a raceway into the next unit is established. For example, incoming water to the

Text Box 9.1   Calculations of Optimum Loading Capacity

Assume an aquaculture facility with two raceways in series. Total flow to the facility is 28.32 Lps (1 cfs). Incoming DO level is 9.0 mg/L. Desired levels are 7.0 mg/L at the effluent of raceway 1 and 6.0 mg/L at the bottom of raceway 2. Average oxygen recharge is 0.7 mg/L between raceway 1 and 2. Oxygen-consumption rate is estimated at 200 mg/hr of oxygen per kg fish. Calculate the fish biomass for each raceway that will provide these effluent levels.

Step 1: Determine the available oxygen/hour for each section.

Raceway 1:
(9 mg/L O_2 incoming - 7.0 mg/L O_2 effluent) × 28.32 Lps × 60 s/m × 60 m/hr = 203,904 mg/hr available O_2

Raceway 2:
(7.0 mg/L O_2 incoming - 6 mg/L O_2 effluent + 0.7 mg/L O_2 recharge) × 28.32 Lps × 60 s/m × 60 m/hr = 173,318 mg/hr available O_2

Step 2: Divide available oxygen by oxygen-consumption rate/kg fish. Convert to pounds if necessary.

Raceway 1:
203,904 mg/L O_2/200 mg/hr O_2 consumed/kg fish = 1020 kg of fish carrying capacity

Raceway 2:
173,318 mg/L O_2/200 mg/hr O_2 consumed/kg fish = 867 kg of fish carrying capacity

Note: Oxygen consumption rates vary by species, water temperature, size of fish, and feeding rate. Consumption estimates can easily be generated for specific facilities by monitoring influent and effluent concentrations for known biomass of fish, feed amounts, and water inflow rate over a period of time or may be found in the literature.

Colt and Orwicz (1991) also describe methods for calculating carrying capacity in culture systems where oxygen is not a limitation, such as when pure oxygen is added in flow-through raceway system.

first unit is at 100% oxygen saturation and the predetermined limit of outgoing water is set at 70% of saturation. The available oxygen is calculated based on the water flow and the incoming concentration of dissolved oxygen. The allowable biomass is calculated by dividing the available oxygen by the metabolic oxygen consumption of the fish. An example of this method is given in text box 9.1.

Table 9.1 Facilities characterization of five model systems.

Facilities	1	2	3	4	5
Raceway flow (Lpm)	946	6,000	5,400	10,000	10,200
Farm flow (m^3/s)	0.09	0.97	3.2	2.7	8.5
Load (kg/Lpm)	1.6	4.3	1.4	1.8	1.5
Fish density (kg/m^3)	38	27	32	51	37
Production (kg/Lpm) (5.34)	4.25	7.81	5.32	6.73	6.23
Water use (m^3/kg) (98)	123	67.4	98.8	78.2	84
Annual production capacity (MT)	23	454	1,021	1,089	3,175

Notes: Facility characterization from True *et al.* (2004); numbers in parentheses represent industry averages.

9.4.5 Production

Much of the commercial aquaculture industry that is raceway-based still operates within the range of productivity outlined by Westers and Pratt (1977), normally carrying between 20 and 80 kg of fish per cubic meter of water volume (approximately 2 to 5 pounds per cubic foot) with water exchange rates per tank of three to six times per hour. The water is reused serially four to six times or until ammonia becomes limiting—then it is discharged. For example, in the major trout producing region of the United States (Idaho), the carrying capacity for trout averages 1.8 kg/Lpm of water flow at densities of 27 to 51 kg/m^3 (table 9.1). Annual production per unit averages 5.3 kg/Lpm of flow (cumulative of multiple crops). When all of the raceways in a series are combined, production may reach 9.66 kg/Lpm of flow (Brannon & Klontz 1989). In areas where the waters are slightly acidic, ten or more uses are typical and other factors such as carbon dioxide or suspended solids may become limiting before ammonia. In a region with lower pH water, such as the southern Appalachian Mountains in the United States, production has reached as high as 17.2 kg/Lpm of water flow (Hinshaw *et al.* 2004, trout yield verification study; unpublished data).

These indices are useful tools to be used in production planning for stocking into and production rates out of most raceway production systems. If properly designed with sufficient water exchange (four to six turnovers per hour) the estimates of carrying capacity obtained from the Flow Index and Density Index should be nearly equal (Hinshaw 2000). For systems that add pure oxygen, the Flow Index is typically higher and other factors other than oxygen will limit production, such as an accumulation of carbon dioxide from the fish. The farm's oxygenation and aeration systems should be designed to also remove carbon dioxide from the water for efficient production.

9.5 Water consumption and waste management

9.5.1 Water budgets

In comparison with ponds, water budgets for flow-through aquaculture systems are simple. The budget is controlled by a regulated water inflow, and the outflow rate equals the inflow rate. The water is retained in the system for only a matter of minutes or hours. Rainfall contributions and evaporation losses are negligible. Flow-through systems do not allow runoff to directly enter the rearing units in an uncontrolled fashion. Farms that rely on surface waters use waters only from protected or underdeveloped watersheds. Most raceways are constructed of impermeable materials so seepage into or out of the system does not occur. Water flows per raceway vary in proportion to raceway size, fish density, and fish metabolic needs as previously described. In practice, water flow rates per raceway are based on desired water exchange rates and, to a lesser extent, water velocity. For example, a 3-m-wide raceway will typically receive 85 to 100 Lps.

Flow-through aquaculture systems are a nonconsumptive use of water. Since evaporation and other water losses from tanks are minimal, the only regular removal of water is in the form of harvested fish or in the removal of fish wastes. Goldburg and Triplett (1997) mischaracterized water use in the Idaho trout industry as consumptive by implying that aquaculture was responsible for declines in groundwater levels in the Eastern Snake Plain Aquifer. In reality, naturally occurring flows are temporarily diverted through production units and are then returned to the receiving stream with little or no actual loss of water.

Although water is not consumed during use, flow-through systems do require the use of large volumes of water per unit of fish production compared to other aquaculture production systems. For example, raceway production of trout in the United States uses 98 m^3 of water per kg of fish produced compared to 1.25 to 1.75 m^3/kg for channel catfish in undrained levee ponds and 6.5 to 10 m^3/kg for channel catfish in watershed ponds (Hargreaves *et al.* 2002). While the classic concept of water conservation is meaningless in flow-through aquaculture systems (Hargreaves *et al.* 2002), their water use efficiency can be improved by using technologies that increase fish production per unit of water flow. Examples of such technologies include rigorous water quality management, high-quality feeds, and using improved fish stocks.

9.5.2   Management of solids

As flow-through systems, raceway-based farms do not retain water long enough for significant biological processes to develop for in situ decomposition and solids can accumulate. Most of the solids are from fish fecal matter and uneaten feed (Fornshell & Hinshaw 2008) and if allowed to accumulate for long periods of time, can degrade environmental conditions in the raceway and stress the fish (Cripps & Berghein 2000).

Just as adequate water flow is important to fish growth, it is also important to solids management. Raceways are designed to promote "plug flow," which is equal water velocity in all parts of the water column. Westers (2001) reported that for a raceway to be self-cleaning, it needs a water velocity of 10 cm/s. However, this rate is higher than can normally be achieved in raceways on a

Table 9.2 Water flow rates for raceways of various dimensions needed to achieve a water velocity of 0.03 m/s (3 cm/s).

Raceway Width (m)	Water Depth (m)	Water Velocity (m/s)	Water Flow Rate (m³/s)
1.8	0.76	0.03	0.041
3.0	0.76	0.03	0.068
3.7	0.76	0.03	0.084
5.5	0.76	0.03	0.125
5.5	0.91	0.03	0.150

practical basis. A water velocity of 3 cm/s is a good compromise flow rate between efficiently managing fish production and still maintaining self-cleaning characteristics of the raceways. Also, in a production raceway system the swimming action of the fish helps limit settling of solids.

To try to achieve the water velocity of 3 cm/s needed for movement of solids, several modifications have been made in raceway design. Raceways are being constructed that are shorter and narrower to increase water velocities (Fornshell & Hinshaw 2008). Table 9.2 gives a matrix of raceway dimensions and flow rates required to achieve a water velocity of 3 cm/s. Another approach has been to install baffles in the water column to constrict the flow to slots near the bottom. This can increase velocities to 10 to 40 cm/s below the baffles and progressively "sweeps" the solids toward the discharge end (Boersen & Westers 1986).

A number of different methods and technologies have been developed and evaluated for the removal of solids from flow-through systems. However, the majority of treatment options are not suitable for commercial aquaculture because they are either impractical to implement due to interference with other aspects of fish management, too costly, or both (Fornshell & Hinshaw 2008). Currently, sedimentation or settling appears to be the only readily adoptable and cost-effective technology. In some designs, a settling basin receives the full flow and has enough volume to slow the water for solids settling (Fornshell & Hinshaw 2008). In others, solids are settled and collected in quiescent (no fish) zones and then routed to settling basins either next to the raceway (off-line) or at the end of raceway (full flow; fig. 9.5). Stechey and Trudell (1990) recommended having quiescent zones in each raceway. These fish-free zones allow the solids to be concentrated and for off-line settling basins to function while handling less than 2% of total facility flow (Fornshell & Hinshaw 2008).

9.5.3 Settling basins

There are three types of settling basins used in flow-through systems to manage solid waste. They are quiescent zones, full-flowing settling basins, and off-line settling basins. Quiescent zones are screened-off areas below the rearing area at

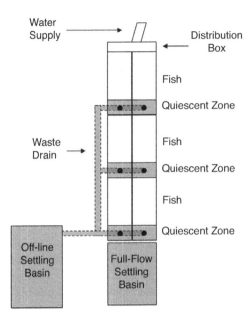

Figure 9.5 A schematic of a solids capture system for raceways utilizing quiescent zones.

the tail end of raceways. The screen, constructed with PVC or aluminum pipe mounted on a wood frame, prohibits fish from entering the quiescent zone. This allows the solids to settle undisturbed. Full-flow settling occurs where the entire flow of a facility passes through a settling basin prior to discharge into receiving waters. Full-flow settling basins are usually used on smaller facilities where total flow rates are less than 283 Lps. Off-line settling basins receive the solids that are removed from quiescent zones. Generally, the combination of quiescent zones in each raceway and off-line settling basins is the most commonly used waste management system to capture and remove solids.

9.6   Feeding and inventory management

9.6.1   Feed quantity and quality

Culture of fish in raceways offers several advantages for farmers in terms of feeding, management of inventory, and harvest. The relatively clear water and high fish densities allow culturists to observe feeding activity and to use a variety of feeding approaches including demand feeders (fig. 9.6). When properly deployed, demand feeders can reduce labor costs, allow all fish access to feed (reducing size variation), and spread out the oxygen demand on the system caused by feeding activities and digestion (Fornshell & Hinshaw 2008). For rainbow trout in raceways, Klontz (1991) recommended six demand feeders (three per side) in rearing units 3 m wide by 30 m long. Since no natural foods

Figure 9.6 A drained raceway showing proper placement of demand feeders.

are produced within the culture unit, any feeds used in flow-through raceway systems must be nutritionally complete, containing all of the macronutrients and micronutrients required by the animal being cultured.

The relatively clear, shallow water allows fish to be observed for behavior, condition, and for any signs of fish health problems. Mortalities are moved with the water flow and collect at the downstream barrier of the tank, allowing rapid removal. Another advantage of fish production in raceway systems derives from the regular shape of the tanks, which allow fish to be crowded into a small section of the tank (fig. 9.7) for self-grading using bar graders or for harvest using lift nets or fish pumps (fig. 9.8).

9.7 Summary

Flow-through raceways are extremely efficient technologies for producing large volumes of fish in a relatively small area. They are very efficient in terms of labor per unit of fish produced and have attributes that allow for very efficient feed utilization. Costs to build raceway systems generally will be intermediate between the cost of building earthen pond systems and the high capital costs of recirculating culture systems. Due to the requirement for large quantities of high quality water, suitable sites and water supplies for raceway production systems are becoming increasingly hard to identify.

Figure 9.7 Crowding of fish in a raceway to allow grading and/or harvest by pump.

Figure 9.8 Harvesting live rainbow trout by fish pump.

9.8   References

Boersen, G. & Westers, H. (1986) Waste solids control in hatchery raceways. *Progressive Fish-Culturist* **48**(2):151–54.

Brannon, E. & Klontz, G. (1989) The Idaho aquaculture industry. *Northwest Environmental Journal* **5**:23–35.

Colt, J. & Orwicz, K. (1991). Modeling production capacity of aquatic culture systems under freshwater conditions. *Aquacultural Engineering* **10**(1):1–29.

Colt, J., Watten, B. & Rust, M. (2009) Modeling carbon dioxide, pH, and un-ionized ammonia relationships in serial reuse systems. *Aquacultural Engineering* **40**(1): 28–44.

Cripps, S.J. & Bergheim, A. (2000) Solids management and removal for intensive land-based aquaculture production systems. *Aquacultural Engineering* **22**:33–56.

D'Souza, G., Miller, D., Semmens, K. & Smith, D. (2004) Mine water aquaculture as an economic strategy: Linking coal mining, fish farming, water conservation and recreation. *Journal of Applied Aquaculture* **15**(1–2):159–72.

Fornshell, G. (2002) Rainbow trout—challenges and solutions. *Reviews in Fisheries Science* **10**(3–4):545–57.

Fornshell, G. & Hinshaw, J.M. (2008) Better management practices for flow-through systems. In *Environmental Best Management Practices for Aquaculture* (Ed. by C.S. Tucker & J.A. Hargreaves), pp. 331–88. Wiley-Blackwell, Ames, IA.

Goldburg, R. & Triplett, T. (1997) *Murky Waters: Environmental Effects of Aquaculture in the United States*. Environmental Defense Fund, Washington, DC.

Hargreaves, J.A., Boyd, C.E. & Tucker, C.S. (2002) Water budgets for aquaculture production. In *Aquaculture and the Environment in the United States* (Ed. by J.R. Tomasso), pp. 9–34. United States Aquaculture Society, Baton Rouge, LA.

Haskell, D.C. (1955) Weight of fish per cubic foot of water in hatchery troughs and ponds. *Progressive Fish-Culturist* **17**:117–8.

Hinshaw, J.M. (2000) Trout farming: Carrying capacity and inventory management. *Southern Regional Aquaculture Center Publication No. 222*.

Hinshaw, J.M, Fornshell, G. & Kinnunen, R. (2004) *A Profile of the Aquaculture of Trout in the United States*. USDA Risk Management Agency, Federal Crop Insurance Corporation, through Mississippi State University. www.agecon.msstate.edu/Aquaculture/pubs/Trout_Profile.pdf.

Klontz, G.W. (1991) *A Manual for Rainbow Trout Production on the Family-Owned Farm*. Nelson and Sons, Inc., Murray, UT.

Landau, M. (1992) *Introduction to Aquaculture*. John Wiley & Sons, Inc. New York, NY.

Mazik, P.M. & Parker, N.C. (2001) Semi-controlled systems. In *Fish Hatchery Management*, 2nd Edition (Ed. by G. Wedemeyer), pp. 241–84. American Fisheries Society, Bethesda, MD.

Piper, R.G., McElwain, I.B., Orme, L.E., McCraren, J.P., Fowler, L.G. & Leonard, J.R. (1982) *Fish Hatchery Management*. United States Department of the Interior, Fish and Wildlife Service, Washington, DC.

Soderberg, R.W. (1995) *Flowing Water Fish Culture*. CRC Press, Inc., Boca Raton, FL.

Stechey, D. & Trudell, Y. (1990) *Aquaculture Wastewater Characterization and Development of Appropriate Treatment Technologies for the Ontario Trout Production Industry*. Final Report. Queen's Printer for Ontario, Ontario, Canada.

True, B., Johnson, W. & Chen, S. (2004) Reducing phosphorus discharge from flow-through aquaculture. I. Facility and effluent characterization. *Aquacultural Engineering* 32:129–44.

Westers, H. & Pratt, M. 1977. Rational design of hatcheries for intensive salmonid culture, based on metabolic characteristics. *Progressive Fish-Culturist* 39(4):157–65.

Westers, H. (2001) Production. In *Fish Hatchery Management*, 2nd Edition (Ed. by G.A. Wedemeyer), pp. 91–186. American Fisheries Society, Bethesda, MD.

Willoughby, H. (1968) A method for calculating carrying capacities of hatchery troughs and ponds. *Progressive Fish-Culturist* 30:173–4.

Chapter 10

Ponds

Craig Tucker and John Hargreaves

Ponds are familiar landscape features in all but the most arid regions of the world. In some areas, ponds are used to grow fish or crustaceans to generate profit from large-scale commercial aquaculture. In other places, ponds are used to grow fish for subsistence consumption at a small scale, often as part of farming systems that integrate plant and animal components (van der Zijpp *et al.* 2007). These noncommercial farms play a critical role in increasing food security and employment for the rural poor of developing countries (Edwards *et al.* 2002). Ponds often have multiple purposes, including potable water storage, irrigation water supply, livestock watering, storm-water retention, landscaping, and recreation. Although ponds can be used to grow aquatic plants, especially single-cell algae such as *Spirulina*, most ponds are used to grow aquatic animals, and such ponds will be the focus of this chapter. As well as being the most common aquaculture system in the world, ponds are also the most ancient way of growing fish, with origins before recorded history (McLarney 1984; Landau 1992). Pond aquaculture likely began as a natural extension of rice-field fisheries in Southeast Asia.

Ponds are commonly defined as small, confined bodies of standing water. This simple definition captures an important ecological feature of ponds. The phrase "of standing water" implies a long hydraulic residence time, which is the key functional attribute of a pond as an aquaculture system. An aquaculture pond can be defined more fully as an aquatic animal culture system where, by virtue of a long hydraulic residence time, suitable water quality for animal

Aquaculture Production Systems, First Edition. Edited by James Tidwell.
© 2012 John Wiley & Sons, Inc. Published 2012 by John Wiley & Sons, Inc.

production is controlled primarily by natural physical, chemical, and biological processes that occur within the water body. The significance of this definition is easier to understand by contrasting ponds with other common aquaculture systems. In flow-through (raceway) and net-pen systems, suitable water quality is maintained by continuous inflow of high-quality water; in recirculating aquaculture systems, water is reconditioned for use by pumping water through discrete water-treatment units (such as settling basins, biofilters, and gas reactors) that are separate from the fish-holding tank. The long hydraulic residence time in ponds also provides opportunity for considerable within-pond (autochthonous) food production for cultured animals, although the degree to which cultured animals rely on autochthonous food production for growth varies widely in different types of pond aquaculture.

The functionality of ponds as aquaculture systems is driven by solar energy, which regulates pond temperature and provides light for photosynthesis. Plants growing in the pond use specific wavelengths (colors) of light energy to synthesize cell components. During this complex process, plants assimilate carbon dioxide and mineral nutrients from the water, produce organic matter, and generate oxygen as a byproduct. Plant growth therefore provides three essential support functions in aquaculture ponds: (1) food for cultured animals, (2) oxygen to support life, and (3) treatment of wastes so that they do not accumulate to toxic levels.

At the most basic level, plants growing in ponds can provide all the resources needed to grow aquatic animals, with little reliance on support from ecosystems outside the pond. Millions of ponds and lakes throughout the world function in this way, supporting aquatic communities that thrive and persist without human intervention, powered only by sunlight falling upon the pond surface. However, energy available from sunlight impinging directly on the pond surface has a fixed maximum, which depends on latitude, time of year or day, and transparency of the atmosphere (generally as affected by cloud cover). Plant growth rate is ultimately limited by the finite energy available from sunlight, which in turn limits the capacity of a pond ecosystem to provide essential life-support functions. These limitations progressively shift from supplying sufficient food to supplying sufficient oxygen to removal of nitrogenous waste products as culturists try to achieve progressively greater crop yields (table 10.1).

Overcoming each limiting factor requires energy input from outside the pond to supplement services dependent on incident sunlight. Those inputs may be in the form of solar energy fixed as organic matter in other ecosystems and then transferred to a pond, or they may be direct inputs of industrial energy, mostly from fossil fuels. Examples of solar energy fixed in other ecosystems include production of plant feedstuffs (such as soybeans and corn) and production of marine plankton that are grazed by pelagic marine fish harvested for fish meal and fish oil. These allochthonous (from outside the system) materials can be formulated into feeds that are used to increase aquaculture production. Industrial energy inputs may include, for example, electricity to power aerators and water pumps to overcome limitations associated with oxygen supply and waste treatment.

Table 10.1 Aquaculture production at increasing levels of input for pond-raised channel catfish and tilapia. Modified from Boyd (1990) and Verdegem *et al.* (2006).

Management input	Annual production (kg/ha)		Limiting factor
	Channel catfish	Tilapia	
Stocking only	50–100	200–500	primary productivity
Stocking, fertilization	200–300	1,000–3,000	primary productivity
Stocking, fertilization, supplemental feeding	500–1,000	3,000–4,000	dissolved oxygen
Stocking, feeding	1,000–2,000	3,000–4,000	dissolved oxygen
Stocking, feeding, emergency aeration	4,000–6,000	4,000–6,000	dissolved oxygen
Stocking, feeding, continuous aeration	6,000–10,000	6,000–10,000	dissolved oxygen, metabolites
Stocking, feeding, continuous aeration, water exchange	10,000–20,000	15,000–35,000	metabolites
Stocking, feeding, continuous aeration, intensive mixing		20,000–100,000	metabolites, suspended solids

Aquaculture ponds may be filled with fresh, brackish, or salt water, and are usually constructed of soil, although some may be lined with plastic or other impervious materials to reduce water loss from seepage. Earthen ponds are typically shallow (<2 m), so interactions between soil and water have important effects on water quality. Most other aquaculture systems are constructed of inert materials (plastics or metals, for example) or water flows through the system so quickly that little time is afforded for interactions between soil and water. This attribute, where the container for water and cultured animals is an active biological and chemical component of the system, is unique to ponds among aquaculture systems.

Most ponds function simultaneously to confine cultured animals and treat wastes. In other words, animals roam free within a pond, usually dispersed at low densities compared to most other culture systems, and wastes produced by animals are treated or transformed by processes occurring in the same space. The animal confinement and waste treatment functions of ponds can be physically separated to achieve greater control over production. In such systems, space allocated to animal confinement is much less than space dedicated to waste treatment. Two "partitioned" variants of traditional ponds are described in other chapters of this book. Partitioned aquaculture systems are discussed in chapter 13 and in-pond raceways are discussed in chapter 15.

10.1 Species cultured

Nearly all types of aquatic animals can be grown in ponds but they are best suited for animals that tolerate a relatively wide range of environmental conditions (Teichert-Coddington *et al.* 1997; LeFrancois *et al.* 2009). This is most obvious in the context of water temperature, the most important environmental factor affecting aquaculture production. Aquaculture ponds have a large ratio of surface area to volume and pond water has a long hydraulic residence time. Pond water temperature therefore closely follows local air temperature, which varies daily and seasonally. Further, because water has a large specific heat capacity, it is almost always impractical to heat or cool ponds to maintain optimum water temperature for culture. Lack of water temperature control has three important implications: (1) pond aquaculture is most common in tropical or subtropical regions that are characterized by long growing seasons; (2) species grown in temperate-region ponds must tolerate wide annual changes in water temperature, often ranging from near freezing to over 35°C; and (3) ponds are uncommon as culture systems in cold regions characterized by a very short growing season.

In addition to water temperature, other pond environmental variables also change over time because biological activity, which greatly influences pond water quality, varies in seasonal and diel cycles. Metabolic wastes are transformed and eliminated from pond water by natural microbial and physical processes, and the rates of those processes vary seasonally. Diel changes in water quality are especially evident for variables affected by photosynthesis and respiration. Dissolved oxygen concentration normally increases during the day and decreases at night; carbon dioxide concentration normally decreases during the day and increases at night. Carbon dioxide concentration controls pH, with greater values during the day and lower values at night. The amplitude of these fluctuations increases as a function of phytoplankton density, which changes in response to culture intensity. In aquaculture ponds operated at a high intensity, environmental conditions may vary from optimum to nearly lethal over time frames as short as hours. As with water temperature, most water quality variables in ponds are difficult, impractical, or expensive to control. One notable exception is dissolved oxygen supply, which is supplemented in some ponds with mechanical aerators. Because water quality varies greatly over short time frames in most ponds, the signature characteristic of animals most commonly cultured in ponds—such as carps, catfishes, tilapias, and penaeid shrimp—is that they are among the hardier species with respect to environmental tolerance.

Global aquaculture production data collected by the United Nations Food and Agriculture Organization are not classified by production system so it is difficult to precisely estimate production from ponds. Based upon the most likely culture system used for various species (FAO 2006b), it appears that approximately 32 million tonnes of fish and crustaceans were produced in ponds throughout the world in 2006. This represents about 60% of total aquaculture production (including plants), 70% of animal aquaculture, and 80% of finfish aquaculture. Most pond production in 2004 was attributable to freshwater cyprinid culture in China, India, and Southeast Asia (18 million tonnes), brackishwater culture of penaeid shrimp (2.5 million tonnes), freshwater tilapia culture (1.8 million tonnes), and brackishwater culture of milkfish (0.5 million tonnes).

In 2004, more than 310,000 tonnes of fish, crawfish, and marine shrimp were produced in 140,000 ha of ponds in the United States, which was about 65% of United States aquaculture production (USDA-NASS 2006). Almost 90% of the finfish produced in the United States were grown in ponds and nearly all were inland, freshwater ponds used to grow channel catfish (*Ictalurus punctatus*), crayfish (*Procambarus* spp.), bait and ornamental fish, and hybrid striped bass (*Morone saxatilis* × *M. chrysops*). A few hundred hectares of ponds located on the Texas coast are used to produce marine shrimp.

10.2   Pond types

Pond aquaculture comprises a wider range of culture species, environments, water sources, resource input levels, and environmental impacts than any other aquaculture system. Several classification systems can be used to simplify discussion and allow comparison among pond types. Some obvious classifications include ultimate use of product (foodfish, ornamental fish, sportfish, baitfish), species of animal (carp, catfish, shrimp), life stage (larval, fingerling, foodfish), salinity preference (freshwater, brackishwater, saltwater), and climatic or temperature regime (tropical, warmwater, coolwater, coldwater).

The type of plant that dominates the food web in a pond is another basis of classification. By far the most common are ponds where phytoplankton is the main plant type. Phytoplankton is an easily digestible form of plant protein and is an effective base for aquatic food chains. Phytoplankton also do not interfere with harvest of fish or crustaceans as do some other plant types. Some ponds are managed to favor attached algae or periphyton, which serves as the preferred natural food for fish such as tilapia (Azim *et al.* 2005). In parts of Southeast Asia, benthic attached algae growth is encouraged by filling brackishwater ponds to a shallow depth and then adding fertilizer. The community that develops with a mat of benthic algae serves as a food resource base for shrimp post-larvae following stocking. Intentional cultivation of aquatic macrophytes in aquaculture ponds is not common because they interfere with harvesting of cultured animals. However, in Louisiana, rice or naturally occurring aquatic macrophytes serve as the forage base of a detrital food web for the production of crayfish, which are harvested by trapping (McClain *et al.* 2007). Although aquaculture ponds are simple agroecosystems, they can be manipulated to expand the number of food niches available for exploitation by cultured animals.

Pond classification by hydrological type is instructive because water use, effluent volume and, to lesser degrees, water quality and effluent quality are impacted by pond hydrology. Various classifications and names are used to describe pond hydrological type; we will use the classification of Boyd and Tucker (1998).

Levee, or embankment, ponds (fig. 10.1) are the most common type used in aquaculture. Levee ponds are built on flat land by excavating a thin layer of soil from the pond bottom and using that soil to form levees or embankments around the pond perimeter. Sometimes fill is obtained by excavating a shallow trench

Figure 10.1 Embankment ponds at a large aquaculture research facility in Mississippi. Pond sizes range widely within this facility because the best size depends on research needs. Ponds in the foreground are approximately 4 ha, which is the average size of embankment ponds used in commercial catfish farming. Photograph courtesy of Danny Oberle, Mississippi State University.

from the internal periphery of the pond bottom, which results in the deepest part of the pond being just inside the toe of the levee. Catchment areas for rainfall and runoff are small, consisting only of the pond surface and the inside embankment slopes, so there must be a source of pumped water to fill ponds and maintain water levels during droughts. Water can be pumped from underground aquifers or nearby surface waters. In ponds sited within the intertidal zone of coastal areas, water can be added without pumping during high tides. Levee ponds can be built in almost any size or shape to accommodate landscape features. Large contiguous tracts of standard-shape (usually rectangular) ponds facilitate farm management.

Excavated ponds (fig. 10.2) are similar to embankment ponds, with respect to general construction and primary water source, but are usually smaller and the elevation of the pond bottom is further below the original ground level of embankment ponds. In areas with high water tables, excavated ponds may extend below the water table and be partially filled with groundwater inflow. Emptying excavated ponds may require pumping to lift water into drainage canals.

Watershed ponds (fig. 10.3) are built in hilly terrain by damming a temporary or permanent stream. The major source of water is runoff from the catchment basin above the dam, which can vary from constant inflow when a permanent stream is dammed to highly variable inflow when the watershed is small. Water level in ponds where the main source of water is runoff may vary greatly through the year and fall to critically low levels during dry seasons or droughts unless a supplemental source of pumped water is available.

The fourth pond type is a hybrid between embankment and watershed ponds. These ponds may have two or three sides consisting of embankments (actually

Figure 10.2 Excavated ponds (approximately 200 m^2) used to grow ornamental fish in Florida. Note the bird-exclusion netting and continuous diffused (bubbler) aeration to protect the valuable crop. Photograph courtesy of Dr. Jimmy Avery, Mississippi State University.

Figure 10.3 Watershed ponds used to grow channel catfish in Alabama. The pond in the foreground is approximately 8 ha. Photograph courtesy of Auburn University Department of Fisheries and Allied Aquacultures.

low dams) across a relatively small, shallow drainage basin. A significant amount of water may be obtained from runoff, but a source of pumped water also must be available because the catchment area above the pond is relatively small. Hybrid watershed-embankment ponds are built on land with gently rolling topography that is not ideally suited for embankment ponds or watershed ponds.

Embankment and excavated ponds have much less overflow than watershed ponds, with the overflow volume from hybrid ponds being intermediate. Also, the quality of effluents from watershed ponds may be affected (either positively or negatively) by upstream water quality, which is in turn affected by land-use practices in the watershed.

10.3   Water use

Compared to other culture systems, pond aquaculture uses relatively large volumes of water, and water availability constrains the location and size of pond aquaculture facilities. Adequate water must be available to fill ponds quickly and maintain the water level during prolonged periods of dry weather. Water use is also inextricably tied to waste discharge because water input and discharge are variables on opposite sides of the hydrological equation: increasing water input volume will increase water volume discharged.

Water budgets quantify inflows and outflows from ponds. Budgets include some or all of the following: inflows include precipitation, runoff, stream inflow, groundwater inflow, and regulated inflow; outflows are made up of evaporation and evapotranspiration, seepage, overflow, and regulated discharge. In the following section we briefly describe water budgets for ponds. Furthermore, Yoo and Boyd (1994) and Hargreaves *et al.* (2002) provide thorough discussions of the subject.

10.3.1   Sources

The main water sources for aquaculture ponds are precipitation, runoff from watersheds, and regulated additions of water from groundwater aquifers or surface water bodies (e.g., streams, reservoirs, and estuaries). Precipitation and runoff are the most important water sources for watershed ponds; precipitation and regulated inflows from groundwater or surface water are the primary water sources for embankment and excavated ponds.

Precipitation falling directly into a pond can be a significant source of water in some regions, although the frequency distribution of precipitation events varies greatly, even in humid climates, with many more precipitation events of low amount than of high amount. In the United States, the range of monthly rainfall among years is great, particularly during winter and early spring. Most tropical areas are characterized by distinct rainy and dry seasons, which have profound impacts on pond water budgets.

Watershed ponds are constructed to capture overland flow from the surrounding watershed, especially if ground or surface water availability is not reliable or predictable. Runoff volume depends on watershed characteristics, particularly slope, type and extent of vegetative cover, soil type, and antecedent soil moisture. Runoff is also a function of rainfall amount and intensity. Across the United States, runoff averages 28% of precipitation. In the southeast United States, overland flow is about 21 to 29% of rainfall from November to March, but very little runoff occurs from April to October (Yoo & Boyd 1994).

Embankment ponds have little area other than the pond surface to capture rainfall and therefore rely primarily on regulated inflows to initially fill ponds, to replace losses from evaporation and seepage, and to manage water quality with water exchange. The main sources of regulated inflow are surface water supplies and groundwater aquifers. Water from streams, rivers, or estuaries can be added by pumping or gravity flow as the sole water source or to supplement groundwater and runoff sources. Seasonal and year-to-year availability of some surface waters can be variable and availability often does not correspond to demand for pond aquaculture. For example, in the southern United States, peak stream discharge occurs during winter and early spring, yet the demand for water is greatest during summer. Surface waters are also undesirable because they are sources of contamination by wild fish, potential disease pathogens, agricultural pesticides, fertilizers, and sediment. Surface waters may also be regulated to limit withdrawals. Overall, groundwater is the preferred source of water for freshwater pond aquaculture. Groundwater is usually free of chemical and biological contaminants that may be present in surface waters and is usually not subject to the same fluctuations in availability as surface waters. However, groundwater aquifers recharge slowly and withdrawals that exceed recharge can reduce groundwater availability and lead to conflicts among users.

10.3.2 Losses

Water is lost from ponds by evaporation, seepage, and overflow that occur when inflows exceed pond water storage capacity. Losses from evaporation represent the only consumptive water use because this water is not available for downstream uses. In theory, water lost by seepage is available as groundwater and thus does not represent a consumptive use. Water is also lost when ponds are intentionally drawn down or drained for harvest, pond renovation, or in the case of crayfish culture, to stimulate crayfish burrowing and prepare ponds for planting vegetation. Water that overflows or is intentionally discharged is available for other purposes and thus does not represent consumptive use.

Evaporation is a function of atmospheric conditions that are beyond control of the culturist. However, seepage can be reduced by good site selection and pond construction practices. Overflow can also be minimized by maintaining water storage capacity and by not managing water quality with water exchange. Losses from intentional discharge can be minimized by reducing frequency of drawdowns or by pond draining.

10.3.3   Pond water use

Water use in aquaculture may be classified as either total or consumptive. Total water use is the sum of all inflows (precipitation, runoff, seepage inflow, and intentional additions) to production facilities. Depending on the type of culture system, a portion of the water entering the facility passes downstream in overflow or intentional discharge and is available for other purposes and thus is not consumed. For example, essentially all water flowing into flow-through systems is discharged, whereas some ponds have no discharge. Consumptive use is the water volume used in aquaculture that is subsequently unavailable for other purposes. Consumptive use includes water lost to evaporation from an aquaculture facility, and water removed in biomass of aquatic animals at harvest. Water in harvest biomass averages about 0.75 L/kg of animal produced, an insignificant quantity compared to losses from evaporation. Although groundwater withdrawals do not necessarily contribute to consumptive water use, groundwater withdrawals may contribute to aquifer depletion if withdrawals exceed recharge.

Total water use in ponds varies over a wider range than for any other aquaculture system, depending primarily on the frequency of intentional water exchange (flushing) and pond drawdowns or draining for harvest. Total water use ranges from less than 2 m³/kg of aquaculture production in undrained ponds operated with no water exchange up to 80 m³/kg for marine shrimp ponds operated with high (10 to 20% of pond volume) daily water exchange (table 10.2). For semi-intensive pond aquaculture, total water use typically ranges from 3 to 10 m³/kg. To put these values in context, total water use is around 100 m³/kg for trout raceways and 0.1 m³/kg for recirculating systems.

Note that total water use to produce fish in extensively managed ponds is similar to that required in intensively managed ponds operated with water exchange (the top two entries in table 10.2, for example). Intensively managed

Table 10.2 Total water use in pond aquaculture. Conditions and values adapted from Phillips *et al.* (1991), Verdegem *et al.* (2006), Boyd (2005), and Tucker *et al.* (2008). Note the effects of culture intensity, species, and hydrological pond type.

Culture conditions	Total water use (m³/kg)
Intensive shrimp, 20% daily water exchange	40–80
Extensive, warmwater fish	45
Intensive shrimp, Taiwan	29–43
Intensive tilapia, Taiwan	21
Semi-intensive shrimp, Taiwan	11–21
Carp polyculture, intensive, Israel	12
Channel catfish, watershed ponds, southeastern US	11
Warmwater fish, pellet-fed	9
Carp polyculture, semi-intensive, Israel	5
Channel catfish, embankment ponds, southeastern US	3–10
Warmwater fish, nighttime aeration	3–6
Low water exchange, sewage-fed, Thailand	1.5–2
Warmwater fish, intensive mixed ponds	0.4–1.6
Air-breathing catfish, Thailand	0.05–0.2

ponds with water exchange use large volumes of water to produce large crops. Extensive ponds use less water but produce proportionately smaller crops. This illustrates a common effect of culture intensity on resource use efficiency, with systems operating at an intermediate level of culture intensity using resources most efficiently.

Ponds in general have the greatest consumptive water use of all aquaculture systems because they have large surface areas for water loss from evaporation. Consumptive water use is typically less than 0.1 m^3/kg of aquaculture production in flow-through and water recirculating systems, but is 1 to 5 m^3/kg or more for aquaculture production in ponds, depending on climate. Water used for water exchange in shrimp ponds is pumped from bays or estuaries and discharged into the same water body, so only water lost to evaporation is used consumptively, but the supply of brackishwater is essentially limitless. In the past, fresh water was added to seawater to adjust salinity in some marine shrimp ponds operated with water exchange. In this case, the added fresh water was used consumptively because it was discharged to the marine environment and not available for other uses.

Although the consumptive use of water in pond aquaculture is relatively large, that use has high economic value as measured by the consumptive water value index (Boyd *et al.* 2007), which compares consumptive water use and gross economic value for crops per unit area. For example, in the southeastern United States, channel catfish aquaculture requires more water than irrigated cotton, corn, and soybeans, but an amount comparable to rice (50 to 100 cm/year). However, the economic value of water used for catfish aquaculture (about US$1.00/$m^3$ of consumptive water use) is much greater than the value of water applied to other crops (less than US$0.10/$m^3$). The consumptive use of water in other aquaculture systems is much less and the corresponding economic value of water use is much higher than for ponds. As examples, the consumptive water value index for trout grown in raceways is approximately US$50/$m^3$ and for tilapia in cages is about US$650/$m^3$ (Boyd *et al.* 2007).

10.4 Pond culture intensity and ecological services

The term *culture intensity* is used to classify agriculture systems based on crop yield per unit land or water area. When defined in that manner, culture intensity varies over at least three orders of magnitude for commonly used aquaculture production systems. At one extreme, water-recirculating systems are capable of annually producing 1 to 2 million kg of fish per hectare of culture unit. The same fish production in flow-through raceways requires about ten times more surface area, and production in ponds may require 1,000 times the area of recirculating systems. Fish production in recirculating systems is therefore considered more intensive than production in ponds. Culture intensity can also describe production characteristics within a production system type, which is especially common in describing production from ponds.

The concept of culture intensity not only includes consideration of relative crop yield but also the broader context of resource inputs. Combining contexts of crop yield and resource use into the overall concept of culture intensity highlights the fact that differences in crop yields among aquaculture systems are not simply a function of systems with greater yields having inherently greater value than systems with lesser yields. Rather, a broader measure of culture intensity considers the extent to which the primary natural resources and ecological services needed to grow aquatic animals—food, life support, and waste treatment—are obtained from outside the culture facility.

The intensity of pond aquaculture—and therefore the degree to which pond aquaculture depends on external resources and services—varies over a wider range than other production systems. Yields in pond aquaculture vary over at least two orders of magnitude, from 200 kg/ha in lightly fertilized recreational sportfish ponds to more than 20,000 kg/ha in intensively managed cultures with feeding and continuous aeration. At the low end of the intensity spectrum, all food and oxygen needed to support fish growth is produced in the pond and most of the wastes produced during culture are retained within the pond, where they are removed or transformed by natural processes. At the other extreme, nearly all resources needed to grow animals in intensive pond systems are provided from ecosystems external to the pond. Animal growth is supported by large inputs of feeds manufactured from ingredients obtained from ecosystems outside the pond. Wastes resulting from feed inputs stimulate high rates of biological activity within the pond and natural supplies of dissolved oxygen are inadequate to meet the overall oxygen demand. At the upper limits of intensity, rates of natural processes within the pond become inadequate to remove all the wastes produced during culture and water exchange is used to eliminate excess organic matter or potentially toxic metabolites. When water is flushed from ponds to remove wastes, adjacent water bodies or land areas are used to treat or assimilate those wastes. Viewed in this manner, culture intensity is closely associated with the concept of ecological footprint, which is a tool for assessing the dependence of an activity on resources or services appropriated from other ecosystems. That is, the overall ecological footprint (which equals the facility area plus support area) of extensive pond aquaculture is small relative to intensive pond culture because land and water area outside the pond is used to provide the ecological services that support high levels of production.

In the next two sections, we describe food production and life-support processes in pond aquaculture. The important life-support processes in ponds are provision of dissolved oxygen and waste treatment. In each of the following sections, note how subsidies of external resources are increasingly needed to make high levels of production possible as culture intensity increases. This underlying theme will then be summarized in the section 10.7 "Land use and the ecological footprint of pond aquaculture."

10.5   Food in pond aquaculture

The fundamental goal of pond aquaculture is to manage a water body so that it will produce more aquatic animals than it would without management. To enhance crop production, providing more food is essential. Food availability can be increased either by enhancing natural productivity within a pond or by providing supplemental food, often in the form of manufactured feed, from sources external to the pond.

10.5.1 Pond fertilization

Primary productivity (plant growth) is the base of the food web in extensive pond systems, but natural levels of primary productivity are usually inadequate to support high aquaculture yields. Primary productivity can be increased by fertilizing ponds with essential plant nutrients—predominantly nitrogen and phosphorus. The goal of a fertilization program is to increase production of natural foods that can be consumed directly or indirectly by cultured aquatic animals. These natural foods include phytoplankton, periphyton (attached algae), phytoplankton-derived detritus, algal-bacterial aggregates, zooplankton, and benthic invertebrates.

Greater food availability in fertilized ponds can increase aquaculture production by factors of five to ten, depending on the food habits of cultured animals and the efficiency of natural food use (table 10.1). For example, Nile tilapia (*Oreochromis niloticus*) responds well to pond fertilization because the fish grazes directly on phytoplankton and periphyton, as well as other natural foods such as detritus and zooplankton. Net fish yields of 10 to 20 kg/ha per day in fertilized tilapia ponds in the tropics are possible. In contrast, fish such as channel catfish are not able to feed directly on algae or other plants, feeding instead on secondary production of zooplankton, insects, and other herbivorous animals. The lower thermodynamic efficiency of animals feeding at higher trophic levels is a well-known ecological principle, implying that production of channel catfish, for example, will not respond to pond fertilization to the same degree as production of strictly herbivorous or omnivorous aquatic animals.

Aquaculture production in fertilized ponds can also be optimized by culturing in the same pond two or more species of animals with different food habits. This strategy, called polyculture, makes more efficient use of the variety of natural foods available in fertilized ponds. For example, a possible combination of fish might be silver carp (*Hypopthalmichthys molotrix*), grass carp (*Ctenopharyngodon idella*), and common carp (*Cyprinus carpio*). Silver carp feed in the water column by filtering plankton; grass carp are adapted to feed on aquatic macrophytes and added green fodders; and common carp are benthic omnivores that feed on detritus and benthic invertebrates. This combination makes efficient use of the different foods available in fertilized ponds. Net fish yields of 10 to 40 kg/ha per day are possible in fertilized polyculture ponds (Lin *et al.* 1997).

The best fertilization strategy maximizes profit, which may not necessarily correspond to maximum yield. The economics of fertilization (like that of other inputs to pond aquaculture systems) is subject to the Law of Diminishing Returns where, past some intermediate input level, each successive increment of an input results in progressively less output. The costs of a fertilization strategy include unit costs of nutrients in a particular fertilizer and costs associated with transport, handling, and application. Furthermore, the opportunity cost of farm capital and labor must be considered.

Algae growth in aquaculture ponds is affected by temperature, light availability, and the rate of nutrient supply. Culturists have no practical control over temperature in ponds, but selecting sites in warm or tropical climates will favor algal growth. Typically, algal growth will double for every 10°C increase in temperature between 10 and 30°C. Fertilization when temperature is too low for phytoplankton growth can be counterproductive and favor growth of undesirable filamentous algae or rooted aquatic plants.

Ultimately, the availability of light limits primary productivity in fertilized ponds. Fertilization can increase algal density to the point where self-shading shifts the limitation of photosynthesis from nutrient supply to light. Availability of light for photosynthesis is reduced by algal and mineral turbidity and water color (staining). The efficiency of light use for photosynthesis can be improved by mixing ponds, which increases average light exposure experienced by phytoplankton. Mixing also increases the availability of nutrients recycled from the sediment.

Given the limited control over temperature and light availability in pond aquaculture, culturists justifiably focus on increasing the supply of nutrients required for plant growth. Algal growth is assumed to be limited by supply of the nutrient—or more broadly, the resource—that is most scarce relative to the requirement, a principle known as Liebig's Law of the Minimum. Nutrient limitation of algal growth must be seen in the context of nutrient requirements relative to the supply of that nutrient. When nutrient demand exceeds supply, nutrient limitation occurs. Of the approximately twenty elements needed for growth by aquatic plants and algae, carbon, nitrogen, and phosphorus are most likely to be in short supply, and, as a practical matter, it is only necessary to meet the needs for those three nutrients in a fertilization program.

Although carbon is the element required by algae in the greatest amount, its supply limits algal growth only in highly productive waters enriched with nitrogen and phosphorus or in waters of very low alkalinity. Dissolved carbon dioxide gas is the primary source of carbon for algal photosynthesis. The gas enters the water by diffusion from the atmosphere or is produced by respiration of pond organisms, including algae, microorganisms, and cultured animals. Many algae also use bicarbonate as a carbon source, either directly as the bicarbonate anion or indirectly when bicarbonate dehydrates to produce carbon dioxide. The water's carbonate-bicarbonate alkalinity system is therefore an important potential source of carbon for photosynthesis. In some cases, lack of a response to fertilization can be attributed to carbon limitation, and in those cases the response to fertilization can be improved with liming to maintain alkalinity greater than 20 to 30 mg/L as $CaCO_3$ (Boyd & Tucker 1998).

To describe nutrient dynamics in aquaculture ponds in the simplest terms, nitrogen cycling is dominated by biological processes and phosphorus cycling is dominated by physicochemical processes. Proportionally more nitrogen than phosphorus is recycled to the water column during the mineralization of nutrients in algal detritus. The preferred form of inorganic nitrogen for algae is ammonia, although algae can take up nitrate if ammonia is absent. Ammonia is the form of nitrogen that increases in pond water with the addition of many common nitrogenous chemical fertilizers (e.g., urea, diammonium phosphate).

In comparison to inorganic nitrogen, inorganic phosphorus quickly becomes incorporated into the sediment in forms of variable availability, with most being tightly bound. Phosphorus solubility is affected by pH, especially in compounds with aluminum and calcium ions, and oxidation-reduction potential, which affects the solubility of phosphorus compounds with iron. Anaerobic conditions at the pond bottom can promote the release of iron-bound phosphorus from the sediment, a process called internal fertilization. Nutrient availability in the water column is enhanced by pond mixing or destratification and is restricted by thermal stratification. In general, ponds are sinks for nutrients, especially phosphorus, but nutrients can be recycled from the sediment to water from organic matter decomposition, diffusion from anaerobic sediment, and desorption during sediment resuspension.

Chemical fertilizers are manufactured in forms—including solid, liquid, instantly soluble, and controlled-release—that vary in solubility, nutrient availability, and cost. Organic fertilizers also vary widely in nutrient content and availability but are characteristically dilute sources of nutrients. Thus, much more organic fertilizers are required than chemical fertilizers to obtain a favorable algal growth response. The cost of organic fertilizers per unit nutrient is much greater than that of chemical fertilizers.

Organic fertilizers encompass a wide range of materials, including animal manure, green manure, fodders, composts, cereal grains, and seed meals. Organic fertilizers are especially important as a source of particulate organic matter for the production of zooplankton in fish nursery ponds (Anderson & Tave 1993). Organic fertilizers act to stimulate autotrophic (algae-based) and heterotrophic (detritus-based) food webs in aquaculture ponds.

Organic fertilizers have very different effects on water quality than do chemical fertilizers. Unlike chemical fertilizers, organic fertilizers decompose and exert an oxygen demand on pond water and also provide carbon dioxide for photosynthesis. Excessive organic loading can lead to oxygen depletion and organic matter accumulation on pond bottoms. Decomposition of organic fertilizers can also result in staining of water, thereby reducing light availability in the water column. In cases where visibility is limited by clay turbidity, organic fertilizers can increase light availability by serving as a coagulating agent.

A fixed-input fertilization strategy based on research is the most common fertilization practice, probably because it is the simplest approach. The selected rate is the input level that maximizes profit in controlled fertilization studies or field trials. Fixed fertilization rates are often adequate but do not account for variation in soils, water quality, and other site-specific characteristics. Fixed rates also do not account for changes in conditions during production and differences in pond-to-pond response, even for adjacent ponds. The differential response of ponds treated similarly is common. Understanding the pond characteristics and environmental factors that cause differential response to fertilization can improve the efficiency of fertilizer use. Thus, as an alternative to a fixed-input fertilization strategy, fertilization rate can be based on pond-specific measurements made during the production cycle, including water and sediment analysis, computer modeling, or an algal bioassay that evaluates the response to nutrient enrichment of water samples (Knud-Hansen 1998). Fertilization programs based on pond-specific measurements of nutrients are complex, labor-intensive, difficult to interpret, and often not practical, cost-effective, or necessary.

Fertilization rates will depend on soil and water type but ultimately are a matter of cost-effectiveness. Recommended fixed fertilizer rates range from 10 to 30 kg N/ha per week, with lower rates in this range used when Secchi disk visibility is less than 20 cm. It is difficult to make specific recommendations about fertilization with phosphorus because availability is strongly affected by fertilization history. Recommended phosphorus fertilization rates range from 1 to 4 kg P/ha per week, with greater rates in this range appropriate for ponds with water of high hardness (>100 mg/L as $CaCO_3$). Phosphorus fertilization rates of 7 to 10 kg P/ha per week can be applied to the water of newly constructed ponds or ponds with clear water. Phosphorus fertilization rates can decrease over time as the capacity of the sediment to adsorb phosphorus decreases. Over time, the N:P ratio of fertilizer can be increased from one or two up to eight by reducing phosphorus input.

From the perspective algal growth requirements, providing frequent doses of fertilizer nutrients is best. The decision of fertilization frequency is based on fertilizer type and the opportunity cost of farm labor. For chemical fertilizers, a fertilization frequency of one application every one to two weeks is sufficient to obtain good response. For manures, daily application of low amounts of fertilizer is best to minimize the risk of oxygen depletion. Typically, chemical fertilizers are broadcast over the pond surface and the poor solubility of many chemical fertilizers can be improved by dissolving them in water before application. Chemical fertilizers can also be applied on platforms or in porous bags that cause the slow leaching of nutrients. Liquid fertilizers are denser than water and should be diluted in water before application.

Ultimately, decisions about fertilization are governed by economics, with costs of the material, transport, and handling part of the decision. Furthermore, the opportunity cost of farm capital and labor must be considered. Finally, the opportunity cost of applying an organic fertilizer to a pond or another crop in the farming system must be evaluated. Fertilizers and fertilization practices are described by Lin *et al.* (1997), Boyd and Tucker (1998), and Knud-Hansen (1998).

10.5.2   Ponds with feeding

Aquaculture yield in fertilized ponds is limited by primary productivity (which is ultimately limited by solar radiation) and by how efficiently the cultured animal uses primary productivity for growth. To increase aquaculture yield past that attainable in fertilized ponds, feed ingredients from outside the pond must be obtained, formulated into a palatable and nutritious feed, and fed to cultured animals. Examples of commercial pond aquaculture relying on manufactured feeds are production of channel catfish in the southeastern United States and penaeid shrimp in tropical coastal zones. Carp aquaculture in Asian ponds, traditionally practiced as extensive culture in fertilized ponds, is increasingly dependent on manufactured feeds to support high levels of production.

Although carbon in ingredients used in manufactured feeds is ultimately traceable to carbon fixed in photosynthesis (as is the case with all foods), the difference between fed aquaculture systems and chemically fertilized ponds is, of course, that the plant growth that supports food webs in fertilized ponds occurs inside the pond, whereas the plant growth that supports animal growth in fed cultures occurs outside the pond. For example, most manufactured feeds are made primarily of terrestrial oilseeds or grains (soybeans, wheat, corn) and fish meal from small pelagic fish harvested from the sea.

The quality of feeds added to ponds varies greatly depending on availability of resources, the nutritional requirements of the cultured animal, and the economics of production. Inputs to some culture ponds consist of low-quality organic matter that might otherwise be considered a waste product, such as animal manures, bedding (litter), and the byproducts of processing agricultural plants. For the most part, little of the aquatic animal production in these ponds is derived from direct consumption of the added organic material because it is generally nutritionally inadequate, unpalatable, or too difficult to ingest to support high aquatic animal production through direct consumption. A more significant source of food in waste-fed ponds is derived via autotrophic and heterotrophic food webs stimulated by addition of organic materials. As such, ponds receiving low-quality organic materials are more properly classified as fertilized ponds rather than fed cultures.

Some animals can, however, efficiently use low-quality organic matter for growth and direct consumption of those materials may account for a large portion of the food consumed. Grass carp is the best example of a commonly cultured animal that grows efficiently on inputs of forages and grasses. Grass carp is the third most commonly cultured aquaculture species in the world, with nearly 4 million tonnes produced in 2004 (FAO 2006b). China is the leading producer of the fish. Most grass carp production relies on inputs of terrestrial or aquatic plants harvested from nearby land or water bodies and then processed to some degree (often by simply chopping into smaller pieces) and fed directly to fish in ponds or cages. Agricultural byproducts, such as leaves of corn, soybean, or other crops, are also used as food in grass carp cultures. Fecal wastes from grass carp then serve as a green manure that stimulates primary productivity when nutrients are released upon decomposition. Silver carp and common carp are often stocked in such ponds to exploit natural productivity enhanced by grass carp manure.

As culture intensity increases, from a nutritional standpoint, growth of cultured animals becomes limited first by the availability of dietary energy (De Silva & Anderson 1995). Therefore, providing supplemental inputs such as energy-rich cereal grains can complement natural foods and increase aquaculture production beyond that obtained by fertilization. As culture intensity increases further, protein becomes the next nutritional factor to limit production.

Most fish and crustaceans grown in aquaculture cannot efficiently make direct use of low-quality organic matter or raw grains, so feeds are manufactured to be a source of concentrated nutrients with high-quality ingredients that are nutritious and highly digestible. High-quality manufactured feeds are expensive, and feed costs may constitute 30 to 60% or more of the overall variable costs of production.

The food web in ponds with feeding is more complicated than simply "animal eats feed." Even when cultured animals are fed a nutritionally complete feed to satiety, they also incidentally consume some of the variety of natural food organisms produced within the pond. Although consumption of natural food organisms is usually not expected to contribute to a significant proportion of the total yield of animals fed manufactured feeds, natural foods can supply vitamins, essential fatty acids, or trace minerals that are important to animal growth, especially when manufactured feeds are not nutritionally complete. In the case of fed penaeid shrimp ponds, direct consumption of feed by shrimp is inefficient and a surprisingly large fraction of added feed functions as an organic fertilizer, stimulating the production of various natural food items that are subsequently consumed by shrimp.

10.6   Life support in pond aquaculture

To maximize production potential and profitability, cultured animals must be grown in an environment conducive to good growth and health. In practice, the most important aspects of maintaining environmental quality in aquaculture are providing adequate dissolved oxygen and removing or transforming metabolic wastes produced during culture. Ponds have a surprisingly large inherent capacity for providing these life-support functions. In fact, the capacity of ponds to provide oxygen and assimilate wastes is a key part of the definition of "ponds," as explained in the introduction to this chapter.

The capacity of pond ecosystems to provide life-support services is, however, limited (Hargreaves & Tucker 2003). As culture intensity increases, so too does dependence on external subsidies of energy and other resources to maintain adequate environmental quality. Whether it is economically justified to supplement the natural life-support capacity of ponds depends on production goals and economics. Typically, the capacity of the pond ecosystem to provide adequate dissolved oxygen is the first life-support function to be exceeded, and it may be profitable to supplement the pond's naturally produced oxygen supply with mechanical aeration.

Further intensification of production will eventually be constrained by the waste assimilation capacity of the pond ecosystem. At that point, excess organic matter and nutrients must be removed by other means, such as exchanging degraded water with high-quality water, in a manner analogous to water exchange in flow-through systems and net pens, or by additional treatment. Water exchange or additional treatment are rarely justified economically and, in the case of water exchange, may be regulated by laws. With the exception of supplementing dissolved oxygen supplies with mechanical aeration, a fundamental goal of most pond aquaculture is to operate within the pond's inherent capacity to provide life-support functions.

10.6.1 Dissolved oxygen

Of resources under the control of the culturist, availability of dissolved oxygen is the next factor to limit production in ponds after meeting food requirements of cultured animals. Two complimentary aspects of dissolved oxygen are important in pond aquaculture: (1) maintaining dissolved oxygen concentration above a minimum threshold for the cultured animal and (2) providing sufficient oxygen supply to meet overall respiratory demand of pond biota. The importance of the first goal is obvious because animals will die or grow poorly if the concentration falls to critical levels for a long duration. Oxygen supply is important because providing adequate oxygen to maintain aerobic conditions throughout the pond maximizes the waste treatment capacity of ponds. Supplying oxygen to the sediment is especially important because many of the waste-treatment processes in ponds occur at or near the sediment-water interface.

Oxygen is not very soluble in water and relatively small changes in its supply or demand can cause large differences in dissolved oxygen concentration. Non-managed ponds seldom experience episodes of low dissolved oxygen concentration unless the pond is built in an area with unusually fertile soils or receives runoff enriched with organic matter or nutrients. But when ponds are fertilized or receive additions of feed to increase aquaculture production, overall rates of plant growth (usually in the form of phytoplankton) and other biological processes increase. As phytoplankton biomass increases in response to greater inputs of plant nutrients, water column gross oxygen production during photosynthesis increases. However, oxygen uptake by overall community respiration also increases because the biomass of plants and bacteria in water and sediment is also greater. This causes wide fluctuations in dissolved oxygen concentration over a 24-hour period as oxygen is produced only during daylight. The magnitude of fluctuations in dissolved oxygen concentration increases as phytoplankton density increases in response to nutrient loading. If the biomass of phytoplankton and other organisms is too high, dissolved oxygen concentration often becomes critically low at night. Maintenance of dissolved oxygen concentration above a critical threshold, which is a characteristic of the species cultured, is best accomplished by mechanical aeration. Using water exchange is an ineffective and inefficient means of adding oxygen to pond water.

Note the relationship between pond culture intensity and source of dissolved oxygen: in low-intensity culture, dissolved oxygen is provided entirely by natural processes (passive diffusion from the atmosphere and photosynthesis in the water column). As culture intensity increases, rates of oxygen supply from natural sources become inadequate to meet overall respiratory demand and natural supplies must be supplemented with mechanical aeration powered by energy from external sources. In the extreme, hard-bottom or lined ponds are aerated continuously with multiple aerators, which produces well-mixed conditions and provides nearly all the dissolved oxygen needed to support respiration of cultured animals and other pond biota.

10.6.1.1   Processes affecting dissolved oxygen concentration

The dissolved oxygen concentration measured at a particular time and place in a pond is the result of many simultaneous biological, chemical, and physical processes that produce or consume oxygen. The primary oxygen sources in most aquaculture ponds are photosynthesis by phytoplankton and gas transfer (diffusion) from the atmosphere. The primary sinks for oxygen include phytoplankton, zooplankton, and bacterial respiration in the water column, respiration of cultured animals, sediment oxygen uptake (which includes respiration of organisms in the sediment plus oxygen-consuming chemical reactions), and gas transfer from water to the atmosphere. Details of these processes, briefly summarized below, are discussed in Boyd and Tucker (1998).

Daily dissolved oxygen budgets in most aquaculture ponds are dominated by photosynthesis and respiration of phytoplankton. At low phytoplankton standing crops (typical of non-fertilized ponds, for example), adequate oxygen is produced during daytime photosynthesis to meet overall pond respiratory demand and dissolved oxygen concentration remains relatively high. As culture intensity increases, average phytoplankton standing crops increase in response to greater loading rates of nitrogen, phosphorus, and other plant nutrients from fertilization or feeding. The effect of increasing phytoplankton biomass on daily net oxygen production is shown in figure 10.4 for a hypothetical aquaculture pond.

The oxygen budget in figure 10.4 uses equations from Smith and Piedrahita (1988) for phytoplankton photosynthesis and respiration. Fish and sediment respiration were calculated for conditions typical of semi-intensive catfish ponds in midsummer. The most important point shown by this general model is that daily dissolved oxygen surpluses occur only at intermediate phytoplankton standing crops. When phytoplankton biomass is very low or very high, daily net oxygen deficits occur that must be offset with mechanical aeration to provide dissolved oxygen to keep the aquaculture crop alive. Daily oxygen deficits develop at low phytoplankton standing crops because inadequate plant biomass is present to produce oxygen in photosynthesis to offset community respiration. Deficits develop at high phytoplankton standing crops because the restricted availability of light caused by algal self-shading, and possibly other factors, increasingly constrains photosynthesis as phytoplankton standing crops increase, but phytoplankton respiration continues to increase as a direct function of algal biomass. Eventually, phytoplankton standing crops can become so great that solar radiation cannot support gross photosynthesis at rates that exceed overall community respiration. Past that point, daily net oxygen deficits occur.

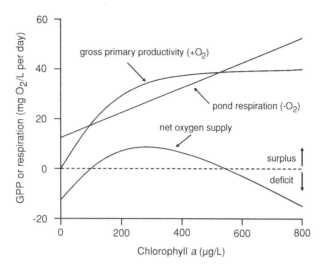

Figure 10.4 Dissolved oxygen budget for a hypothetical aquaculture pond as a function of phytoplankton biomass (chlorophyll *a* concentration is a common surrogate measure of algal biomass). In this budget, photosynthesis (gross primary productivity, GPP) by phytoplankton is the sole source of oxygen; respiration includes oxygen consumed by phytoplankton, fish, and sediment. The lower curved line is the net oxygen surplus or deficit calculated by subtracting respiration from gross primary productivity.

The relationship between phytoplankton density and net daily oxygen production shown in figure 10.4 assumes optimum conditions for algal photosynthesis. Conditions that reduce photosynthesis will reduce net oxygen production and may lead to critically low dissolved oxygen concentrations under conditions that otherwise would be safe. Examples of conditions that may reduce gross oxygen production and contribute to oxygen supply deficits include reduced solar radiation (prolonged cloudy weather, for example), reduced light penetration into the water from non-algal turbidity (from suspended clay particles, for example), and poisoning of phytoplankton with herbicides (Smith & Piedrahita 1988).

After plankton respiration, sediment respiration is the next largest sink for oxygen in most semi-intensive aquaculture ponds. Oxygen is consumed by the respiration of microorganisms and invertebrates living on or in the sediment and by chemical oxidation of reduced substances that diffuse from deeper sediment layers to the sediment surface. The sediment surface is the primary location for mineralization of organic matter, consisting mostly of detritus derived from phytoplankton and fish fecal solids that settle to the bottom. Nitrification of ammonia, a process that consumes oxygen, also occurs at the sediment surface.

Respiration of cultured animals is a function of biomass, animal size, water temperature, ambient dissolved oxygen concentration, time after feeding, and animal activity. Contrary to intuition, cultured animal respiration usually accounts for a relatively low proportion of overall oxygen use in ponds. For example, in semi-intensive catfish ponds, total respiration in ponds is partitioned among the water column (65%), sediment (20%), and fish (15%) (Steeby *et al.* 2004).

Gas transfer across the air-water interface can result in a gain or loss of dissolved oxygen from pond water, the direction of diffusion dependent on the difference between the dissolved oxygen concentration in pond water and the equilibrium (saturation) concentration. Diffusion can be an important source of oxygen when the deficit of dissolved oxygen concentration in pond water is large. Oxygen transfer across the air-water interface is a function of the difference in the partial pressure of oxygen in pond water and the atmosphere, turbulent mixing, and the pond surface area to volume ratio. The rate and direction of gas transfer therefore varies during the day and is at its maximum when the dissolved oxygen deficit (or surplus) is at maximum and when wind speed is high. Air to water oxygen transfer becomes of great practical importance when dissolved oxygen concentration falls to dangerously low levels because increasing oxygen transfer rate is the basis for managing dissolved oxygen supply with mechanical aerators. Mechanical aerators augment overall oxygen transfer rates by greatly increasing the air-water interfacial area and increasing turbulence within the pond.

10.6.1.2 Mechanical aeration

As pond aquaculture intensity increases, the imbalance between daily oxygen production and oxygen consumption increases, and aeration is needed more often and in greater amounts. This trend is illustrated by changes in aeration practices over the last fifty years of channel catfish farming. In the 1950s and early 1960s, catfish farming was practiced mainly to supply local markets as a more dependable and profitable alternative to wild-caught fish. Ponds without management were too unproductive to be profitable, even for these early low-intensity endeavors, and farmers sought to increase fish yield by fertilizing ponds or adding crude feeds. The risk of dissolved oxygen depletion increased as feeding and fertilization rates increased, and an early goal of pond management was to improve catfish production without degrading the environment to the point where aeration was needed. One of the first recommendations for channel catfish farming in the southeastern United States was to restrict maximum daily feeding rate to about 35 kg/ha to minimize the risk of oxygen depletion and the need for aeration.

In the 1960s and early 1970s, catfish farmers increased production by increasing fish stocking rates and feeding at correspondingly higher levels. They reached a point where supplemental aeration was occasionally needed to keep fish alive. Problems usually occurred late at night or around dawn, when nighttime respiration had depleted the water of oxygen produced during daytime photosynthesis. Episodes of critically low dissolved oxygen concentrations were relatively rare, occurring only under unusual conditions such as prolonged periods of warm, cloudy weather or after sudden, unpredictable phytoplankton die-offs. Fish farmers often resorted to using irrigation pumps or other rudimentary equipment to oxygenate pond water.

Through the 1970s, catfish farmers continued to intensify production. Nighttime oxygen depletions became more frequent and equipment was developed specifically for pond aeration. Aeration of large ponds required a relatively

Figure 10.5 A tractor-powered paddlewheel aerator. This paddlewheel directs water parallel to the pond embankment. Other designs push aerated water away and perpendicular to the embankment. Photograph courtesy of E. L. Torrans, United States Department of Agriculture.

large energy input, and early catfish pond aerators were powered by farm tractors (fig. 10.5). Farmers typically monitored dissolved oxygen concentration throughout the night and moved tractor-powered aerators to ponds where dissolved oxygen concentration was expected to decline to critical levels (usually 2 to 4 mg/L). This approach, called emergency aeration, made no attempt to manage dissolved oxygen levels throughout the pond. Rather, the goal was to provide a zone of well-oxygenated water near the aerator to support the respiratory needs of the fish biomass and operate the aerator until pond-wide dissolved oxygen concentration increased in daytime from algal photosynthesis.

Further intensification of catfish farming through the late 1970s and early 1980s was accompanied by a dramatic increase in the use of aeration. When average daily feeding rates exceeded 50 kg/ha through the growing season, often exceeding 100 kg/ha for long periods during summer, supplemental aeration was needed every night in most ponds. Initially, tractors were a convenient power source for aeration, but were expensive and inefficient. In the early 1980s, commercial catfish farmers shifted to permanently installed electric aerators (fig. 10.6). Electric aerators are much more efficient and currently the most common aeration device in most commercial pond aquaculture.

The evolution of catfish pond aeration practices illustrates the relationship among nutrient loading (intensification), oxygen budgets, and the need to supplement natural oxygen supplies. The frequency and amount of supplemental aeration also depends on cultured species. For example, penaeid shrimp live

Figure 10.6 A floating, electric paddlewheel aerator. Note the splash pattern and the zone of oxygenated water adjacent to the aerator. Photograph courtesy of E. L. Torrans, United States Department of Agriculture.

near the pond bottom and cannot quickly move large distances to find zones of aerated water. Shrimp ponds are therefore aerated for longer periods of time (many are aerated continuously) and with more aerator power per unit pond area than catfish ponds at equivalent nutrient loading rates. Aerators used in shrimp culture not only oxygenate water but also prevent stratification by mixing the water column to assure that oxygen supplied by aeration reaches the pond bottom where shrimp forage.

Aeration practices also depend on the goals of pond management and relative crop value. In some types of low-intensity aquaculture, such as sportfish ponds, the goal is to increase fish production through fertilization, but not to the point where phytoplankton and animal biomass becomes so great that additional management is needed to prevent oxygen-related fish kills. On the other hand, ornamental fish may be grown under relatively low-intensity conditions in ponds where episodes of critically low dissolved oxygen concentration are rare. However, the fish crop is so valuable that farmers may aerate ponds continuously to remove any risk of oxygen depletion from natural causes.

Aerators used in pond aquaculture are based on designs for similar equipment used in wastewater treatment. Aerators either splash water into the air (surface aerators) or release air bubbles under water to increase gas-water contact area. Surface aerators are most common in pond aquaculture. Aerators for large ponds are usually based on the paddlewheel design whereas impeller-type vertical pump aerators (fig. 10.7) are often used in small ponds. Aeration equipment and practices are reviewed by Boyd and Tucker (1998) and Boyd (1998).

Figure 10.7 An impeller-type vertical pump aerator. A submerged electric motor drives an impeller that throws water into the air. This type of aerator is especially suited for small ponds. Photograph courtesy of E. L. Torrans, United States Department of Agriculture.

Threshold minimum dissolved oxygen concentrations for warmwater fish (2 to 4 mg/L) are similar for good fish growth and survival and the maintenance of waste assimilation capacity. As such, pond aeration also serves to enhance other life support functions, such as waste nitrogen assimilation, in addition to the direct effect of providing dissolved oxygen to cultured animals. The direct value of aeration in preventing oxygen-related mortalities is obvious; however, the economic benefit of aeration to enhance the pond waste assimilation rate is unknown but likely not cost-effective. The most common aeration strategy in finfish ponds is to maintain a zone of aerated water near the aerator and not attempt to maintain dissolved oxygen concentration above a critical threshold throughout the pond to enhance waste assimilation capacity.

10.6.2 Waste treatment

Fish and crustaceans produce wastes as a byproduct of growth, and the amount of waste produced is proportional to animal biomass and feeding rate. Fertilization or feeding practices used to increase aquaculture production therefore increase the amounts of organic matter, solids, nitrogen, phosphorus, and other substances added to pond water.

Waste treatment in aquaculture has two distinct, but highly interrelated, functions. First, removal of metabolic wastes prevents degradation of the environment to the point where aquatic animal growth or health is negatively impacted. For example, bacteria that decompose organic materials added to ponds or produced during culture express an oxygen demand and compete with cultured animals for oxygen. Some substances, such as ammonia or nitrite, may accumulate and directly affect growth and health of cultured animals. Second, wastes produced during aquaculture also represent potential pollution if culture water is discharged from the facility. Regulatory constraints or negative impacts on the downstream environment can limit aquaculture production and profitability. As such, waste-treatment processes in aquaculture, whether inherent in the production system or as an adjunct to culture, reduce the pollution potential of aquaculture effluents.

10.6.2.1   Sources of nutrients and organic matter

The major nutrients applied to ponds in fertilizers are nitrogen and phosphorus. A portion of the fertilizer nutrients added to ponds is assimilated by plants, which then produce organic matter in photosynthesis. Plants are the base for the food web, eventually culminating in growth of cultured fish or crustaceans. Wastes are produced at each step in the food web because ecological efficiencies of nutrient use are not high, especially with natural foods. Even for fertilized ponds that are carefully managed, less than half the nitrogen and less than a quarter of the phosphorus applied as chemical fertilizer are eventually recovered in fish harvested. Efficiencies of carbon use are even lower, with less than 10% of the carbon fixed in photosynthesis recovered in fish harvested from fertilized ponds (Boyd & Tucker 1998). Differences between nutrient inputs (or production in the case of carbon) and amounts recovered in fish represent the waste load added to ponds.

Waste loads for ponds with feeding can be calculated by subtracting amounts in fish harvest from amounts in feed. Waste loads as percentages of feed inputs vary depending on species cultured, feed quality, and other factors, but using channel catfish as an example, about 80 to 90% of the dry matter and carbon, and 70 to 80% of the nitrogen and phosphorus in feeds is released to ponds as waste. Nutrients that are not assimilated stimulate high levels of organic matter production in photosynthesis, so overall loading of organic matter in ponds with feeding is much greater than indicated by simple mass balance. In one study (Boyd 1985), nutrients released during production of 1,000 kg of channel catfish supported production of 2,000 kg of dry matter in the form of phytoplankton.

Because large fractions of nutrient and organic matter input are not recovered in harvested animals, considerable quantities of waste are added to water in intensive pond aquaculture. Most waste nitrogen is excreted as ammonia through the gills, but some is excreted as organic nitrogen in fecal solids that are rapidly mineralized by pond bacteria to ammonia. In channel catfish ponds with a typical midsummer daily feeding rate of 100 kg/ha, more than 500 mg/m^2 of nitrogen are added to ponds in the form of feed protein. Assuming 70% of nitrogen added to feed is excreted to the pond environment as ammonia, approximately 350 mg/m^2 per day (or 0.35 mg/L per day in a 1-m deep pond) is excreted by fish as waste. At that daily loading rate, ammonia should rapidly increase to levels that kill fish. However, total ammonia in catfish ponds is usually less than 1 to 2 mg N/L, even after many months of feeding. This illustrates a phenomenon common to all waste budgets in pond aquaculture; that is, observed concentration is always less than expected based on simple mass balance. The difference between observed levels of nutrients and organic matter and the amount expected based on mass balance is attributable to biological, chemical, and physical processes that transform and remove wastes from pond water. These processes are important because they maintain, at no direct economic cost to the farmer (other than that for supplemental aeration), adequate water quality for animal production and they reduce the pond's pollution potential. In fact, the widespread economic success of pond aquaculture can be attributed in large part to low production costs associated with the inherent waste assimilation capacity of ponds, despite the internalization of those costs, unlike more open production systems.

Qualitative fates and cycling of organic matter and nutrients in aquaculture ponds do not differ greatly from other shallow aquatic ecosystems, and detailed discussions of these processes can be found in any good limnology textbook (Scheffer 1998). Details of aquaculture pond ecology and nutrient cycling are also presented in Delincé (1992), Tucker (1996), and Boyd and Tucker (1998). The following briefly summarizes fates of organic matter, nitrogen, and phosphorus in pond water. These three variables have important implications for pond water quality and the environmental impacts of pond effluents.

10.6.2.2   Organic matter

In ponds with fertilization or feeding, the creation of "new" organic matter by algal photosynthesis far exceeds the amount added from solid fecal wastes or intentional additions of organic fertilizers. Organic matter loading from photosynthesis can be very large, with annual production of several thousand kilograms of organic solids per hectare in intensively managed ponds. In fact, sustainable operation of pond aquaculture is predicated in large measure on the capacity of the pond to remove the large amounts of organic matter produced during culture.

Some phytoplankton and other organic solids are consumed by zooplankton or other invertebrates, but most of the organic matter from phytoplankton biomass and solid fecal wastes is decomposed in the water column and at the sediment-water interface. Aquaculture ponds are shallow so most organic matter settles to the sediment before it is decomposed. The settled organic matter is readily decomposed because it consists of easily biodegradable organic matter in the form of dead phytoplankton, detritus of phytoplankton origin, or fecal solids. Water temperature and oxygen availability are the two major factors affecting rates of organic matter decomposition in ponds. Decomposition converts (mineralizes) the organic matter into inorganic components such as carbon dioxide, ammonia, and orthophosphate, which may be recycled into new organic matter production when assimilated by phytoplankton. Organic matter decomposition also produces dissolved organic compounds that may serve as substrate for bacterial or fungal growth. All these processes occur continuously in a complex cycle of input, uptake, decomposition, reuse, and removal. When water is discharged from ponds, some fraction of the various forms of solid and dissolved organic matter is released to the downstream environment.

10.6.2.3   Nitrogen

Nitrogen is a major nutrient affecting productivity of aquatic ecosystems because it is an essential component of protein and other constituents of cellular protoplasm. Nitrogen is important in fertilized ponds because it is a key plant nutrient that may be in short supply relative to the amount needed for plant growth. In culture systems provided with manufactured feed, nitrogen is important as a constituent of feed protein and as a waste product of animal metabolism. Nitrogen in animal wastes may contribute to excessive phytoplankton abundance and may also lead to the accumulation of ammonia and nitrite, which can be toxic to aquatic animals. Waters discharged from ponds may also be enriched with inorganic and organic combined nitrogen and contribute to nutrient enrichment of receiving waters.

Nitrogen occurs in water as various inorganic compounds and myriad organic compounds. Transformations among the various forms of nitrogen constitute the nitrogen cycle. Most transformations are biochemical oxidation-reduction reactions that are strongly interdependent, with rates of one process sometimes limited by the rate of substrate formation in a preceding process. For example, rates of denitrification in aquaculture ponds are often constrained by the rate at which nitrate (the substrate for denitrification) is produced by nitrification.

Variable but relatively small amounts of nitrogen enter the aquatic nitrogen cycle via biological nitrogen fixation by aquatic bacteria or blue-green algae. Fertilizers or feed accounts for most of the nitrogen input to aquaculture ponds. In fertilized ponds, nitrogen is added as chemical fertilizers or in nitrogen-rich organic materials. In fertilized ponds managed for high fish production from algal-based food webs, daily nitrogen input may exceed 400 mg/m^2. In ponds with feeding, almost all nitrogen originates from feed protein. As a typical value, about 25% of the feed nitrogen is recovered in fish production or, conversely, about 75% of the nitrogen in feed is released to pond water as waste from fish. Roughly 80% of the waste nitrogen in ponds with feeding enters the pond as ammonia, the primary nitrogenous waste product excreted by aquatic animals from the gills. Nitrogen also enters the water in fecal solids and uneaten feeds, but ammonia is quickly liberated when those materials decompose. Assuming a 25% retention of feed nitrogen and 30% protein in the feed, total waste nitrogen input to the pond water will be about 36 mg/kg of feed. At a feeding rate of 100 kg of feed/ha per day, a typical summertime feeding rate used in channel catfish culture in the southeastern United States, daily nitrogen loading would be 360 mg/m^2.

Decomposition of dead phytoplankton cells, fish fecal solids, uneaten feed, and other organic material releases ammonia. Some decomposition occurs in the water column but most takes place in the surface layer of pond sediment. The rate of ammonia production from organic matter mineralization depends on temperature, oxygen availability, pH, and the quantity and quality of organic matter. Some nitrogen-containing organic material accumulates in sediments but most is quickly decomposed and nitrogen is recycled to the water column. Ammonia recycled from organic matter decomposition in pond sediments accounts for 25 to 33% of the total ammonia input to water of catfish ponds (Hargreaves 1997), with fish excretion accounting for the balance.

Uptake by phytoplankton is the main sink for most combined inorganic nitrogen in aquaculture ponds. Rates of algal nitrogen assimilation are roughly proportional to rates of net photosynthesis, so factors affecting plant growth directly affect nitrogen assimilation. For instance, nitrogen uptake is reduced when water temperature is low, during periods of low sunlight intensity, or when the availability of some other nutrient limits plant growth. The magnitude of average rates of nitrogen assimilation by phytoplankton can be estimated by assuming that phytoplankton assimilate nitrogen in a fixed proportion to carbon. The so-called Redfield ratio of C:N:P in "average" phytoplankton tissue is 42:7:1 by weight. Daily net carbon fixation rates range from less than 0.5 to over 5 g/m^2 in semi-intensive aquaculture ponds, which corresponds to daily nitrogen assimilation rates of less than 80 to over 800 mg/m^2. Most of the ammonia produced from fish excretion and organic matter mineralization is assimilated by phytoplankton. The continuous recycling of nitrogen through the processes of phytoplankton uptake, cell death, mineralization of organic nitrogen, and re-assimilation by phytoplankton is an important feature of nitrogen dynamics in aquaculture ponds.

Ammonia from fish excretion or organic matter mineralization may also be adsorbed as ammonium ion by clays and organic colloids in pond bottom soils, volatilized into the atmosphere as ammonia gas, or nitrified by bacteria to nitrate. Quantitatively nitrification is second to phytoplankton uptake as a major sink for ammonia. Nitrification is the two-step transformation of ammonia to nitrite and then nitrate by chemoautotrophic bacteria. Nitrification rates are controlled by water temperature, oxygen availability, pH, and ammonia concentration. Nitrifying bacteria are strictly aerobic and rates of nitrification drop dramatically when dissolved oxygen level falls below 1 to 2 mg/L. Nitrification occurs over a wide temperature and pH range but optimum conditions are 25 to 35°C and pH 7 to 8.5. Nitrifiers tend to colonize surfaces, and highest rates occur at aerobic, sediment-water interfaces. Few direct measurements of nitrification in ponds have been made, but estimated rates range from 0 to 150 mg N/m^2 per day, with most estimates ranging from 25 to 50 mg N/m^2 per day (Hargreaves 1998).

Denitrification of the nitrate produced in nitrification is the main process by which nitrogen is lost from aquaculture ponds. Denitrification is the production of nitrogen gas by common heterotrophic bacteria that use nitrate or other oxidized forms of nitrogen, rather than oxygen, as terminal electron acceptors in respiration. Denitrification is an anaerobic process so it occurs primarily in pond sediments just below the thin layer of oxidized surface sediment. The denitrification potential of aquaculture ponds is great because the volume of anaerobic sediment is large. The rate of denitrification depends on the availability of nitrate, so the process is tightly coupled with nitrification, which is the process by which nitrate is generated in aquaculture ponds. The coupling of these two processes—one aerobic and the other anaerobic—is ultimately dependent on oxygen supply because the rate-limiting step is the production of nitrate from nitrification.

Un-ionized ammonia and nitrite are potentially toxic to some aquatic animals and the risk associated with these two toxicants increases as culture intensity increases. In low-intensity pond aquaculture, nitrogen loading rates are low and problems with accumulation of ammonia or nitrite are extremely rare. In intensive culture, nitrogen loading rates from manufactured feeds are high, and at any one time large quantities of nitrogenous compounds are in flux among the various nitrogen cycle components. Under these conditions, relatively minor changes in process rates can result in accumulation of potentially toxic intermediate products (Hargreaves and Tucker 1996; 2005). For example, most of the ammonia excreted by fish is quickly detoxified when phytoplankton assimilate and incorporate ammonia into algal proteins. Ammonia will accumulate when rates of phytoplankton uptake are reduced, which may occur during prolonged cloudy weather, with herbicide use, or following naturally occurring algal die-offs. Likewise, nitrite, which is an intermediate in the two-step process of nitrification, may accumulate when the rate of the first nitrification step exceeds that of the second step. In both examples, the amount of reaction substrate (ammonia in the first example, nitrite in the second) that may accumulate is greater when the overall nitrogen loading rate to the pond is high.

Treating pond water to remove ammonia is difficult because large volumes of water are involved. Various treatments, some based on accelerating natural processes, have been proposed but most either do not work or are too expensive (Hargreaves & Tucker 2004). Ammonia can be removed by exchanging water and this is commonly practiced in some pond cultures. Water exchange may be continuous or occasional, such as whenever the pond manager feels that fish health is threatened. The need for water exchange and the efficacy of the practice as an ammonia-removal strategy is debatable in most pond aquaculture, particularly if ponds are large and water exchange rates are low (less than 10 to 20% per day). Moreover, if frequent water exchange is required to maintain ammonia concentration at a nonthreatening level, it is a clear indicator that the waste assimilation capacity of the pond has been exceeded. However, practicing water exchange to reduce phytoplankton density to the intermediate levels that maximize algal growth and ammonia uptake may be an effective control strategy.

In pond aquaculture, as defined here, the key to ammonia management is to minimize the probability of accumulation to toxic levels by operating within the assimilative capacity of the pond ecosystem. If ammonia concentration is routinely high in fertilized ponds, nitrogen is being supplied in excess of requirements for algal growth, which is wasteful of fertilizer and money. In ponds receiving feed, the risk of ammonia toxicity is reduced by using moderate stocking and feeding rates and using high-quality feeds and good feeding practices to maximize nitrogen retention by the cultured animals. Other than these general rules and using experience to empirically determine the pond's assimilative capacity, it is difficult to precisely define nitrogen loading limits for ponds. As a rough estimate, however, the nitrogen assimilative capacity of a pond is roughly equivalent to the average rate of nitrogen uptake by phytoplankton. Nitrogen uptake rates vary from day to day, but a reasonable average daily uptake rate for warm, shallow, unmixed ponds is in the range of 300 to 500 mg/m^2. This corresponds to the amount of nitrogen excreted by fish when daily feeding rates for high-quality feed are in the range of 80 to 140 kg/ha. Algal nitrogen assimilative capacity must be increased (as it is in the partitioned aquaculture system; see chapter 13) if greater feeding rates are sustained over long periods.

Nitrite causes methemoglobinemia in some fish, which is a condition where nitrite enters the bloodstream across the gills and oxidizes iron in hemoglobin to the ferric state. Oxidized hemoglobin, called methemoglobin, is incapable of reversibly binding with oxygen and causes a functional anemia that puts fish at risk of hypoxia. Ictalurid catfish, salmonids, carps, and tilapias are especially sensitive to nitrite toxicosis (Tomasso 1994).

Nitrite accumulation in ponds is highly sporadic and appears to be most common in warm-temperate regions with pronounced seasonal changes in water temperature and solar radiation. In those regions, nitrite concentration in intensive aquaculture ponds shows a bimodal pattern, with maximum concentrations in spring and autumn. A model of nitrogen dynamics in catfish ponds (Hargreaves 1997) indicates that nitrification rate (and therefore the rate at which nitrite is produced) is a function of ammonia concentration and water temperature. During summer, nitrification rate is low despite favorable temperature because phytoplankton are better competitors for ammonia than nitrifying bacteria, and the low ammonia substrate concentration limits nitrification. In winter, nitrification substrate is not limiting because phytoplankton growth is slow under conditions of low light and cold water temperature, and ammonia concentration is correspondingly high. However, wintertime nitrification rates are slow because low temperature restricts microbial activity. Nitrite therefore accumulates when nitrification rates are maximum, which occurs at intermediate values of temperature and ammonia concentration, during spring and autumn.

Similar to considerations associated with the control of ammonia, removal of nitrite from large ponds is not practical. Fortunately there is an inexpensive treatment that protects nitrite-sensitive fish from toxicosis. Chloride is a competitive inhibitor of gill lamellar uptake of nitrite in nitrite-sensitive fish. Adding common salt to pond water to increase chloride concentration can prevent nitrite toxicosis. The required amount of chloride depends on the highest expected nitrite concentration and the fish species cultured. In channel catfish culture, where nitrite toxicosis is a significant potential problem, salt is routinely added so that the chloride concentration is at least thirty times the highest expected nitrite-nitrogen concentration. Typically salt is added to catfish ponds to maintain a chloride concentration of about 100 mg/L, which protects fish from all but the most extreme episodes of nitrite accumulation. Salt treatment is long-lasting because chloride is lost from ponds only during overflow associated with excessive rains, intentional water exchange, or pond draining.

10.6.2.4 Phosphorus

Phosphorus is a key metabolic nutrient that is available in relatively small amounts in most natural surface waters. The supply of phosphorus therefore regulates primary productivity in many natural waters. It is the key nutrient, with nitrogen, in pond fertilizers. Phosphorus is also important because aquaculture pond water is often enriched with phosphorus relative to natural waters and the discharge of pond water can increase the phosphorus concentration of receiving waters, potentially leading to eutrophication.

Fertilizer or feed accounts for virtually all phosphorus inputs to aquaculture ponds. Phosphorus may be added to ponds in water used for filling or water exchange, but amounts added are usually small unless the water supply is polluted or otherwise enriched with phosphorus. In ponds receiving applications of manufactured feeds, about 65 to 75% of ingested phosphorus is excreted, primarily as particulate fecal solids.

Unlike the nitrogen cycle, with many forms of inorganic nitrogen, the phosphorus cycle has one main form of inorganic phosphorus, called orthophosphate. Orthophosphate is formed when phosphorus fertilizers dissolve in pond water and released when organic compounds such as fish fecal solids and dead phytoplankton cells decompose. Much of the phosphorus in phytoplankton cells and organic detritus settles to the pond bottom before the opportunity for bacterial degradation and release of orthophosphate into the water column.

The ultimate fate of most phosphorus added to ponds is accumulation in pond sediment through chemical reactions with sediment minerals that render phosphorus unavailable for algal growth. The net movement of phosphorus from water column to sediment is, in fact, the major feature of the phosphorus cycle in aquaculture ponds. Depending on soil type and soil pH, phosphorus may be strongly bound as iron and aluminum phosphates in low pH pond sediments or calcium phosphates in sediments of higher pH. The continuous loss of phosphorus from the water column and accumulation in the sediment is the primary reason why phosphorus is usually the first limiting factor for phytoplankton growth, necessitating multiple ongoing additions of phosphorus to maintain high rates of primary productivity. Phosphorus accumulation in sediment is also important as a mechanism that reduces the potential impact of pond effluent on receiving streams. Water released from ponds always contains far less phosphorus than the amount added in feeds or fertilizer because nearly all the phosphorus input is retained in pond sediment. Detailed and quantitative discussion of this topic can be found in Boyd (1995).

10.7 Land use and the ecological footprint of pond aquaculture

Ponds use large areas of land to produce aquaculture crops, requiring orders of magnitude more facility area than flow-through or recirculating aquaculture systems. In that regard, ponds might be perceived as inefficient systems for producing food. However, as discussed in the section on culture intensity, land use in aquaculture is more complicated than simply accounting for facility area.

Aquaculture, as with agriculture generally, usually relies on additional area to provide ecosystem support or service functions. The overall land or sea area occupied by the facility plus that needed to provide all resources to grow a crop is called the ecological footprint. In aquaculture, the overall footprint includes areas for the facility, food production, and life support (which includes oxygen supply and waste treatment). Footprints may also include ecosystem requirements for other services, such as freshwater supply, forest or ocean area for carbon dioxide sequestration, and, for some systems, nursery areas used to produce seedstock or brood animals. As the term implies, the concept of ecological footprint focuses on requirements for land area, but the analysis can also apply to requirements and flows of energy and materials through the larger system that includes fish production and ecosystem support.

Fundamental thermodynamic principles dictate that the need for external ecosystem support increases as culture intensity (expressed as crop yield per unit area) increases. That is, low intensity culture systems have a low ratio of total support area to facility area whereas high-intensity systems require large areas external to the facility to support high levels of aquaculture production. For example, much of the ecosystem support for extensive pond aquaculture is inherent in the system. Traditional aquaculture ponds used in Chinese carp polyculture function not only to confine fish, but also to provide an internal area for food production, oxygen production, and waste treatment. Such systems may have ratios of support area to facility area that approach one. Or, stated another way, the ecological footprint is approximately equal to facility area. On the other hand, some aquaculture systems function only to confine the crop. Food for the confined animals is produced in external ecosystems and wastes are exported and treated outside the facility. In other words, the footprint of those systems is much larger than the facility area.

When considering the total ecosystem area needed to support a certain level of aquaculture production, the apparent differences in culture intensity based on yield per unit area nearly disappear. Although ponds require hundreds or thousands of times the facility area as other systems, those other systems rely on hundreds or thousands of times the external ecosystem area to provide food and waste treatment compared with ponds.

10.7.1 The footprint of food production

Most global aquaculture production depends on plant photosynthesis either within the culture unit (ponds) or in adjacent waters (open-water molluscan shellfish culture) to produce most of the food that supports animal growth. In ponds managed for autochthonous food production, aquaculture yield is limited by primary productivity, which in turn is ultimately limited by the amount of solar radiation that the culture unit receives. Such ponds are usually fertilized with plant nutrients to make the best use of incident solar radiation and are often stocked with two or more herbivorous and omnivorous species to make efficient use of the variety of available natural foods.

Annual yield of herbivorous or omnivorous fish can be quite impressive (3,000 to 10,000 kg/ha, or more) in fertilized ponds. To increase aquaculture productivity past that achievable in systems dependent only on autochthonous primary productivity, food produced outside the culture system must be imported. In some systems, that food may be low-quality organic matter that might otherwise be considered a waste, such as agricultural byproducts or livestock manures. In many aquaculture systems worldwide—and in nearly all commercial systems in the United States—allochthonous organic matter added to increase productivity consists of manufactured feeds made from plant (soybean and corn, for example) and animal (usually fish) meals. Production of terrestrial feedstuffs requires agricultural land, and production of fish meal and oil requires an area of marine ecosystems to produce small pelagic fish.

The ratio of supporting food production area to aquaculture facility area varies over four orders of magnitude depending on culture system, type of food, and feed composition. Carp, tilapia, and other fish at lower trophic levels can be grown in ponds fertilized with agricultural wastes or byproducts and do not require external ecosystem area for food production. Those systems have a value of around one for the ratio of ecosystem support to facility area. Growing omnivorous species like channel catfish in ponds appropriates an area for procurement of feedstuffs that is approximately ten times the area of the pond (recalculated from Boyd *et al.* 2007 using additional data on ecosystem area needed to procure fish meal). Intensive net-pen culture of carnivorous fish, such as salmon, requires an ecosystem support area for food production that is more than 10,000 times the area of the culture system (Folke *et al.* 1998). For aquatic animals that are fed, the ecological footprint for food production dominates the total footprint and the total footprint for a given level of fish production is similar among culture systems.

10.7.2   The footprint of life support

Every culture system must provide an environment conducive to growth and survival of cultured animals. Maintaining environmental quality within tolerance limits has two interrelated goals: providing oxygen and providing a mechanism to maintain concentrations of potentially toxic metabolites below some threshold. Providing oxygen sustains metabolism of cultured animals and enhances waste assimilation in ponds. Depending on culture intensity, life support services are provided to a variable degree by natural processes within the pond. As culture intensity increases, life support is increasingly dependent on biophysical resources and ecological services from outside the pond.

10.7.2.1   Dissolved oxygen

Solar radiation entering the water column provides energy to drive photosynthesis, producing oxygen as a byproduct. In extensive pond aquaculture, dissolved oxygen needed to sustain the aquaculture crop is produced within the pond.

The footprint of oxygen production therefore equals the facility footprint. Past that point, expressing external ecosystem support on an areal basis is inconvenient when speaking of operating inputs such as aeration, which are more intuitively (and commonly) expressed in energy units, such as kilowatt-hours (or kilojoules) of industrial energy input per kilogram of production. Expressed in this manner, no input of industrial energy for aeration is necessary for extensive systems where photosynthesis and gas transfer from the atmosphere supply all the dissolved oxygen.

Intensification of pond aquaculture results in greater biological activity, and gross daily oxygen production is often less than overall community respiration. The daily net oxygen deficit must be offset by mechanical aeration, which is an energy-intensive process. Typically, as culture intensity increases, aeration is needed for longer periods each day and higher-power aerators (or more aerator units) are needed to provide enough oxygen to offset the supply deficit. Overall energy inputs for aeration increase dramatically as production intensity increases or for animals such as shrimp that require more aeration than finfish. For example, catfish ponds in the southeastern United States may be aerated for 1,200 hours per year (about 6 hours per night over the 200-day warm-weather growing season). The total nominal power rating of the aerator motors divided by pond area averages about 1 kW/ha. If net fish production is 5,000 kg/ha, the industrial energy input for aeration is approximately 0.24 kW-hours/kg of fish produced (860 kJ/kg). Marine shrimp ponds are often aerated continuously and with more aerator power per unit pond area than catfish ponds. In contrast to catfish ponds, which may be aerated with 1 to 2 kW/ha, shrimp ponds may be aerated with as much as 10 to 20 kW/ha (Boyd 1998). If 10 kW/ha of continuously operated aeration will support a standing crop of 6,500 kg/ha of shrimp (Peterson *et al.* 2001) and two 150-day crops of shrimp are produced annually, then the industrial energy input for aeration is approximately 5.5 kW-hours/kg of shrimp produced (20,000 kJ/kg), or more than twenty times the energy input per unit production for catfish.

Although energy expended for aeration represents a significant operating cost in intensive pond aquaculture, it is much less than the energy required to produce or obtain feedstuffs and process them into feeds. The energy input associated with catfish feed is about 86,000 kJ/kg of fish produced (Troell *et al.* 2004) or more than 100 times the industrial energy input for aeration calculated above. However, direct energy input captures only a part of the energy subsidies for pond aeration. Energy is also required to obtain raw materials, construct, and maintain aeration equipment. These costs are difficult to calculate, but still represent only a small fraction of the overall industrial energy subsidies in intensive pond aquaculture because of the overwhelming energy inputs associated with manufactured feeds.

10.7.2.2 Waste treatment

All aquaculture produces waste and water or land area is required to assimilate or otherwise treat those wastes. It is instructive in this regard to consider the degree to which aquaculture production systems are open to the environment. In relatively closed production systems, like recirculating aquaculture systems and ponds operated with long hydraulic residence times, significant quantities of waste produced during culture are treated within the facility and there is little external area needed for waste treatment. In contrast, much of the waste produced in flow-through and net-pen culture systems is discharged directly to the environment. The capacity of the external ecosystem to assimilate those wastes may limit aquaculture production either by polluting the surrounding water to the point where animal welfare inside facilities dependent on that water source is endangered ("self-pollution") or by imposed regulatory constraints on the amount of waste that can be discharged. In addition to effects on aquaculture production, waste discharge into public waters may create problems such as transfer of pathogenic organisms, degraded water quality that limits options for use, water treatment costs, and other downstream impacts.

The ecosystem area needed for waste assimilation, expressed as waste treatment area per unit production facility area, varies over at least two orders of magnitude, depending on the type of production system. At one extreme are aquaculture ponds with low to moderate stocking and feeding rates that can, on the basis of inherent waste assimilation capacity, be operated for many years without draining or intentional water exchange to remove wastes. Natural processes described above remove or transform wastes at rates adequate to prevent long-term accumulation of potential pollutants. In theory, no outside ecosystem support area is needed to treat wastes produced during culture and the pond functions as its own waste treatment facility. In practice, ponds must be occasionally drained for renovation and inventory adjustment and some overflow is inevitable during periods of heavy precipitation. Nevertheless, for ponds operated with long hydraulic residence times, more than 90% of the waste organic matter, nitrogen, and phosphorus produced during culture is assimilated inside the pond before water is discharged (Tucker *et al.* 1996).

Internalizing waste treatment imposes a relatively high direct land cost to pond aquaculture and this cost is evident in the large size of pond facilities. Extensive land use is the result of the pond functioning as both an animal confinement area as well as a waste treatment facility. In theory, less than 5% of the pond area is needed to confine fish, meaning that more than 95% of the land and construction costs for ponds can be assigned to waste treatment functions. Effectively this means that the relatively large land area occupied by a pond aquaculture facility is a price the culturist pays for treating wastes on site rather than discharging them to public waters. Engle and Valderrama (2002) explored other costs (such as aeration, labor, etc.) associated with internalizing waste treatment in channel catfish ponds and calculated that almost 30% of the total cost of producing channel catfish can be ascribed to internal waste treatment processes. Land area requirements and other costs resulting from internalizing waste treatment limit profitable pond aquaculture to areas where large tracts of flat land are available at a reasonable price.

Aquaculture production in ponds is therefore limited by the finite capacity of the pond ecosystem to treat wastes produced during culture (Hargreaves &

Tucker 2003). Further intensification of production is possible only if wastes are treated outside the culture unit, usually by discharging wastes to public waters. At that point, the culture system is no longer a pond as defined here, but represents a system that is intermediate between a pond and a flow-through system. Displacing wastes from the pond requires an ecosystem area outside the culture facility for treatment. In that case the cost of treating wastes discharged to public waters is borne by society, not the aquaculturist.

The external ecosystem area needed to treat aquaculture wastes depends on the quality and amount of waste discharged and the hydrology and biology of the ecosystem into which wastes are discharged. Based on the few studies conducted, it appears that areas between 100 and 300 times the facility area are needed to treat wastes produced from intensive cage and net-pen fish culture (see, for example, Berg *et al.* 1996; Kautsky *et al.* 1997; Folke *et al.* 1998; Brummett 1999). Note that these values for the ratio of waste assimilation area to fish-holding area are somewhat larger, but within the same order of magnitude, as the ratio of waste treatment area to fish-holding area in intensive catfish ponds described previously. This is not a coincidence. The same biological and physicochemical processes are responsible for waste treatment whether the water is inside a pond or has been discharged to a lake, stream, or estuary. Therefore, all other things being equal, the area required for treatment of a unit of waste should be of similar magnitude regardless of the degree to which a particular culture system is open to the environment. The difference, of course, is that the ecosystem area needed for waste treatment in ponds is, for the most part, inherent in the production system whereas waste treatment for flow-through systems and net pens is external to the system.

10.8 Consequences of unregulated algal growth

An important characteristic of pond aquaculture is the low level of control over many of the important factors of production. Phytoplankton communities are the dominant ecological feature of aquaculture ponds because their metabolic activities directly or indirectly affect the suitability of the pond environment for growing the crop. Some ponds are fertilized to increase phytoplankton production as the food base for cultured fish. In ponds with feeding, phytoplankton communities develop as the unintentional consequence of applying nutrient-dense feeds. As a practical matter, the abundance and species composition of phytoplankton communities cannot be controlled.

Phytoplankton are simultaneously desirable and undesirable in aquaculture ponds. Some type of plant community is inevitable in ponds and, of the various communities that may develop, phytoplankton does not interfere with fish harvest as do filamentous algae and emergent or floating aquatic plants. Phytoplankton also produce oxygen and play a key role in waste assimilation. However, phytoplankton can be undesirable when biomass becomes excessive or when the community contains species that produce odorous metabolites or toxins. Odorous algal metabolites impart undesirable off-flavors to the edible portions of cultured animals, thereby affecting product quality and marketability. Algal toxins may kill the cultured animal.

Communities of odorous or toxic algae are called "harmful algal blooms" or HABs. Harmful algal blooms are usually transient in ponds and they develop and then disappear in a seemingly random manner. Environmental factors regulating phytoplankton community composition are understood only at the most basic level (Reynolds 1984). For example, HABs tend to be more frequently encountered in ecosystems with high nutrient loading. Past that generality, predicting the occurrence of odorous or toxic species is impossible and managing the environment to prevent harmful blooms is very difficult.

10.8.1 Off-flavors

The most common off-flavors in pond-raised animals are caused by either geosmin or 2-methylisoborneol (MIB). Geosmin has an earthy odor redolent of a damp basement. MIB has a unique, musty-medicinal odor, somewhat like camphor. Both compounds are essentially nontoxic to plants or animals. They are among the most highly odorous compounds found in nature and are common in aquatic and terrestrial environments—they give soil its characteristic odor, for example. Geosmin and MIB are synthesized by a variety of fungi, actinomycete bacteria, and blue-green algae (cyanobacteria) but in ponds they are nearly always produced by blue-green algae. Off-flavors in aquaculture are reviewed by Tucker (2000).

Relatively few species of blue-green algae synthesize geosmin or MIB, and flavor problems in cultured animals coincide with the development and disappearance of odor-producing species. When odorous species are present, waterborne geosmin or MIB is absorbed by fish or crustaceans across the gills and deposited in tissues throughout the animal. Geosmin and MIB are extremely fat soluble and strongly concentrated from the water into animal tissues. Further, the human sensory threshold for these compounds occurs at low tissue concentrations. Because geosmin and MIB are intensely odorous and highly bioconcentrated from water, they can cause flavor problems in fish or crustaceans when present in water even in trace amounts.

When odor-producing algae disappear from the phytoplankton community, geosmin and MIB in fish or crustacean tissues are purged or metabolized. Rates of depuration depend primarily on water temperature and tissue fat content (Johnsen *et al.* 1996). Lean fish in warm water may completely depurate intense off-flavors within days whereas fatty fish in cold water may remain off-flavor for months.

Off-flavors caused by geosmin or MIB occur in pond-cultured animals throughout the world. Earthy-musty off-flavors have been reported from such widely different species and geographical regions as pond-raised channel catfish in the southeastern United States, rainbow trout (*Oncorhynchus mykiss*) raised in prairie "pothole" lakes in central Canada, common carp (*Cyprinus carpio*) from intensive fish culture ponds in Israel, Atlantic salmon from net-pens in lakes in Scotland, and penaeid shrimp in low-salinity ponds in Ecuador (summarized in Tucker 2000).

There are strong ecological similarities among waters that routinely produce tainted fish. Nearly all waters commonly producing fish tainted with geosmin or MIB are static, nutrient-enriched freshwaters. In these systems (as exemplified by warmwater aquaculture ponds), blue-green algal blooms are a natural consequence of prevailing environmental conditions and off-flavors are relatively common. Odorous blue-green algae do not thrive in pond water with salinity greater than about 5 ppt, and shrimp or fish cultured in such water seldom, if ever, have taints associated with geosmin or MIB.

The occurrence of odorous blue-green algal blooms is unpredictable and, in pond cultures where off-flavors are common, fish are usually "taste-tested" before harvest. If off-flavors are detected, the crop is not harvested. Development of off-flavors can be a severe economic burden for aquaculturists because it prevents timely crop harvest, increases production costs, disrupts cash flows, and interrupts the orderly flow of fish from farm to processor. Furthermore, if off-flavored fish or crustaceans are inadvertently marketed, the negative reaction of consumers may adversely affect market demand.

Management of off-flavors is difficult and relies upon natural elimination of odorous compounds from flesh to improve flavor quality once fish are no longer in the presence of the organism producing the compound. Fish with off-flavors may be simply held in the pond until odor-producing algae disappear. At times, farmers attempt to expedite the process by treating ponds with algicides to eliminate or reduce the density of odor-producing algae.

10.8.2 Algal toxins

Many species of algae produce toxins that directly or indirectly affect aquaculture. Some toxins reduce the growth of or possibly kill the cultured animal. Other toxin-producing algae indirectly affect aquaculture production when toxins accumulate in the tissues of animals that feed upon the algae and are transmitted through the food chain. Toxins that are passed along the food chain may represent a health threat to humans who consume the product. Toxin-producing algae come from diverse taxonomic groups, including blue-green algae, diatoms, and a variety of flagellated marine and freshwater species of prymnesiophytes, euglenoids, dinoflagellates, and chloromonads. Most toxin-producing algae are marine and are important in shellfish farming or net-pen fish culture in bays and estuaries. Fewer species are important in pond aquaculture (Boyd & Tucker 1998).

Algae-related toxic events in ponds may result when toxin-producing algae are present in the water supply or when toxin-producing populations develop naturally as part of the pond phytoplankton community. Ponds filled with water from estuaries, bays, or other saline water bodies are particularly susceptible to toxic algae events when blooms of toxic dinoflagellates are present in the water supply. The infamous "golden alga" *Prymnesium parvum* occurs worldwide in warm estuaries and fresh waters with a relatively high mineral content. Populations of *P. parvum* may be pumped into ponds from a water source or populations may develop naturally in ponds of the proper salinity. Certain blue-green algae produce hepatotoxins (liver toxins) or neurotoxins that are toxic to a wide range of vertebrates and invertebrates, especially warm-blooded animals. Populations of toxic blue-green algae develop in freshwater ponds as part of the normal succession of phytoplankton communities.

Fish kills caused by *Prymnesium parvum* have been documented throughout the world and considerable research has been conducted on mechanisms of toxicity and control measures. A website maintained by the Texas Parks and Wildlife Department (www.tpwd.state.tx.us/landwater/water/environconcerns/hab) contains abundant information on the ecology, impacts, and management of *Prymnesium parvum* and other toxic algae. Quite unexpectedly, aside from toxic events related to *Prymnesium,* relatively few fish kills have proven to be caused by algal toxins in aquaculture ponds, despite the high potential for such problems to occur. In many instances, fish kills are caused by dissolved oxygen depletions associated with excessive algal abundance but are mistakenly attributed to algal toxins because of the coincident occurrence of suspected toxin-producing algae. The lack of widespread algae-related toxic events in aquaculture is especially puzzling for blue-green algae because toxin-producing blue-green algal strains are very common in warm, freshwater aquaculture ponds. Further, neurotoxins and hepatotoxins from blue-green algae are quite toxic to fish when injected directly. There is no good explanation for this paradox although several factors may contribute to the apparent tolerance of freshwater fish to blooms of potentially toxic blue-green algae (Boyd & Tucker 1998).

Managing toxic algal blooms is difficult in ponds. When the water supply is the source of toxic algae, careful monitoring of source water may allow the aquaculturist to avoid pumping from the water supply until the toxic bloom disappears. Selecting sites and water sources that are not historically prone to toxic algae blooms may also reduce risk. Once blooms are established in ponds, attempting to eliminate toxin-producing algae by killing them with algicides may make conditions worse because large amounts of toxin will be released into the water when algae die and cells lyse. The dilemma of managing toxic blooms is especially problematic for blue-green algae because toxin-producing strains are common in aquaculture ponds across the world, yet there is little direct evidence of harmful consequences related to toxin production. Furthermore, managing the taxonomic composition of phytoplankton communities in ponds is effectively impossible. The ecology of toxin-producing algae and their potential impacts and management in aquaculture are complex subjects, and additional information can be found in Boyd and Tucker (1998) or at the Texas Parks and Wildlife Department website previously referenced.

10.9   Practical constraints on pond aquaculture production

This chapter's recurrent theme has been that pond aquaculture production is limited by the capacity of ecosystems to provide essential life-support functions.

Overcoming these limiting factors—which progressively shift from food supply to oxygen production to waste removal—requires energy input from outside the pond. At some point, waste treatment capacity is exceeded and production can be increased only by changing the system so that it is functionally no longer a pond—at least as ponds are defined in this chapter. That is, when production is intensified to the point where the pond's waste-treatment capacity is exceeded, water must either be rapidly exchanged (at which point it becomes a flow-through system) or treated in separate units or filters (at which point it becomes a water-recirculating aquaculture system).

In reality, potential aquaculture production is affected by numerous factors unrelated to ecosystem services. At the most basic level, production is affected by factors that affect fish growth and survival. Some of these factors can be controlled by the culturist, but many are not amenable to management control. Those factors act as constraints and result in a situation where potential production is seldom achieved under commercial conditions. In fact, actual production usually falls far short of potential production. This section explains why fish production in ponds always falls below potential by examining the effects of seedstock availability, cropping system selection, and crop losses as practical, production-limiting constraints.

Most factors affecting production are manageable to some degree, although overcoming these constraints may not be practical or economically justified. For example, water temperature is the most important variable affecting animal metabolism and pond ecology, but, as a practical matter, temperature cannot be controlled by the culturist (other than through site selection). Another common example involves the decision to treat infectious diseases by adding a chemical therapeutant to pond water. Therapeutants are greatly diluted when added to ponds, and large amounts of chemical may be required to achieve an effective concentration. In some instances, treating the disease may not be justified economically for commodity fish production.

10.9.1   Seedstock availability

Aquaculture production and profitability can be optimized only when the culturist has complete control of the production cycle. Production should begin with healthy juveniles of the desired size and the production cycle should be initiated when conditions are best for early growth. Most commercial aquaculture relies on controlled reproduction of captive animals. There are, however, important examples of aquaculture relying on wild-caught juveniles for seedstock (Hair *et al.* 2002), although the practice is increasingly discouraged due to potential adverse effects on natural populations.

Timing of aquatic animal reproduction depends on species-specific sets of environmental cues, such as changes in temperature, photoperiod, tides, or moon stages. For some species, such as channel catfish, captive broodstock are allowed to reproduce under conditions that resemble those encountered in the wild. This practice is simple and inexpensive, but supplies of juveniles are seasonal and reproductive success may vary from year to year depending on changes in weather or other factors. Seasonal availability of seedstock and an uncertain supply can mean that growout strategies (such as initiation of the production cycle) depend more on seedstock availability than on optimizing the temperature-dependent growout process.

Greater control of reproduction can be gained by artificially manipulating the environment to stimulate gametogenesis, often followed by administration of hormones to induce spawning. Complete control over reproduction allows a consistent, year-round supply of seedstock, which is especially important for pond systems with a relatively constant water temperature in which multiple crops can be produced annually or cropping cycles can be staggered. Tilapia are particularly exemplary as the kind of species amenable to this approach. Examples of culture systems that function best when seedstock are available year-round include recirculating systems, flow-through systems with a constant-temperature water supply, and ponds in tropical regions. Year-round spawning is less important and may not be desirable for commercial pond aquaculture in temperate regions. For example, out-of-season spawning is easily induced in brood channel catfish, carps, or other seasonal spawners by manipulating water temperatures in indoor systems. However, for large-scale fingerling production, fish must be moved into ponds for further growth and little advantage is gained by moving fish to ponds when water temperatures are suboptimal.

10.9.2   Cropping system and market demand

For purposes of this discussion, cropping system is defined as the manipulation of stocking density and fish biomass in ponds to maximize the temperature-dependent waste assimilation capacity, often indicated by maximum daily feeding rate. The most common cropping system in ponds is the single batch where one cohort of fish is stocked as fingerlings and grown out to market size. Assuming favorable temperatures for growth at stocking, the production potential of the pond is not achieved at this time because most of the waste assimilation capacity is not used when fish are small. In fact, the carrying capacity of the pond is only achieved for a short period just before harvest. To make better use of production capacity, culturists may manipulate density by splitting the stock periodically during the culture period. The best frequency of stock splitting is a trade-off between gains associated with maximizing carrying capacity and additional labor costs for handling fish. The production-to-capacity (P/C) ratio is the metric that is used to describe use of the pond carrying capacity. It is calculated as production per unit time (usually one year) divided by the theoretical capacity for production. In recirculating aquaculture systems, P/C of 2 to 3 is possible, and indeed are necessary for economic sustainability. In ponds, P/C is always less than one.

In channel catfish farming, the multiple batch production system is used where multiple cohorts of fish are present in the pond at any one time and ponds are not drained for harvest. The system has evolved as a response to the need to meet regular supplier demand for market-size fish and as a hedge against the risk of off-flavor that prevents timely harvest. Harvest timing in commercial aquaculture is usually dictated by market demands rather than being set to optimize production. These market demands might require a constant supply or an elevated supply during certain times of the year or holidays. Market demand for the products of pond aquaculture rarely corresponds to the end of the growing season, when a single-batch pond would normally be harvested, especially in temperate regions. It is difficult to match market demand with fish production cycles.

Channel catfish processors in the southeastern United States require a year-round fish supply that is met by holding some food-sized fish in ponds throughout the year. In effect, pond space that could be used to produce a new fish crop is used indirectly by the processors to hold fully grown fish in inventory. Farmers attempt to address this processor-imposed harvest schedule by staggering production start dates so that multiple populations reach market size at different times throughout the year. Even if production is staggered to meet processor demands, some harvest-sized fish must be held over winter to meet processor demands for fish in winter and spring. Overwintering harvest-sized catfish to meet winter and early-spring processor demand exposes fish to increased risk of crop loss from infectious disease.

Furthermore, as described in a previous section, algae-related off-flavors are common in pond aquaculture and off-flavored fish are usually not accepted for processing. Similarly fish may not be accepted for harvest and processing during periods of temporary oversupply. In either case, fish must be held past optimum harvest time, which increases risk of loss and reduces long-term productivity.

Because specific growth rate and feed conversion decrease as fish get larger, the production cycle should end when fish reach the desired market size. Growing fish past that point extends production cycle duration, worsens feed conversion, and increases the risk of crop loss to disease or predators. Production cycle planning should also minimize intervals between crops when the pond is empty and unproductive.

10.9.3 Crop loss

Crop loss to disease and predation affects production and profitability in all types of agriculture. Losses can be particularly troublesome in pond aquaculture because it is exceedingly difficult to make ponds biosecure. Commercial aquaculture ponds are relatively large outdoor systems that are difficult or practically impossible to disinfect between crops. Ponds are accessible to, or easily colonized by, a variety of disease vectors including mammals, birds, and invertebrates. Pond facilities are often located in rural or isolated areas that support diverse and abundant wildlife and, relative to other aquaculture production systems, ponds are especially vulnerable to wildlife depredations. The isolated nature of pond facilities also makes them an attractive target for poaching.

Morbidity and mortality from disease epizootics and crop loss to predation and poaching are manageable to varying degrees. The degree of management possible depends on culture species, types of diseases encountered, the availability of management tools (such as vaccines or medicated feeds), facility size, hydrology, and other considerations. The degree to which crop loss can (or should) be managed has an obvious economic component. Ornamental fish and other high-value crops may justify extreme measures such exclusion structures to prevent predator access, surveillance and security measures to prevent poaching, disinfection of source water, disinfection of pond muds between crops, and use of expensive therapeutants or drugs. Such measures may not be cost-effective—or even possible—for lower-value commodity fish crops.

10.9.4 An example using channel catfish pond aquaculture

The impact of practical constraints on the productivity of pond aquaculture has been estimated using data from commercial catfish aquaculture in the southeastern United States (Tucker 2005). Assuming that adequate aeration is available to overcome oxygen supply limitations, potential production is defined by the effects of water temperature on catfish metabolism when water temperatures are below about 20°C and by the nitrogen assimilation capacity of the pond when water temperatures are warmer. Based on those two initial constraints, maximum annual net fish production under climatic conditions in the region is approximately 19,000 kg/ha, which corresponds to the range for maximum production in table 10.1.

The simple model used by Tucker (2005) estimated the crop yield reduction due to the combined effects of market constraints (off-flavors and the need for a year-round fish supply), infectious disease losses, and bird predation. The model also estimated the effects of dissolved oxygen limitation on fish growth. Although an initial assumption of the model was that adequate aeration is available to overcome oxygen supply limitations, commercial catfish ponds seldom have adequate aeration to provide non-limiting oxygen supplies. Combined, these factors reduce annual yields by 60% to 7,500 kg/ha. Although the model is a vast oversimplification of actual conditions, the estimated annual yield under commercial conditions is in the range of yields reported by farmers in the southeastern United States.

This exercise has significant implications. Catfish production could be increased 25 to 50% by improving fish health management, reducing predator loss, and managing fish flavor quality so that market-sized fish are promptly removed from ponds. Significant progress toward that goal can be made by properly implementing current technologies. Greater improvement in yields will depend on overcoming limitations of current aeration technology, such as using some variation of the partitioned aquaculture system (chapter 13).

10.10 Comparative economics of culture systems

Comparing economic performance among culture systems is difficult because economic sustainability is more strongly related to operational characteristics that apply universally to culture systems. Although market demand, product price, and feed cost are critical external aspects of economic sustainability, site selection is the most important operational characteristic affecting potential success. Physical site characteristics have a strong deterministic effect on selection of culture system and ultimate system performance. For example, selecting a pond culture system is a rational economic decision where there are large areas of flat, low-cost land with abundant water resources. For ponds, physical characteristics like soil type, topography, hydrology, and climate are basic aspects of site selection. The environmental variables that affect production vary by species but temperature is universally important as the main factor controlling metabolic rate. Control of temperature at levels optimum for aquatic animal growth is seldom practical or cost-effective, which is especially true in ponds. Climate is a key site selection criterion that determines the maximum growth potential of cultured animals and the productive potential of a site.

For any aquaculture operation, choosing the appropriate scale of production is a decision rivaled in importance only by site selection. A production scale that is too small restricts marketing options and a production scale that is too large often means that fixed costs, especially the cost to borrow capital (indicated by debt-to-equity ratio), are too large. Irrespective of culture system type, a small-scale operation can be profitable with a niche or targeted market for high-value products. A large-scale operation can be profitable by achieving economies of scale on inputs associated with large size and marketing products as a relatively low-value commodity. As a commodity, seafood competes with other sources of animal protein, especially poultry, in the market. Worldwide, nearly all commodity fish for domestic consumption are produced in ponds.

All aquaculture operations must have access to basic natural resources (land, water, and energy), capital resources, and human resources (labor). Selection of appropriate culture systems also depends on the availability of physical infrastructure and access to science, technology, and information infrastructure. Economic performance will depend on how resource inputs and infrastructure are combined and applied in production, the cost of the inputs, and how products are marketed (Shang 1990). Net revenue is generally most sensitive to market price of product and unit feed costs. In general, pond aquaculture is favored where land-use intensity and cost is low and water resources are abundant. Increasing land value and water scarcity favors intensification of pond culture.

With pond aquaculture, unlike other production systems, a wide range of culture system intensity is possible. For example, flow-through and recirculating systems are entirely dependent on external food supplies but ponds are dependent on external food supplies to a variable degree, largely as a function of culture system intensity. As culture system intensity increases, the dependence on externally provided inputs, especially manufactured feed, increases. The proportion of variable costs represented by feed increases with culture intensity. However, variable costs per unit production in intensive systems are potentially lower because of increased control over the factors that affect growth and survival, especially feeding.

Across all culture systems, managing production capacity of the culture unit is critical to good performance. Examples of cropping systems that can increase the ratio of production to carrying capacity include manipulation of stocking density as a function of size, stocking multiple cohorts, and partially harvesting the population followed by another stocking. Growth and survival can be increased by fertilization, feeding, water quality management, and disease and predator control.

A common management goal among production systems is to increase efficiency and reduce waste. Efficiency indicators can be used to evaluate system performance as a function of major resource inputs, such as energy, land, water, and feed. However, comparing efficiency indicators across culture systems can result in misleading conclusions. For example, production per unit land area in pond aquaculture is much less than inherently intensive systems like cages, raceways, or recirculating systems. A better comparison would consider performance in relation to the factors of production that are most limiting, and therefore most valuable, in a particular location. More broadly, considering the additional area required for the production of feed ingredients and waste treatment tends to overwhelm a land-use efficiency indicator based solely on the land area occupied by a particular facility.

Other than comparing specific efficiency measures, broad qualitative statements can be made about the comparative economic performance of different culture systems. In this context it is useful to consider the economic dimensions of culture systems with respect to food supply and life support.

Food supply is the essential service with the greatest economic implications, one that transcends the diversity of culture systems as a concern. Given that feed represents from 30 to 60% of variable costs in semi-intensive or intensive pond aquaculture, economic performance is strongly dependent on improving feed conversion or reducing feed cost per unit production. In some cases the cost-benefit ratio favors pond production systems where cultured animals are fed; in other situations, more traditional, low-input or fertilized systems are favored.

Providing oxygen to cultured animals is an essential life-support service. In ponds, oxygen is mainly provided by algal photosynthesis. Supplementation of the oxygen supply by aeration, circulation, or water exchange has a cost. In the context of the collective magnitude of the three main cost items (feed, seed, and labor), such a cost increase is often justified by the potential benefit. Obviously the availability of reliable power supplies is a prerequisite for the supplementation of oxygen supplies by aeration.

Fundamentally waste from a culture system can be treated within the farm or facility or outside a farm or facility, or sometimes both. Where waste is treated outside the farm or facility, those costs are borne by society and are considered an externality of production. Where waste is treated inside the farm or facility, those costs are internalized and borne by the farm owner. Given that the most important aspects of economic sustainability are independent of the culture system, ponds have two potential economic advantages relative to other culture systems. To a degree dependent on the food habits of cultured species, production in ponds based on fertilization to produce natural food items may have a high benefit to cost ratio. Internalizing waste treatment represents an additional operating cost compared to culture systems where waste treatment takes place external to the farm or facility. However, if ponds are operated within the waste assimilation capacity, the operating cost of waste treatment is essentially free. Furthermore, treating as much waste as possible internal to the farm or facility is responsible aquaculture. The often-lengthy process of permitting aquaculture facilities that release waste to public waters can increase production costs due to attention to regulatory matters or litigation.

10.11   Sustainability issues

Although there is no consensus on a functional definition of sustainability, the concept is understood to encompass the intersection of three overlapping domains: economic, environmental, and social. Sustainability implies continuity of production, efficiency of resource use, and responsibility to the welfare of the supporting environment, cultured animals, farm workers, fish consumers, and the broader society. Many of the important environmental and social impacts in aquaculture are independent of culture system. The previous section discussed elements of economic sustainability; here we describe the interrelated environmental and social aspects of sustainability issues that have particular relevance to pond aquaculture, especially those relating to site selection, land use, and property rights.

An underlying theme of this chapter has been that the size of the ecological footprint (an index of sustainability) in pond aquaculture ranges widely. Pond aquaculture extends from nearly self-sufficient production systems with few external inputs to relatively intensive systems relying on large external subsidies such as manufactured feeds, mechanical aeration, and water exchange for waste management. Water-use efficiency in pond aquaculture also varies greatly, with some types having the highest consumptive water use per unit production of all other forms of aquaculture. On the other hand, most ponds are relatively closed systems with an indirect or intermittent hydrological connection to outside water bodies. When pond aquaculture facilities are properly sited, issues such as waste loading, pathogen transmission, and escape of animals from culture units are usually less important for ponds than for culture systems more open to the environment, such as net-pens and flow-through raceways. Of course, the relative risk of those impacts increases as hydraulic retention time of ponds decreases (that is, as ponds become more open to the environment with respect to hydrological connection).

Overall, ponds have a greater range of potential environmental impacts than other aquaculture systems. The relative importance of specific impacts of pond aquaculture depends upon, among others:

- culture intensity
- production scale
- trophic level of species cultured

- hydrological pond type
- local hydrology and climate
- type and scale of local human activities
- extent or concentration of local aquaculture development
- functional integrity and resilience of downstream aquatic ecosystems.

The extraordinary diversity of potential environmental effects makes it impossible to assess the overall environmental performance of pond aquaculture and compile a simple list of impacts. Boyd *et al.* (2007) propose several quantitative indicators of resource use and waste production for aquaculture and provide examples of relative environmental performance of different aquaculture systems. However, it is not possible to develop an overall index of resource use efficiency and waste production by species, production method, or facility. Further details of environmental impacts of aquaculture and their management are discussed in Tucker and Hargreaves (2008). Many environmental and social impacts of aquaculture can be avoided or minimized by implementing better management practices at the farm level, especially those related to site selection.

Different pond production systems may perform well in one area of potential impact but less well in others. For example, channel catfish ponds in the southeastern United States can be operated for many years with very little discharge, so impacts related to waste loading and pollution are low. However, catfish ponds using groundwater supplies may have significant negative impacts related to aquifer overdraft. Which issue is more important? Do the benefits of one cancel the liabilities of the other? Answers to these two simple questions depend on assessing impacts in a more general context than simply quantifying resource use or waste loading. Answers also depend on values, perceptions, and, to an increasing extent, the outcome of public participation in stakeholder consultations. Further confounding these simple questions, there is no consensus on the most serious impacts of aquaculture. Issues such as water pollution, animal escape, pathogen transmission, excessive use of fish meal in manufactured feeds, and land use have all been proposed as serious concerns for one sector of aquaculture or another.

Pond facilities use large land areas compared with other productions systems, so issues such as habitat conversion and land-use conflicts are more frequently encountered in pond aquaculture than in other forms of aquaculture, especially in the coastal zone. At the core of land-use conflicts are disputes between land owners and others over equitable distribution of the costs and benefits of private aquaculture and the fairness of using public resources for private gain. The benefits of private aquaculture often accrue solely to the entrepreneur but in some situations some of the costs of production (such as waste treatment in ponds operated with water exchange) are borne by society, with the potential to reduce public welfare through ecosystem damage and pollution of water resources. Responsible aquaculture means internalizing such costs to the greatest extent.

Perhaps the most prominent example of land use conflicts in aquaculture is the conversion of mangrove forests to large tracts of coastal aquaculture ponds, primarily for the culture of penaeid shrimp. It is one of the most contentious environmental and social issues in aquaculture (Boyd 2002). The location of aquaculture ponds in the coastal zone can restrict access to the shoreline and to resources that are traditionally considered to be open access or common property. Numerous instances of conflict, rising to the point of violence, between coastal shrimp farmers and small-scale fishers have occurred. Land- and water-use conflicts also occur between shrimp farmers and crop farmers. The discharge of saline water from inland shrimp farming into irrigation canals can cause local soil salination and reduce the capacity of adjacent lands to grow crops. In some areas, pond aquaculture competes for limited supplies of freshwater with other uses, including for irrigated agriculture, industry, and household use. At the most basic level, land-use conflicts are more common for pond aquaculture simply because they use more land. Ponds are the most commonly used aquaculture system and individual facilities occupy a greater land area than more intensive systems. Proper siting, using selection criteria that transcend basic physical attributes, can often prevent most of the problems that lead to land-use conflicts.

Decisions on land use do not necessarily have to favor one use over another, but pond aquaculture may be relatively advantageous in certain sites. For example, large expanses of land are used for catfish aquaculture in the southeastern United States but those soils are poorly suited for most row crops and their use in aquaculture does not therefore compete directly with other agricultural uses. Pond aquaculture is often possible on marginal or degraded land, such as areas with salinized soils that are not suitable for terrestrial agriculture.

Using fish meal derived from pelagic marine fish in feeds for other fish is another contentious issue in aquaculture. Some believe that culture of certain fish or crustaceans using fish meal in feeds does not contribute to a net increase in global fish production (Naylor & Burke 2005). That is, each kilogram of fish produced in aquaculture consumes more than one kilogram of small pelagic marine fish in the form of fish meal. Others argue that this analysis is flawed and overly simplistic (Tidwell & Allan 2002). Regardless, the point here is that feeds used to grow certain species of animals in ponds have higher levels of fish meal than others, and, correspondingly, the impacts (real or perceived) vary depending on the trophic level of the species in question rather than on which culture system is used to raise the animal. The fish feed equivalence (FFE) can be used to determine the net impact on global fish supplies (Boyd *et al.* 2007). The FFE is a unitless ratio of the weight of pelagic marine "feed fish" needed to produce 1 kg of cultured animal. The FFE for pond-raised channel catfish is 0.2, indicating that the weight of fish used in the feed is much less than the weight of fish produced in culture. The FFE for pond-raised black tiger prawns (*Peneaus monodon*) is 1.7, meaning that the weight of feed fish used in shrimp diets is greater than the weight of shrimp produced. Based on these two examples, some argue that culture of tiger prawns has a greater negative impact on world seafood supplies than catfish farming, even though both are grown in ponds. Most fish grown in pond culture worldwide feed at low trophic levels and therefore contribute to a net increase in global fish supplies. However, increasingly more fish meal is being used in diets for omnivorous carps, especially in China, as culturists there intensify production.

Most small-scale aquaculturists in the world produce fish in ponds. Small-scale pond aquaculture has many positive social benefits, including increased employment, protein supply, food security, and economic diversification. Pond aquaculture can function as the nexus for economic development and poverty alleviation in rural areas (Edwards *et al.* 2002). Producing fish in pond aquaculture is more labor intensive than in other production systems. Pond aquaculture, as practiced in most of the world, is technologically basic and simple, thereby providing employment opportunities for participation of poor unskilled laborers. However, this attribute highlights the need to protect workers with labor and safety standards.

There are at least three ways to address the environmental and social impacts of pond aquaculture (FAO 2006a). First, the costs of maintaining environmental quality should be internalized to the production system. In pond aquaculture, these costs are internalized to a variable degree but most ponds are operated as static-water systems, which function to internalize essentially all waste treatment costs. Second, impacts can be addressed by adopting better management practices. Although there is some overlap, there are better management practices for each culture system (Tucker & Hargreaves 2008). In pond aquaculture, practices related to site selection and pond construction are particularly important. Finally, impacts can be addressed by integrating aquaculture into development and land management programs, including coastal zone management. Nearly all rural aquaculture is conducted in ponds, and incorporating aquaculture into planning and implementation of development programs can help reduce rural poverty.

10.12   Trends and research needs

Predicting the future of pond aquaculture is difficult because of rapidly changing global economics that will influence the profitability of aquaculture generally. Also, pond aquaculture, which can be an elegant way to grow food with respect to the ecological footprint for food production and waste treatment, uses large tracts of land for the facility and is a large consumptive user of water, relative to other aquaculture systems. As land and water resources become limiting or find more favored alternative uses, pond aquaculture may become a less desirable way of growing aquatic animals. Pond aquaculture will undoubtedly continue to expand as the most economical method of producing certain species, but it will probably contribute a smaller proportion of overall aquaculture production than it did in the past.

Increasing the fisheries supply from pond aquaculture can occur by increasing pond area or increasing the productivity of existing ponds. Expansion of the area of pond aquaculture will continue to be constrained by the availability of suitable sites and competition for arable land with other uses. Construction of additional ponds is likely to occur on more marginal land and other sites that are not ideally suited for pond aquaculture. Increasing productivity of existing ponds (intensification) is driven by the high levels of human appropriation of existing freshwater resources and chronic water shortages in some aquaculture producing areas. Intensification of pond aquaculture will require additional subsidies of industrial energy to support higher levels of dependence on externally supplied food and enhancement of internal waste treatment processes.

Nutrient budgets for aquaculture systems where animals are fed indicate that 25 to 30% of the nutrients and organic matter in the form of feed is removed at harvest. Given the importance of manufactured feed as the principal cost item, the 70 to 75% of nutrients and organic matter released to the pond environment represents a potential source of revenue if they can be recovered in cultured animals or other economically valuable products. Unfortunately, ponds function to "treat" wastes and present few opportunities for waste recovery and recycling. New pond designs, such as the partitioned aquaculture system or other compartmentalized pond systems, offer opportunities to recover waste nutrients that conventional ponds do not have. Another approach is to recover waste nutrients in microbial biomass or detrital aggregates by adding a source of organic carbon to pond water (Hargreaves 2006; Crab *et al.* 2007). One drawback of this approach is that effectiveness depends on the inputs of industrial energy in the form of aeration to maintain well-mixed conditions.

Small-scale pond aquaculture for the production of fish at low trophic levels for direct consumption or local marketing is widespread in developing countries, especially in Asia. Many are operated as components of traditional farming systems where wastes from one subsystem are a resource for another subsystem. These integrated production systems are well adapted to local conditions and define a benchmark for sustainable aquaculture. Increasing pressures on land and water resources to be more productive suggests that some intensification of traditional systems in lesser-developed countries is also necessary. The goal will be to combine traditional and scientific knowledge, and the judicious use of the most limiting biophysical resources, to increase the productivity and production efficiency of traditional pond systems.

Finally, one recent trend worth noting is the proliferation of product certification efforts. In the context of pond aquaculture, there is great interest in developing ecolabeling programs for sustainable or responsible production of penaeid shrimp and certain fish species in ponds. Many in the environmental community view ecolabeling as an approach to achieve environmental protection and conservation goals, particularly with respect to coastal wetlands. Producers view the label as an opportunity to obtain a price premium for certified shrimp and fish and reduce the environmental and social costs associated with production. Ecolabels have been proliferating but it remains to be seen if consumer demand for these products will be sufficient to change producer behavior and warrant continuing certification efforts. It is also unclear if small-scale producers in developing countries will be able to access the potential benefits associated with ecolabeling.

10.13　References

Anderson, R.O. & Tave, D. (1993) *Strategies and Tactics for Management of Fertilized Hatchery Ponds*. Food Products Press, New York.

Azim, M.E., Verdegem, M.C.J., van Dam, A.A., & Beveridge, M.C.M. (2005) *Periphyton: Ecology, Exploitation, and Management*. CABI Publishing, Massachusetts.

Berg, H., Michelson, P., Folke, C., Kautsky, N. & Troell, M. (1996) Managing aquaculture for sustainability in tropical Lake Kariba, Zimbabwe. *Ecological Economics* 18:141–59.

Boyd, C.E. (1985) Chemical budgets for channel catfish ponds. *Transactions of the American Fisheries Society* 114:291–8.

Boyd, C.E. (1990) *Water Quality in Ponds for Aquaculture*. Alabama Agricultural Experiment Station, Auburn.

Boyd, C.E. (1995) *Bottom Soils, Sediment, and Pond Aquaculture*. Chapman and Hall, New York.

Boyd, C.E. (1998) Pond water aeration systems. *Aquacultural Engineering* 18:9–40.

Boyd, C.E. (2005) Water use in aquaculture. *World Aquaculture* 36(3):12–3, 70.

Boyd, C.E. (2002) Mangroves and coastal aquaculture. In *Responsible Marine Aquaculture* (Ed. by R.R. Stickney and J.P. McVey), pp. 145–58. CABI Publishing, New York.

Boyd, C.E. & Tucker, C.S. (1998) *Pond Aquaculture Water Quality Management*. Kluwer Academic Publishers Boston.

Boyd, C.E., Tucker, C.S., McNevin, A., Bostick, K. & Clay, J. (2007) Indicators of resource use efficiency and environmental performance in fish and crustacean aquaculture. *Reviews in Fisheries Science* 15:327–60.

Brummett, R.E. (1999) Integrated aquaculture in Sub-Saharan Africa. *Environment, Development, and Sustainability* 1:315–21.

Crab, R., Avnimelech, Y., Deifordt, T., Bossier, P. & Verstraete, W. (2007) Nitrogen removal techniques in aquaculture for sustainable production. *Aquaculture* 270: 1–14.

Delincé, G. (1992) *The Ecology of the Fish Pond Ecosystem: With Special Reference to Africa*. Kluwer Academic Publishers, Boston.

De Silva, S.S. & Anderson, T.A. (1995) *Fish Nutrition in Aquaculture*. Chapman and Hall, New York.

Edwards, P., Little, D.C. & Demaine, H. (2002) *Rural Aquaculture*. CABI Publishing, New York.

Engle, C.R. & Valderrama, D. (2002) The economics of environmental impacts in the United States. In *Aquaculture and the Environment in the United States* (Ed. by J.R. Tomasso), pp. 240–70. United States Aquaculture Society, Baton Rouge.

FAO (Food and Agriculture Organization of the United Nations) (2006a) *State of World Aquaculture: 2006*. FAO Fisheries Technical Paper 500. FAO, Rome.

FAO (Food and Agriculture Organization of the United Nations) (2006b) *Yearbook of Fishery Statistics, Commodities 2004*. FAO, Rome.

Folke, C., Kautsky, N., Berg, H., Jansson, A. & Troell, M. (1998) The ecological footprint concept for sustainable seafood production: A review. *Ecological Applications* 8(1):63–71.

Hair, C., Bell, J. & Doherty, P. (2002) The use of wild-caught juveniles in coastal aquaculture and its application to reef fishes. In *Responsible Marine Aquaculture* (Ed. by R.R. Stickney & J.P. McVey), pp. 327–54. CABI Publishing, New York.

Hargreaves, J.A. (1997) A simulation model of ammonia dynamics in commercial catfish ponds in the southeastern United States. *Aquacultural Engineering* **16**:27–43.

Hargreaves, J.A. (1998) Nitrogen biogeochemistry of aquaculture ponds. *Aquaculture* **166**:181–212.

Hargreaves, J.A. (2006) Photosynthetic suspended-growth systems in aquaculture. *Aquacultural Engineering* **34**:344–63.

Hargreaves, J.A. & Tucker, C.S. (1996) Evidence of control of water quality in channel catfish (*Ictalurus punctatus*) ponds by phytoplankton biomass and sediment oxidation. *Journal of the World Aquaculture Society* **27**:21–9.

Hargreaves, J.A. & Tucker, C.S. (2003) Defining loading limits of static ponds for catfish aquaculture. *Aquacultural Engineering* **28**:47–63.

Hargreaves, J.A. & Tucker, C.S. (2004) *Management of Ammonia in Fish Ponds*. Southern Regional Aquaculture Center Publication 4603, Southern Regional Aquaculture Center, Stoneville.

Hargreaves, J.A. & Tucker, C.S. (2005) Conditions associated with sub-lethal ammonia toxicity in warmwater aquaculture ponds. *World Aquaculture* **36**(3):20–4.

Hargreaves, J.A., Boyd, C.E. & Tucker, C.S. (2002) Water budgets for aquaculture production. In *Aquaculture and the Environment in the United States* (Ed. by J.R. Tomasso), pp. 9–33. United States Aquaculture Society, Baton Rouge.

Johnsen, P. B., Lloyd, S.W., Vinyard, B.T. & Dionigi, C.P. (1996) Effects of temperature on the uptake and depuration of 2-methylisoborneol (MIB) in channel catfish *Ictalurus punctatus*. *Journal of the World Aquaculture Society* **27**:15–20.

Kautsky, N., Berg, H., Folke, C., Larsson, J. & Troell, M. (1997) Ecological footprint for assessment of resource use and development limitations in shrimp and tilapia aquaculture. *Aquaculture Research* **28**:753–66.

Knud-Hansen, C.F. (1998) *Pond Fertilization: Ecological Approach and Practical Applications*. Oregon State University Pond Dynamics/Aquaculture Collaborative Research Support Program, Corvallis.

Landau, M. (1992) *Introduction to Aquaculture*. John Wiley and Sons, Inc., New York.

LeFrancois, N.R., Jobling, M., Carter, C. & Blier, P.U. (2009) *Finfish Aquaculture: Species Selection for Diversification*. CABI Publishing, New York.

Lin, C.K., Teichert-Coddington, D.R., Green, B.W. & Veverica, K.L. (1997) Fertilization regimes. In *Dynamics of Pond Aquaculture* (Ed. by C.E. Boyd & H.S. Egna), pp. 73–107. CRC Press, New York.

McClain, W.R., Romaire, R.P., Lutz, C.G. & Shirley, M.G. (2007) *Louisiana Crawfish Production Manual*. Louisiana State University Agricultural Center, Baton Rouge.

McLarney, W. (1984) *The Freshwater Aquaculture Book*. Hartley & Marks, Point Roberts, Washington.

Naylor, R. & Burke, M. (2005) Aquaculture and ocean resources: Raising tigers of the sea. *Annual Reviews of Environment and Resources* **30**:185–218.

Peterson, E.L., Wadhwa, L.C. & Harris, J.A. (2001) Arrangement of aerators in an intensive shrimp growout pond having a rectangular shape. *Aquacultural Engineering* **25**:51–65.

Phillips, M.J., Beveridge, M.C.M. & Clarke, R.M. (1991) Impact of aquaculture on water resources. In *Aquaculture and Water Quality* (Ed. by D.E. Brune & J.R. Tomasso), pp. 506–33. United States Aquaculture Society, Baton Rouge.

Reynolds, C.S. (1984) *The Ecology of Freshwater Phytoplankton*. Cambridge University Press, Cambridge.

Scheffer, M. (1998) *Ecology of Shallow Lakes*. Kluwer Academic Publishers, Boston.

Shang, Y.C. (1990) *Aquaculture Economic Analysis: An Introduction. Advances in World Aquaculture*, Volume 2. World Aquaculture Society, Baton Rouge.

Smith, D.W. & Piedrahita, R. (1988) The relation between phytoplankton and dissolved oxygen in fish ponds. *Aquaculture* 68:249–65.

Steeby, J.A., Hargreaves, J.A., Tucker, C.S. & Cathcart, T.P. (2004) Modeling industry-wide sediment oxygen demand and estimation of the contribution of sediment to total respiration in commercial channel catfish ponds. *Aquacultural Engineering* 31:247–62.

Teichert-Coddington, D.R., Popma, T.J. & Lovshin, L.L. (1997) Attributes of tropical pond-cultured fish. In *Dynamics of Pond Aquaculture* (Ed. by C.E. Boyd & H.S. Egna), pp. 183–98. CRC Press, New York.

Tidwell, J.H. & Allan, G.L. (2002) Fish as food: aquaculture's contribution. *EMBO (European Molecular Biology Organization) Reports* 21:958–63.

Tomasso, J.R. (1994) The toxicity of nitrogenous wastes to aquaculture animals. *Reviews in Fisheries Science* 2:291–314.

Troell, M., Tyedmers, P., Kautsky, N. & Rönnbäck, P. (2004) Aquaculture and energy use. In *Encyclopedia of Energy* (Ed. by C. Cleveland), pp. 97–108. Elsevier, Amsterdam.

Tucker, C.S. (1996) The ecology of channel catfish culture ponds in northwest Mississippi. *Reviews in Fisheries Science* 4:1–55.

Tucker, C.S. (2000) Off-flavor problems in aquaculture. *Reviews in Fisheries Science* 8:1–44.

Tucker, C.S. (2005). Limits of catfish production in ponds. *Global Aquaculture Advocate* 8(6):59–60.

Tucker, C.S. & Hargreaves, J.A. (2008) *Environmental Best Management Practices for Aquaculture*. Wiley-Blackwell, Ames.

Tucker, C.S., Hargreaves, J.A. & Boyd, C.E. (2008) Aquaculture and the environment in the United States. In *Environmental Best Management Practices for Aquaculture* (Ed. by C.S. Tucker and J.A. Hargreaves), pp. 3–54. Wiley-Blackwell, Ames.

Tucker, C.S., Kingsbury, S.K., Pote, J.W. & Wax, C.W. (1996) Effects of water management practices on discharge of nutrients and organic matter from channel catfish ponds. *Aquaculture* 147:57–69.

USDA-NASS (United States Department of Agriculture, National Agricultural Statistics Service) (2006) *Census of Aquaculture (2005), Volume 3, Special Studies Part 2*. Publication AC-02-SP-2. United States Department of Agriculture, Washington, DC.

van der Zijpp, A.J., Verreth, J.A.J., Tri, L.Q., van Mensvoort, M.E.F., Bosma, R.H. & Beveridge, M.C.M. (2007). *Fishponds in Farming Systems*. Wageningen Academic Publishers, Wageningen.

Verdegem, M.C.J., Bosma, R.H. & Verreth, J.A.J. (2006) Reducing water use for animal production through aquaculture. *Water Resources Development* 22:101–13.

Yoo, K.H. & Boyd, C.E. (1994). *Hydrology and Water Supply for Pond Aquaculture*. Chapman and Hall, New York.

Chapter 11

Recirculating Aquaculture Systems

James M. Ebeling and Michael B. Timmons

The oceans of the world were long considered to be an unlimited source of fishery products. Current estimates are that the maximum sustainable yield of many species through harvest of wild stock has been or will soon be reached, and many species are overfished. Yet due to the rapid increase of commercial scale aquaculture, production and per capita consumption of fishery products has been continuously increasing. Two examples of extremely successful aquaculture products that employ intensive recirculating systems at some stage of their growout are the salmon industry and tilapia. The Chilean salmon industry has grown from US$159 million industry in 1991 to exporting over US$2.3 billion in 2007 (498,000 tonnes). The production of tilapia has been exponential in the last several years to the point that the US market demand for tilapia has gone from essentially nothing to importing the equivalent whole fish of 554,000 tonnes in 2007.

Aquaculture is the most probable and feasible solution to providing the aquatic products for an ever-increasing market demand. It provides a consistent and reliable source of high-quality, fresh seafood that is nutritious, safe to eat, and reasonably priced. The basic thesis of this chapter is that Recirculating Aquaculture Systems (RAS) are the key technology that will allow the world aquaculture community to supply the world per capita needs for aquatic species over the coming decades and will do so in an environmentally friendly manner.

Aquaculture Production Systems, First Edition. Edited by James Tidwell.
© 2012 John Wiley & Sons, Inc. Published 2012 by John Wiley & Sons, Inc.

In this chapter, we will review the basic unit operations that make up a RAS (e.g., oxygen supply, carbon dioxide removal, nitrogenous waste management including biological filters for ammonia removal, and solid waste removal). The main advantage of a RAS is that water quality can be managed to create the desired target environments for the fish being cultured, instead of the environment defining what fish can be grown. There are numerous ways to design a RAS, and most people who design such systems think that their design is the best. We do not attempt to define "best," but after reviewing this chapter, the reader should have a much better idea of how to go about designing such a system. There is no single system that will address all needs. Designs are often defined by economic constraints and availability of resources to farm owners. We have written a 948-page book on the design and management of RAS, and the reader is encouraged to read this book for further details on the various unit processes and management of such systems (see www.c-a-v.net).

11.1   Positive attributes

As a comparison, conventional aquaculture methods, such as outdoor pond systems and net pen systems, are *not* likely to be sustainable in the long term, due to significant environmental issues and their inability to guarantee the safety of their products to the consumer. Conversely, indoor fish production employing RAS is sustainable, infinitely expandable, environmentally compatible, and has the ability to guarantee both the safety and the quality of the fish produced throughout the year. Indoor RAS offer the advantage of raising fish in a controlled environment, permitting controlled product growth rates and predictable harvesting schedules. RAS conserve heat and water through water reuse after reconditioning by biological filtration using biofilters.

RAS allow effective economies of scale, which results in the highest production per unit area and per unit worker of any aquaculture system. RAS are environmentally sustainable: They use 90 to 99% less water than conventional aquaculture systems, less than 1% of the land area, and they provide for environmentally safe waste management treatment. RAS allow year-round production of consistent volumes of product, and complete climate control of the rearing environment. Because RAS can be set up to produce the same volume of fish every week, week in and week out, they have a competitive advantage over outdoor tanks, pond systems, or wild catch, which are seasonal and sporadic in harvest. Widespread usage of RAS has not yet been adapted for food fish production, mostly due to the high capital costs associated with most designs and the generally higher production costs for these systems, as water must be pumped. However, RAS are almost universally employed to produce salmon smolts, and designs continue to improve that are more cost competitive with conventional aquaculture systems. Recent economic studies on land-based salmon systems show that salmon can actually be produced more cheaply in RAS, but the capital costs are still too high to justify major changes in the way salmon are currently grown in net-pen systems.

RAS-designed aquaculture systems are infinitely scalable. There are no environmental limitations to the size of the intended fish farm to be built because waste streams are controllable in environmentally sustainable ways. Indoor aquaculture is probably the only potential method that could be used to ensure a 100% safe source of seafood, free from all chemicals and heavy metals. With increasing consumer concerns about food safety, aquaculture producers using RAS have an unprecedented opportunity to meet the demands for safe seafood. Attributes of fresher, safer, and locally raised products are clear advantages for RAS-produced seafood.

Over the past several decades, numerous recirculating system designs have been proposed and researched. Numerous commercial systems have opened with fanfare and have quietly gone out of business. Many of these early systems were designed using traditional wastewater treatment concepts and engineering, or by a simple trial and error approach. More recently, aquaculture engineering has come of age, and systems have been engineered specifically for aquaculture and the unique needs of an aquatic/biological system (Timmons & Ebeling 2010). In addition, numerous commercial sources of equipment and supplies are now available and are specifically designed and marketed for aquaculture. Species successfully being cultivated in intensive recirculating systems include tilapia, striped bass, cobia, pompano, barramundi, and marine shrimp.

What follows is an overview of the engineering aspects of intensive recirculating systems and system components as they have emerged over the past three decades. First, the concept of a "unit process" is introduced, where a specific treatment process is used to treat the water as it is recirculated through the system (i.e., solids removed, ammonia converted into nitrate, oxygen added, carbon dioxide removed, and, in some cases if necessary, disinfected). Then, examples of treatment technologies currently used are described and linked together in a complete recirculating system design. Finally, a design example uses the engineering concept of mass balance to show how to estimate water flow requirements for oxygen, ammonia removal, waste solid removal, and carbon dioxide removal. The majority of this chapter has been taken from Timmons and Ebeling (2010), and we recommend that the reader refer to this work for further explanation of the various concepts presented in the following sections.

11.2   Overview of system engineering

Engineers like to divide complicated systems into small parts, called unit processes, which correspond to a specific treatment process. Figure 11.1 shows how a recirculation system can be subdivided into several individual unit processes that may correspond to separate systems or be linked together in a process stream. There are numerous solutions to each "unit operation" and although some are more efficient or more cost effective than others, there is no right or wrong technology. Some work better in large-scale applications, some in small scale, but it usually is the case that all work to some extent. What final choice

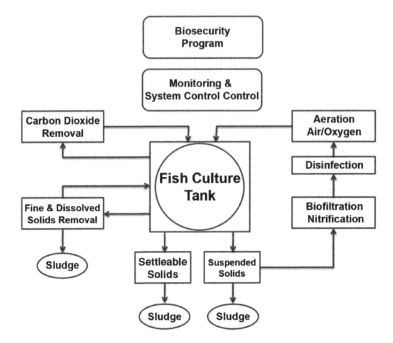

Figure 11.1 Unit processes used on a recirculating system.

is made should be based on sound engineering analysis of the options available for the species to be grown and what is appropriate technology or management for the region of the world in which the application will be implemented. And of course, as any good engineer knows, these choices must be made in the context of economic viability. A fish farmer must be profitable to stay in business, and an elegant design that produces beautiful fish but loses money is of no practical value to a fish farmer.

Returning to the process flow diagram in figure 11.1, the water moves from the central fish culture tank and flows through systems that remove the settleable and suspended waste solids, remove the fine and dissolved solids, convert the ammonia to nitrate, remove carbon dioxide and add oxygen, and finally, when required, disinfect the water before returning it to the culture tank. The monitoring and control system oversees all of these processes and controls the set points for water quality and sounds an alarm if they move outside of acceptable ranges. A biosecurity program and process must be absolutely superimposed on the whole process and farm to prevent losses due to disease introduction from the outside.

In a real world system, as many as possible of the individual unit processes are usually linked together as the water flows through each process (circulation). Usually 10 to 20% of the discharge flow from the culture tank is removed from the center drain to take advantage of the "tea cup" effect, which concentrates the solids in the central drain flow. Some form of settleable solids removal device (swirl separator, radial flow clarifier, settling basin) pretreats this flow stream, which is then combined with the remaining 80 to 90% of the discharge from a side outlet. This combination of center drain and sidewall drain is usually referred to as a Cornell dual-drain system. The remaining suspended solids are then removed usually by either a rotating microscreen filter or floating bead filter.

The water then flows to some form of biofiltration, such as a trickling tower, bead filter, fluidized sand filter, or moving bed bioreactor, where the ammonia is converted to nitrate by autotrophic bacteria. At high loading densities, a carbon dioxide stripping column is then required to remove excess CO_2 and aerate the water to saturation. Finally, where high levels of stocking density are used (>60 kg/m^3) in commercial systems, an oxygenation device is employed to supersaturate the flow to provide sufficient oxygen. In some cases, a UV or ozone system is added to disinfect the returning water stream as part of a biosecurity program or where extremely high-quality water is required.

11.3 Culture tanks

Over the years, people have used a wide variety of materials to construct tanks, including just about anything that holds water. Simple molded polyethylene tanks are often used due to their low cost, smooth surface that makes for easy cleaning, and their low weight, which allows for quick set-up and relocation. They work well for the most part, but because they are very soft and malleable, they need to be well supported on the bottom.

Fiberglass is a popular choice for tanks because of its flexibility in both size and shape. Fiberglass tank panels are easy to cut, drill, and modify, and repairs are relatively easy. Large fiberglass tanks come in easily transported components that are field assembled to almost any diameter.

For maximum strength and resistance to abuse, galvanized or epoxy-coated steel modules can be bolted together to form tanks that are extremely large (32 m) and deep (4 to 5 m). The bottoms are usually poured concrete where drains and heating coils are embedded, along with the tank sidewalls. Tanks can be partially buried to conserve heat and make fish observations easier.

Concrete is always an option in making tanks. It is extremely strong, but very permanent. Mixed-cell raceway designs are another option for tank design, combining the advantages of a raceway for grading and sorting and a round tank for solids removal.

To promote mixing and solids removal, it is important that a tank has good hydraulic characteristics. One of the most important rules for RAS is that solids need to be removed from the tank as quickly and efficiently as possible to reduce the solids' impact on water quality. Tanks can be characterized by diameter (D) to depth (d) ratio (D/d). This ratio should be maintained greater than 3, less than 10, and preferably less than 5 (as in a deeper tank given the same diameter). Using this range of D/d will minimize the problems of solids settling before reaching a center drain (indicating that a tank is too shallow; that is, an excessive D/d ratio) or not settling fast enough to reach the drain by being continually resuspended (indicating a silo-type tank that is too deep; that is, too small of a D/d ratio).

11.4　Waste solids removal

Waste solids are produced in an aquaculture system as uneaten feed, feed fines, fish fecal matter, algae, and sloughed biofilm cell mass from biological filters. Waste solids influence the efficiency of all other unit processes in the recirculating system. They are a major source of carbonaceous oxygen demand and nutrient input into the water and can directly affect fish health within recirculating systems. Therefore, solids removal is considered one of the most critical processes in aquaculture systems. Optimally, solids need to be removed from the fish culture tank as soon as possible, with minimum turbulence and mechanical shearing.

Solids are generally classified into three categories: settleable, suspended, and fine or dissolved solids. In recirculation systems, the first two are the primary concern, while dissolved organic solids can become a problem in systems with very little water exchange. Waste solids can be removed by either settling within the culture unit or through the use of a solids removal unit following the rearing tank. Several unit process options are currently being used in aquaculture: settling basins, radial flow clarifiers, mechanical filters, granular media filters, and floatation or foam fractionation.

11.5　Cornell dual-drain system

The Cornell dual-drain system uses the culture tank itself as a swirl separator ("tea-cup" effect) and removes most of the settleable solids from a small percentage of discharge from the center drain. This design system is applicable to round tanks and is very effective in removing the majority of large, quickly settleable solids such as uneaten feed and fecal matter using only a small portion of the total water flow (10 to 25%). This waste stream can then be further concentrated using either settling basins or additional swirl separators. Remember, the objective is to create a small, concentrated flow that is easily treated, rather than a large, diluted flow of waste.

Solids that concentrate at the bottom center can be removed in a small flow stream by using a bottom-drawing center drain, while the majority of flow is withdrawn at an elevated drain. The dual-drain system uses a center drain that removes from 10 to 25% of the flow and a higher sidewall drain for the majority of the flow (fig. 11.2). The location of the two tank drains has typically been at the center of the tank, which then takes advantage of both the tea cup effect and the strength of the overall flow when it drains through the tank center. The

Figure 11.2 Sidewall discharge box with standpipe for water level control (75 to 90% of total discharge flow from rearing tank).

design criteria for the percentage of flow that should minimally be removed by the center drain are:

- 6 Lpm/m^2 (0.15 gpm/ft^2) of tank floor area
- Hydraulic retention time (HRT) center drain <200 minutes
- 15 to 25% of total tank flow rate.

After calculating these three values, the largest of the three values should be used.

A final criterion that should be applied to the drain design outlet is that the tank outlet should be approximately 10% of the tank diameter. This is because if the D/d ratio recommendation is followed, then solids should settle on the floor within about 5% (of tank diameter) of the tank center drain and be removed. For example, if a tank were 10 m in diameter, the center drain should have an effective diameter of 1 m (by using a flat plate slightly elevated above the tank floor) and the solids will settle within 0.5 m of the center point of the tank floor, thus these solids would be drawn into the center drain by high velocity created by the flow out the center drain. There are imaginative ways of achieving this besides making a large outlet drain: for example, flat plates on a center standpipe correctly spaced off the tank floor to create a desired capture velocity—at least 0.3 m/s—for the settling solids.

The alternative Cornell dual-drain approach has significant economic implications. When using this approach, solids removal costs are controlled more by the volume of flow that is treated rather than the solids concentration of the effluent that is treated. By concentrating the majority of the solid wastes in only 10 to 25% of the flow leaving a tank via the center low-flow drain treatment costs are proportionately reduced as well.

11.6   Settling basins and tanks

A settling basin is simply a tank that provides a quiet, nonturbulent area, where the flow rate is slowed and the solids are allowed to settle out of suspension by gravity. In an attempt to increase sedimentation, tube or plate settlers—consisting of a sequence of inclined tubes or plates that are stacked several centimeters apart—are often used. This increases the effective settling area per unit volume and reduces the depth to which a particle must settle to contact a surface. Advantages of settling basins are the simplicity of operation, low energy requirements, and their low construction costs. Disadvantages include the relatively large size of settling basins, their low removal efficiencies of small or low density particles, and leaching of nutrients from the settled solids back into the system while the solids are stored in the settling basin.

Another method to increase sedimentation rate is through the use of swirl separators or hydrocyclones. The effluent water from the culture tank is injected at the outer radius of a conical tank, so that the water spins around the tank's center axis. The spinning creates a centrifugal force that moves the particulates toward the wall, where they settle and can be removed continuously. These have most often been used in aquaculture recirculating systems to concentrate the solids from the dual-drain culture tank systems.

11.7   Mechanical filters

Two types of mechanical filters are commonly used in aquaculture to remove suspended solids: screen filters and expandable granular media filters. Suspended solids from an aquaculture viewpoint are within the fraction of total solids that will not settle out of the water column in a reasonable amount of time (30 to 60 minutes). This fraction of the solid waste needs to be removed from the culture tank water because of its potentially high oxygen demand and mineralization (the increased rate of ammonia-nitrogen being added to the water column as the protein and urea in the feces is broken down by bacteria into ammonia).

Microscreens are currently being widely used to remove suspended solids in both research and commercial recirculating systems. Microscreen filters act as a form of sieve that retains suspended particles larger than the fine mesh filter screen openings.

Rotating microscreen filters are available in a variety of sizes and flow rate capacities (fig. 11.3). They have numerous advantages, the primary one being that they are easy to install and operate. Almost all microscreen filters work on the principle of the physical interception of particles on a screen and their removal by means of a water spray. The screens are interchangeable and mesh size is usually selected based on the characteristics of the water to be treated, the required discharge water quality, and the trade-offs of size, cost, and waste discharge volumes. Microscreens are especially attractive when used to remove

Figure 11.3 Rotating microscreen filter and spray bars for cleansing.

solids from large flow streams. Additionally, they are compact in size and cause minimal head loss. The disadvantages of these systems are the high maintenance requirements and their relative high capital and operating costs. The performance of the microscreen filter is largely dependent on the size of the filter screen openings, influencing the filter's hydraulic capacity, the fraction of particles removed, the sludge-water production rate and concentration, and the filter backwash frequency that is generally discharged from the system (water loss that must be replaced).

11.8　Granular media filters

Granular media filters operate by passing water laden with suspended solids through a bed of granular material, which traps the solids through sedimentation, straining, and interception. The bead filters are the most popular form, due to their low head loss and low water requirements for backflushing. Bead filters have replaced the use of sand filters for all practical purposes. Sand filters can do a superb job at removing solids, where typically a sand bed is flooded from the top and the water is allowed to percolate through the sand. Sand filters, such as those used in swimming pools, operate under the same principle but use water pressure to force the solids-laden water through the sand. When the pressure requirements to push the water through the sand become excessive, then the sand filter goes into a backwash mode and in the process discharges large quantities of water. With either sand filter approach, excessive water loss occurs and the management of either sand filter has always been a challenge.

Bead filters employ small 3- to 5-mm low-density, polyethylene beads as the filtering media in pressurized, upflow filters. This type of plastic media is commonly used as feed stock for plastic injection molding process. Filtration is accomplished by trapping suspended solids particles within the bead matrix and then periodically agitating the beads to settle out the trapped solids and biofloc. These filters have been successfully used for both solids capture and biofiltration. Traditionally, the chief disadvantage of this type of filter has been the large volume of water required for back-flushing (although much less is required than in a sand-filter design), although the PolyGeyser class of bead filters have successfully addressed this problem. For more details on bead filters see www.beadfilters.com.

Many of the fine suspended solids and the dissolved organic compounds that build up over time in intensive recirculating systems are not removed by mechanical filtration or granular media filters. These solids can be removed by using a process usually referred to as foam fractionation, air stripping, or protein skimming. In this process, air bubbles rising through a closed-contact column physically adsorb the fine suspended solids and dissolved organic compounds onto the surfaces of the air bubbles. The bubbles create foam at the top of the liquid column and the organic wastes can then be disposed of along with the foam produced.

11.9   Disposal of the solids

The solids generated by these removal methods can have significant impact on the environment if not disposed of appropriately. In general, aquaculture solid wastes are treated as an agriculture waste and considered a nontoxic nutrient source. Several options are utilized for disposal, including agricultural application on land and composting. Land application of waste solids to an on-site agricultural field is usually the cheapest methods of solids disposal. Land application is governed by regulations that limit the amount of pathogens, heavy metals, and other contaminants, nutrient content, soil type and plant nutrient-uptake characteristics to prevent run-off or groundwater contamination.

One of the newest techniques for containing sludge is to use geotextile bags (fig. 11.4). These are porous textile bags that receive the waste stream and capture solids while allowing the water to drain off for recapture. In many cases, a polymer or flocculation agent is added to the influent to improve solids/liquid separation. Alum or ferric chloride can also be added as a coagulation aid, which also helps to sequester dissolved phosphorus. Composting of waste solids using thermophilic bacteria creates a valuable soil amendment and also can be used to dispose of fish mortalities in a safe manner. Anaerobic and aerobic waste lagoons can be used, but will require careful engineering design to be properly sized and managed.

11.10   Biofiltration

Nitrogen is an essential nutrient for all living organisms and is found in proteins, nucleic acids, adenosine phosphates, pyridine nucleotides, and pigments. In the

Figure 11.4 Geotextile bag on gravel bed designed to capture leached water for reuse.

aquaculture environment, there are four primary sources of nitrogenous wastes: (1) urea, uric acid and amino acid excreted by the fish; (2) organic debris from dead and dying organisms; (3) uneaten feed and feces; and (4) nitrogen gas from the atmosphere. In particular, various nitrogenous waste products are expelled by fish through gill diffusion, gill cation exchange, urine, and feces. The decomposition of these nitrogenous compounds is particularly important in intensive recirculating aquaculture systems because of the toxicity of ammonia, nitrite, and to some limited extent, nitrate. The process of bacterial driven ammonia removal in a biological filter is called nitrification, and consists of the successive oxidation of ammonia to nitrite and finally to nitrate. The reverse process is called denitrification and is an anaerobic process where nitrate is converted to nitrogen gas. Although not normally employed in commercial freshwater aquaculture facilities today, the denitrification process is becoming increasingly important, especially in marine systems, as stocking densities increase and water exchange rates are reduced, resulting in excessive levels of nitrate in the culture system.

Ammonia, nitrite, and nitrate are all highly soluble in water. Ammonia exists in two forms, un-ionized NH_3 and ionized, NH_4^+, with the relative concentration primarily a function of pH and temperature. An increase in pH or temperature increases the proportion of the un-ionized form of ammonia nitrogen. For example, at $20°C$ and a pH of 7.0, the mole fraction of un-ionized ammonia is only 0.004, but at the same temperature the mole fraction increases to 0.80 at a pH of 10.0. Un-ionized ammonia is toxic to fish at low concentrations.

Nitrification is a two-step process, where ammonia is first oxidized to nitrite and then nitrite is oxidized to nitrate. The two steps in the reaction are normally carried out sequentially. Since the first step has a higher kinetic reaction rate than the second step, the overall kinetics are usually controlled by ammonia oxidation, and, as a result, there is usually no appreciable amount of nitrite accumulation. Equations 1, 2, and 3 show the basic chemical conversions occurring during oxidation by *Nitrosomonas* and *Nitrobacter* and the overall oxidation reaction (US EPA 1975).

Nitrosomonas:

$$NH_4^+ + 1.5\,O_2 \rightarrow NO_2^- + 2\,H^+ + H_2O \tag{1}$$

Nitrobacter:

$$NO_2^- + 0.5\,O_2 \rightarrow NO_3^- \tag{2}$$

Overall:

$$NH_4^+ + 2\,O_2 \rightarrow NO_3^- + 2\,H^+ + H_2O \tag{3}$$

Based on these relationships, 4.57 g of O_2 and approximately 7.14 g of alkalinity as $CaCO_3$ are needed for the complete oxidation of 1 g of ammonia-nitrogen. Alkalinity (all alkalinity is measured or defined in terms of calcium carbonate, $CaCO_3$) is generally added by using sodium bicarbonate/baking soda ($NaHCO_3$) or calcium carbonate, slaked lime (CaO), or hydrated lime, $Ca(OH)_2$. Sodium carbonate has the advantage of being rapidly dissolved and being very safe to handle (you can eat it if your stomach is upset!). Conversely, the various lime compounds are not easy to dissolve (mixing tanks are required) and they are dangerous to handle. The lime products though are often much less expensive per unit of alkalinity than baking soda.

The ammonia removal capacity of biological filters is largely dependent upon the total surface area available for biological growth of the nitrifying bacteria. For maximum efficiency, the media used must balance a high specific surface area (i.e., surface per unit volume) with appreciable voids ratio (pore space) for adequate hydraulic performance of the system. The media used in the biofilters must be inert, noncompressible, and not biologically degradable. Typical media used in aquaculture biofilters are sand or some form of plastic or ceramic material shaped as small beads, or large spheres, rings, or saddles. Biofilters must be carefully designed to avoid oxygen limitation or excessive loading of solids, biochemical oxygen demand, or ammonia. Several types of biofilters commonly used in commercial intensive recirculating aquaculture systems are: trickling biofilters, floating bead filters, fluidized-bed biofilters, downflow micro-bead biofilters, and moving bed bioreactors (MBBR).

11.10.1   Trickling towers

The trickling tower is a classical biofilter, combining both biofiltration, aeration, and degassing into one unit process. Water cascades over some media on which bacteria grow, oxygen diffuses into the water, and nitrogen and carbon dioxide diffuse out. They can be constructed to any diameter required. Effective distribution of the influent water over all the media both horizontally and vertically is a continual challenge.

11.10.2   Floating bead filters

Floating bead filters use beads that are slightly buoyant. The beads provide surface area for bacteria and also trap solids, thus doing two jobs for the price of one filter. Water is introduced below a bed of packed bead media and travels upward through the filtration chamber where mechanical and biological filtration takes place. Backwashing of the filter is accomplished either mechanically with a motor/propeller or with air bubbles (fig. 11.5). At some predetermined

Figure 11.5 Propeller washed bead filter for biofiltration and solids capture.

rate, a desired backwash or mixing cycle (after mixing and breaking up the static floating bed, the bed is allowed to settle for a minute or so, and then the settled sludge is discharged by opening a valve at the bottom of the filter) will be imposed to clean the beads and remove the resulting sludge. Newer designs for bead filters (PolyGeyser Drop Filters; see www.beadfilters.com) have the capacity to minimize water loss during the cleaning cycle. This is particularly advantageous in marine systems where the loss of saltwater is minimized and thus operating costs are decreased.

11.10.3   Fluidized sand beds

Usually used in large-scale cool or cold-water applications, fluidized sand beds provide large surface area for bacteria in a small footprint. These filters get their name because as the water flows up through the sand bed, the sand becomes suspended in the flow or fluidized. Numerous designs have been investigated and found to perform effectively, particularly for cool water applications that use fine sands. Fluidized sand beds can be relatively more expensive to operate than other filters due to the high pump rates and pressures required to fluidize the sand bed.

11.10.4   Downflow micro-bead biofilter

Downflow micro-bead biofilters have been used for several years due to their simplicity and low cost for media. The filters use small plastic beads (1 to 3 mm) that float in the biofilter as the water flows down through them. The high specific surface area, low head loss, and small footprint makes them a strong competitor to other biofilter designs. The Styrofoam micro-beads are also a fraction of the cost of other bead media.

11.10.5   Moving bed bioreactors

Moving bed bioreactors (MBBR) have been introduced over the last several years and appear to be one of the most competitive of all the biofilter types (fig. 11.6). The media remain in suspension as the water flows through the biofilter, which is actively aerated. The high turbulence and aeration provide good mixing and contract with the media.

11.11   Choice of biofilter

Each biofilter described has advantages and disadvantages that need to be taken in consideration during the early design phase. One of the chief advantages of

Figure 11.6 Media and discharge manifold in a MBBR.

both the trickling biofilter and the MBBR is that they both add oxygen to the water flow during normal operation. In addition, they can provide some carbon dioxide stripping. In contrast, the submerged biofilters, floating bead filters, micro-bead filters, and fluidized-bed biofilters are all net oxygen consumers and must rely solely on the oxygen in the influent flow to maintain aerobic conditions for the biofilm. If, for whatever reason, the influent flow is low in dissolved oxygen or the incoming flow to the biofilter is too low, interrupted anaerobic conditions will be generated within the biofilter.

The application of low specific surface area media is a distinct disadvantage for both the trickling biofilters and the MBBR. Since the capital cost is proportional to the total surface area of the filters, the result is physically large and requires more costly filters. In contrast, floating bead filters, and especially fluidized-bed filters and downflow micro-bead filters, use media with high specific surface area resulting in reduced cost and space requirements for the equivalent surface area.

11.12 Aeration and oxygenation

Dissolved oxygen is the first limiting factor in intensive aquaculture systems. Minimum dissolved oxygen concentrations of from 4 to 6 mg/L are required for optimal growth and survival of most aquaculture species. At densities up to 45 kg/m^3, aeration with atmospheric oxygen is adequate to maintain this level, and is commonly referred to as aeration. At higher stocking densities, pure oxygen is required and is usually referred to as oxygenation.

During aeration, air is brought into contact with water, either by bubbling air through the water or forcing small droplets of water through air. In each case, oxygen is transferred into the water and to some extent carbon dioxide and excess nitrogen are stripped out. In the oxygenation process, pure oxygen is used

Figure 11.7 Speece cone for oxygen transfer.

to increase the transfer rate and yields supersaturated oxygen concentrations, theoretically five times that of aeration with atmospheric oxygen.

Oxygen can be produced on site using pressure swing adsorption (PSA) equipment or purchased in bulk liquid or gas form from commercial sources. Numerous types of oxygen transfer equipment are now available from commercial sources, including U-tubes, multistaged low head oxygenation units, packed columns, pressurized columns, oxygenation cones, oxygen aspirators, bubble diffusers, and enclosed mechanical-surface mixers. The two most often used in large-scale commercial systems are the downflow bubble contactor (Speece cones; fig. 11.7) or the multistage low head oxygenation (LHO) unit.

The downflow bubble contactor or Speece cone consists of a cone shaped column with water and oxygen injected at the top and removed at the bottom. As the water and the oxygen bubbles move down the cone, the downward velocity of the water decreases, due to the increasing cross-sectional area, until it equals the upward velocity of the oxygen bubbles. This yields a long contact time between the oxygen bubbles and the water, resulting in oxygen diffusion efficiencies at or above 90%.

A multistage low head oxygenation unit (LHO) increases oxygen transfer efficiency by reusing the oxygen feed gas through a series of contact chambers. Water flow is distributed equally through a series of chambers via a perforated plate, flows through the chambers and into a pool of water, which seals the lower outlets. Oxygen gas enters at one side of the series of chambers and passes through each chamber due to the pressure variation across the unit. The repeated contact of oxygen and water in each chamber provides a high rate of gas transfer, both oxygen into the water and excess nitrogen out. LHOs provide moderate oxygen transfer efficiency and operate at very low-pressure heads. Gas to liquid ratios should be maintained at less than 1.5% to obtain reasonable transfer efficiencies ($\sim$70%).

11.13 Carbon dioxide removal

Carbon dioxide is a product of respiration of fish and other organisms in the production system, such as the bacteria in the biofilm. At low stocking densities, carbon dioxide is removed by the physical agitation and exposure of the water to air. But advances in recirculation system design and management have resulted in a steadily increasing stocking density of fish and a reduction in water exchange rates. As a result, carbon dioxide often becomes a limiting factor in some intensive production systems.

For every gram of oxygen consumed, 1.38 grams of carbon dioxide is produced. Carbon dioxide obstructs the respiration of fish by reducing the capacity of the blood to transport oxygen through a physiological phenomenon known as the Bohr effect. Carbon dioxide also directly affects the overall system performance by decreasing its pH, which can stress the fish and inhibit the nitrifying bacteria in the biofilters.

Carbon dioxide can be controlled by gas exchange in a downflow trickling tower or packed column aerator, similar to the downflow trickling tower used for biofiltration. As the water containing carbon dioxide flows over the media in the column, air is forced either up or down the column and carbon dioxide is removed by gas transfer across the thin water/air interface on the media surface. Carbon dioxide removal by gas exchange is often limited by the buildup of carbon dioxide in the gas phase, requiring gas to liquid flow ratios substantially higher than those needed for traditional oxygen or nitrogen gas transfers. Gas liquid (G/L) ratios should be maintained above 5 for effective removal conditions. Carbon dioxide concentration in culture waters will also be impacted by the alkalinity concentration of the water (i.e., for a specific pH value, carbon dioxide concentration is proportional to alkalinity concentration). In fact, there is a stoichiometric relationship between pH, carbon dioxide concentration, and alkalinity concentration; any two of these three variables defines the third variable. See Timmons and Ebeling (2010) for a complete review of these relationships.

11.14   Monitoring and control

As the stocking density of fish continues to increase, the need for continuous monitoring and control of critical operations such as water flow, tank water level, and oxygen concentration also becomes essential to prevent catastrophic fish losses. It takes only one mistake to *kill everything in your facility!* For example, if a power failure occurred in a moderately stocked (60 kg/m^3) warmwater system, oxygen levels could drop to stressful levels in approximately 60 minutes—and if highly stocked (120 kg/m^3), in only 10 minutes.

Today, continuous monitoring systems are commercially available and relatively reliable. All the alarms should report to a phone dialer to notify staff both during and after working hours (the authors recommend twenty-four-hour people coverage wherever possible). Monitoring and control systems can be as simple as a solenoid valve that automatically opens an oxygen supply line to aerate during a power failure. For most facilities, simple, fairly inexpensive monitoring/auto dialer systems are commercially available, which will monitor eight or more switch inputs and automatically switch on backup systems and dial out in emergency situations. At the other extreme are sophisticated computer control systems that monitor multiple parameters with redundant backup systems, Internet pages for off-site monitoring, and phone dialers to alert management of system status and alarms.

Regular monitoring of other parameters that change slowly with time requires a well equipped and maintained water-quality lab (e.g., that tests for ammonia, nitrite, nitrate, alkalinity, etc.). It doesn't have to be an elaborate expensive facility, but it needs to be in a dedicated location with all the support equipment for good laboratory practices.

11.15   Current system engineering design

As can be seen from the above descriptions, there are several alternative solutions to each of the unit processes utilized in intensive recirculation aquaculture systems. Sometimes designing recirculation systems seems more like ordering from a Chinese menu, where the designer selects one unit from column A and another from column B and another from column C and then integrates them into the "optimal" design. Important design questions that determine what unit processes to use include: species to be produced, stocking density, production capacity, and sensitivity of species to water quality parameters such as temperature, dissolved oxygen, ammonia-nitrogen, pH, and CO_2. For example, tilapia is a species with excellent market potential and tolerance to poor water quality. On the other hand, many marine species of commercial interest have very high water-quality requirements.

Numerous combination and permutations of the above unit processes have been implemented over the past decade, and although no one system design has become the standard, several combinations seem to work well together. For example, many large-scale commercial recirculation systems use some form of a rotating microscreen filter for solids removal combined with a Cornell dual-drain system. Although microscreen filters have a high capital cost, their ease of installation and operation make them economically attractive. Most biofiltration is now accomplished with either a moving bed bioreactor (MBBR) or some form of granular media filter, such as a downflow bead filter or a fluidized-sand filter. The main advantage of the MBBR is the aeration and degassing they provide, reducing aeration or oxygenation requirements. Fluidized-sand filters have the advantage of lower capital cost and smaller size, but higher operating cost. Aeration and degassing of carbon dioxide is usually accomplished with some form of cascade column with high air:water volumetric ratios. Finally, oxygenation of the recirculated water is usually performed with downflow bubble contactors or LHOs.

11.16 Recirculation system design

11.16.1 Mass balances, loading rates, and fish growth

All system designs should begin with a mathematical analysis of the system loadings. This is how the various system components and unit processes are integrated to create a system that maintains targeted water quality parameters for some specific type of aquatic animal. Water flow is the mechanism by which oxygen is transported into a fish culture vessel and the waste products being generated within are removed. The design of a recirculating aquaculture system (RAS) should ensure that the important parameters affecting water quality and fish productivity (e.g., oxygen, ammonia, carbon dioxide, and suspended solids) are properly balanced. This requires calculating the required flow rates via mass balance equations to maintain the design water quality variables at or below (or above, in the case of oxygen) their maximum tolerable or design target values. Then the system must be operated at the highest flow rate calculated for these four critical water quality parameters. Obviously, the maximum flow rate used to maintain the constraining water quality parameter will be higher than necessary for the others, which simply means these water quality parameters will be at "better" values than design targets. The same mass balance approach can be utilized on any variable affecting water quality.

The mass balance approach simply comes down to balancing the transport in of some parameter, the production of this particular parameter within the culture tank, and the transport out of the parameter. In word equation form:

Transport **in** of X + **production** of X = transport **out** of X

The production term can be the production of oxygen, ammonia, suspended solids, or CO_2. Note that the production term can be *negative,* meaning consumption of a certain component (e.g., oxygen). Note that X does not refer to a concentrations of x, for example, in mg/L, but rather about a mass quantity of some "stuff," referred to as X in the word equation.

This is the control volume approach. Engineers like to depict mass transport across some "imaginary" box (unit processes) that designates the vessel or container that we are trying to analyze. For RAS it can be assumed that the culture tank represents a completely well-mixed tank and that the tank has reached a non-changing condition with respect to time or steady-state conditions. Each of the boxes shown in figure 11.1 represents some treatment device or process that changes the concentration of the noted parameter X. (Note: There could be several treatment devices, each treating a different water quality variable.)

Another way of representing the mass balance equation uses the product of the flow rate (volume/unit time) multiplied by the concentration (mass/volume) resulting in a mass per unit time, as, for example, kgs of oxygen/day, kg of solids per day, etc. Neglecting the effects of the flow-through component (which should be considered if the system discharge per day is one volume or more), the mass balance equation in its simplest form is:

$$Q\ C_{in} + P = Q\ C_{out}$$

C_{in}, C_{out}: Concentrations of parameter X into and out of the culture tank, mass/volume
Q: Water that is recirculated, mass/day
P: Production rate or consumption (negative), mass/day

11.16.2   Selecting tank values

The designer/manager must choose *design* or target operating conditions to calculate the mass balances. These are the C values in the mass balance equation. These design numbers are species dependent and are continually being refined for RAS applications. Calculating the minimum flows required to maintain targeted values for water quality (and then using the largest minimum value found for all the different water quality variables) will show how sensitive the calculated flow rates are to the value selected for the design value. A typical scenario is to select a value, do the calculations, realize that there is no way you could afford to supply such a high flow rate, and then start to make adjustments in the targeted values (for example, 4 mg/L oxygen is probably acceptable instead of the 6 mg/L you originally chose). In the end, one must choose realistic values at the beginning of the design process and then stay with these choices and the ramifications of the resulting flows required to maintain the mass balances. *Do not ever* compromise on the required flow rates.

11.16.3   Water treatment device

Each treatment box (unit process) is a system that "improves" the water quality of the water. The concentration of the particular water quality parameter leaving the treatment device can be calculated by knowing the "absolute" best the treatment device could achieve and the treatment efficiency of the particular device being used. The treatment efficiency should be provided by the device supplier or manufacturer.

Solving the general mass balance equation applied to the treatment box determines the concentration of each particular water quality parameter leaving the treatment device (C_2). The treatment device could be a biofilter, a CO_2 stripper, or a solids settling chamber. Each will have its own treatment efficiency for the particular water quality parameter it is designed to treat.

The water quality concentration C_{out} leaving the treatment box is found by using a mass balance analysis to be:

$$C_{out} = C_{in} + T/100 \, (C_{best} - C_{in})$$

In this equation, T is the treatment efficiency (%) and C_{best} is the absolute best result obtainable by a treatment system (e.g., zero ammonia or saturated oxygen, zero suspended solids). Note that if the device is an oxygen addition unit, the C_{best} term can be increased above atmospheric concentration values for oxygen by increasing the partial pressure above atmospheric oxygen partial pressure in the device. For example, a pure oxygen device will have a C_{best} value of roughly five times the C_{best} value obtained if normal air were used at atmospheric pressure (e.g., trickling tower). C_{best} for most other parameters should be fairly obvious to the reader (e.g., ammonia and TSS are zero, but CO_2 will be around 0.5 mg/L since there is some CO_2 in the air).

11.17   Four major water-treatment variables

11.17.1   Oxygen

The major reason most fish die is from lack of oxygen due to a loss of water flow. This is because oxygen is consumed at a fairly high rate (fish metabolism) and oxygen is transported by water flow. Due to low inherent concentrations of oxygen, "high" flows are required to transport the required oxygen. Flows required to maintain a satisfactory oxygen level are generally the controlling flow rate parameter when solving the series of mass balance equations to determine the most restrictive parameter. Even a *partial* loss of flow will generally result in insufficient oxygen for the fish, resulting in death. Ask yourself if your monitoring system will sense a reduction in flow as opposed to a loss of flow.

It is important to remember that not only do the fish require oxygen, but also the biological filter, which is critically dependent upon adequate oxygen levels to support bacteria metabolism. The dissolved oxygen (DO) concentration within the filter must be maintained at or above 2.0 mg/L to ensure that the rate of nitrification in the filter does not become limited because of oxygen depletion. Always measure the DO coming out of the biofilters and if the concentration starts to approach 2.0 mg/L, then take corrective action (e.g., increase flow rate through the biofilter by increasing the hydraulic loading rate on the filter or add oxygen prior to the flow entering the filter or within the filter itself).

The production term for oxygen (P_{Oxygen}) is:

$$P_{Oxygen} = -0.25 \text{ kg per kg feed consumed by fish}$$
$$-0.12 \text{ kg per kg feed consumed by nitrifying bacteria}$$
$$-0.13 \text{ kg per kg feed consumed by heterotrophic bacteria}$$
$$(\text{can be as high as } 0.5)$$
$$= -0.50 \text{ kg (sum of above) per kg feed for system}$$

These oxygen terms are all negative since they "consume" oxygen from the system as opposed to adding a mass quantity to the water column. The heterotrophic bacteria load can be as much as 0.5 kg of oxygen per kg of feed fed or higher if the system has poor solids removal. The oxygen consumption term is always much higher when solids management in a RAS is not good; a safe number for design purposes might be closer to 1 kg of oxygen per 1 kg of feed used (this would be twice as high as the above suggestion).

11.17.2   Ammonia

There is considerable confusion about design target values for ammonia concentration. Definitive values for the toxic levels of ammonia and the differentiation between the toxic NH_3 form and the supposed nontoxic NH_4^+ have not been precisely determined. The apparent toxicity of ammonia is extremely variable and depends on more than the mean or maximum concentration of ammonia.

The European Inland Fishery Advisory Commission (EIFAC) of FAO has set 0.025 mg/L as the maximum allowable concentration for un-ionized ammonia (NH_3 or ANH_3-N). A good rule of thumb is values of 1 mg/L for total ammonia nitrogen (TAN) for cool water and 2 or 3 mg/L for warmwater fish. You should always check your TAN target value selection by assuming some pH and temperature that you intend to maintain and see if your NH_3 concentrations will exceed the 0.025 mg/L value using readily available ammonia-ammonium pH temperature tables from several texts (Timmons & Ebeling 2010).

The rate of ammonia generation is considered to be a "soft" number. For simplicity, one could simply assume 10% of the protein in the feed becomes the ammonia-N generation rate. For a more precise estimate and one that is affected by protein content, the following equation for the production rate of ammonia

(P_{TAN}) can be used:

$$P_{TAN} = F \times PC \times 0.092$$

Here, F is the daily feeding level and PC is the protein content of the feed being fed. In this equation, the time period used is one day. In RAS, feed can be fed uniformly over a twenty-four-hour period, thus distributing the ammonia load uniformly over the entire day as well. If a uniform twenty-four-hour feeding is not used, then the above equation should be adjusted and the time period should be the time between feedings, or if a single feeding per day is used, then use four hours as the time period as an estimate of the time for the ammonia to be excreted from a feeding event. Note that in the above equation, the feed is assumed to be fed uniformly over a twenty-four-hour period. If the feeding period is concentrated to some fraction of the day, then the production equation for TAN must be accordingly adjusted upward. For example, if you feed over a twelve-hour period, then the P_{TAN} term will be twice as large for design purposes, requiring your biofilters to be twice as large (and cost twice as much!). This is why in RAS, you generally feed the fish over a twenty-four-hour period when possible.

11.17.3 Carbon dioxide

Carbon dioxide is an important, but largely overlooked water quality limiting parameter. This is probably because until recently, most systems were generally low density (less than 40 kg/m^3) and relied on aeration as the main means of supplying oxygen. This type of management also kept CO_2 values at low levels (e.g., less than 20 mg/L). However, loading rates have increased in recent years, and it has become necessary to inject pure oxygen into these systems, instead of using aeration. As result, the natural stripping of CO_2 that occurs when using aeration systems was no longer taking place. CO_2 production can be explicitly related to oxygen consumption as the ratio of molecular weights (44/32), $P_{CO2} = 1.375 \times P_{Oxygen}$.

11.17.4 Suspended solids

The effective control of solids generated in RAS is probably the most important task that must be accomplished to ensure long-term successful operation of a RAS. The quantity of suspended solids or total suspended solids (TSS) generated per unit of feed being fed is estimated as:

$P_{solids} = TSS = 0.25 \times$ kg feed fed (dry matter basis; range of 0.20 to 0.40)

TSS is treated as a dilute waste. TSS design concentrations in RAS will be in the 10 to 100 mg/L range. Even after concentrating TSS with some type of treatment process such as a rotating screen filter, the discharge water will still contain only around 0.5 to 1% solids on a dry matter basis. In comparison, cow manure is 20% solids. TSS captured in a settling basin has a fluffy consistency and will require substantial volumetric space depending upon frequency of cleaning. As a rule of thumb, assume that each kg of dry feed fed will produce approximately 25 liters of liquid waste (1% solids).

11.17.5   Nitrate nitrogen

Typically, nitrate nitrogen is not considered in the mass balance equations to determine maximum required flow rate. In saltwater systems, though, nitrate should be considered. Nitrate nitrogen (NO_3-N) is the end product of the nitrification process. In general, concentrations of nitrate are not extremely adverse to RAS water quality. Nitrogen is essentially conserved throughout the nitrification process. Thus, if 1 kg per day of TAN is being produced, then 1 kg of nitrate-N is being produced. The equilibrium concentration of nitrate will therefore be directly dependent upon the overall water exchange rate throughout the system. Nitrate-nitrogen is relatively nontoxic to freshwater fish and as such will not influence the controlling flow rates in the system. One can choose some value such as 200 mg/L, if you want a number to work with. More recent unpublished information seems to indicate that for various salmonids, the nitrate-N levels should not exceed 40 to 50 mg/L. Design information is very limited relative to this water quality parameter, especially for saltwater systems.

11.18   Summary of four production terms

These four terms are called the production terms, or the "P terms," in mass balance equations. Summarizing the four production terms as related to feed being fed is as follows:

$$P_{Oxygen} = -0.50 \text{ kg per kg feed consumed by fish}$$
$$P_{CO2} = 1.375 \text{ grams produced for each gram } O_2 \text{ consumed}$$
$$P_{TAN} = F \times PC \times 0.092$$
$$P_{solids} = 0.25 \times \text{kg feed fed (dry matter basis; range of 0.20 to 0.40)}$$

11.18.1   Predicting fish growth

The basic premise of RAS engineering is to produce a given volume of market size fish at some predetermined schedule. Based on the growth rate and stocking density, the size and number of production tanks can then be estimated. The production rate also defines the required fish feeding rate, which then in turn determines the waste generation loads (i.e., ammonia-nitrogen, carbon dioxide,

Table 11.1 Temperature growth units in °C (°F) for trout, tilapia, yellow perch, and hybrid striped bass.

	Trout	Tilapia	Yellow Perch	Hybrid Striped Bass
T_{base}	0 (32)	18.3 (65)	10 (50)	10 (50)
TU_{base}	6.1 (28)	3.3 (15)	5.5 (25)	5.5 (25)
T_{max}	22.2 (72)	29.5 (85)	23.9 (75)	23.9 (75)

solids) and also oxygen consumption rate. Based on these data, the most appropriate treatment systems can then be designed based on both water quality demands and economics.

A common way of defining fish growth is based on existing data sets of production trials under commercial stocking densities, feed rates, and water quality conditions. A second way is based upon a temperature unit approach and some defined number of temperature units to create a unit growth rate, such as centimeters per month:

$$Growth = \frac{T - T_{base}}{TU_{base}}$$

Where:

T = water rearing temperature

T_{base} = practical lower temperature where fish growth still is achieved

TU_{base} = monthly temperature units needed to achieve one unit of growth, e.g., 1 cm or 1 inch per month.

The above equation predicts growth, using units of cm/month or inches/month, and the T values are all in either degree Centigrade or Fahrenheit (see table 11.1 for values for various species; Timmons & Ebeling 2010). The growth equation is subject to the limitations that if T is greater than T_{max}, then calculate the growth at T_{max}. Note that excessive temperatures will compromise growth and/or feed conversion.

The weight of fish can be mathematically related to their length by using a term called the condition factor (K or CF); the bigger the condition factor, the more weight per unit length. The condition factor is expressed quantitatively using the equation below for either metric or English units. Each fish species will have an associated K or CF factor value to describe expected or normal body condition. The value of this factor is influenced by age of fish, sex, season, stage of maturation, fullness of gut, type of food consumed, amount of fat reserve and degree of muscular development. Typically, the condition factor will stay relatively constant from a very early fingerling stage (few grams) to near maturity size and, as such, is an excellent management tool for producers to determine over and underfeeding or poor water quality (causing high coefficient of variation in the tank cohort).

$$WT(lbs) = \frac{CF\,(L_{inches})^3}{10^6} \text{ or } WT(g) = \frac{K\,(L_{cm})^3}{10^2}$$

11.19 Stocking density

The final and most important questions that must be addressed in system design is the mass of fish that can be maintained in a tank or, more appropriately, the mass of fish that can be maximally supported. We should probably call this term the harvest density (or the density that when reached will trigger an action by the fish manager to move the fish to their next stage or be taken to market). The number of fish and their individual weight will define the feeding rates from which all other individual engineering components are designed. The mass of fish that can be stocked (or harvested) per unit volume ($D_{density}$) will depend on both the fish species and the fish size. We use an approach that is based upon body length (L) to estimate the number of fish that can be carried per unit volume of tank:

$$D_{density} = \frac{L}{C_{density}}$$

Where:
 $D_{density}$ Density in kg/m^3 (lb/ft^3)
 L Length of fish in cm (inches)
 $C_{density}$: 0.34 for L in cm (2.1 for L in inches) for a trout species (different species have their own $C_{density}$ value).

Maximum allowable or safe densities (D_{fish}) for fish stocking is primarily a function of fish size, species, and the characteristics of the rearing environment and management skill. New growers tend to overestimate their own safe loading densities and assume they can establish and sustain densities from the very beginning that in fact require expert management skills. *Do not fall into this trap.* You will kill fish. New growers should target about half the densities recommended for expert growers. For example, a tilapia system for a new grower should probably be operated at a maximum density of 50 kg/m^3 rather than 100 to 120 kg/m^3.

11.20 Engineering design example

11.20.1 Required minimal flow rates

Let's assume a target production capacity of 500 metric tonnes of tilapia per year with a target market size fish of 750 g (1.65 lb), whole, on ice. Tilapia

Table 11.2 Condition factors for various fish (Timmons & Ebeling 2010).

Species	CF (length in inches, weight in lb)	K (length in cm, weight in g)
Tilapia	750–900	2.08–2.50
Tilapia <1 gm	500	1.39
Rainbow, Brown Trout	400	1.11
Lake Trout	250	0.69
Charr	520	1.45
Hybrid Striped Bass	720	1.99
Perch	490	1.36
Muskellunge	150	0.42
Northern Pike	200	0.56
Largemouth Bass	450	1.25
Walleye	300	0.83

fingerlings are purchased at a mean size of 50 g. A three-stage production system is decided upon (i.e., fingerling, intermediate, and growout).

Step 1: Calculate growth period from stocking to harvest
The following calculation is employed to determine the total growth period required from 50 grams to 750 grams, at 28°C (83°F); assume a K = 2.10 (CF = 760; see table 11.2 for tilapia). Calculate length of fingerlings at 50 g and the final market sized fish at 750 grams:

$$L\ (cm\ at\ 50\ g) = \left[\frac{100 \cdot 50}{2.10}\right]^{1/3} = 13.4 \text{ cm}$$

$$L\ (cm\ at\ 750\ g) = \left[\frac{100 \cdot 750}{2.10}\right]^{1/3} = 33.9 \text{ cm}$$

Total change in length required from 50 to 750 grams:

$$\Delta L = (32.9 - 13.4) = 19.5 \text{ cm}$$

Step 2: Determine time for growout
Calculate growth rate at the specified rearing temperature of 28°C (83°F):

$$Growth_{rate} = \frac{28 - 18.3}{3.28} = 2.96\frac{cm}{month}$$

Total time to achieve the required length:

$$Growth_{period} = \frac{19.5\ cm}{2.96\frac{cm}{month}} = 6.6\ months$$

Step 3: Determine production strategy

Note that this growth could occur in a single tank and the fish remain there for 6.6 months, or the growth could be managed to occur in three tanks that are managed sequentially. If a three-stage scheme is used with equal time at each state, then the fish would reside in each tank for one-third of the total growth time, or 2.2 months/size class, or 9.5 weeks in each tank; in practice, we'd probably round this up to ten weeks.

A very simple concept here that is sometimes missed is that whatever the production cycle is for a stage will determine how many tanks are needed for that stage. If ten weeks are required for the final production stage, and weekly harvests are required, then ten final growout tanks will be required. The other twenty weeks of the growout cycle can be distributed however you choose, but their cumulative growth time must add up to twenty weeks. More commonly, the time the fish stay in stage 3 might be increased (e.g., stage 1 for four weeks, stage 2 for eight weeks, and stage 3 for eighteen weeks) so there is less stress on moving fish from stage 2 to stage 3 (this scheme allows you to move smaller fish, which is always easier).

To keep things simple in this example, assume that the fish will be kept in each stage for an identical time. This will require designing a fingerling to growout production strategy that requires ten tanks (rounding the 9.5 weeks to 10 weeks) per stage of increasing size, plus systems for circulation, solids removal, biofiltration, and gas exchange. During each stage, the change in length of the fish would be the total increase (19.5 cm) divided by the number of stages (three) or 6.5 cm. Using the length/weight relationship, the initial and final weights can be calculated at each stage as shown in table 11.3.

Step 4: Calculate weekly harvest weight

Assume that fifty weekly harvests are conducted with a two-week end of year vacation break:

$$Harvest_{weekly} = \frac{500{,}000 \ kg/year}{50\dfrac{weeks}{year}} = 10{,}000 \ kg \ per \ week$$

Step 5: Determine the number of fish per tank at harvest

Note that no allowance was made for mortalities, but they are a part of life and are best estimated based on actual production experience.

$$Fish_{Tank} = \frac{10{,}000 \ kg}{750\dfrac{g}{fish}} * \frac{1000 \ g}{1 \ kg} = 13{,}333 \ fish \ per \ Tank$$

Step 6: Estimate final tank biomass at each stage

$$Biomass_{Tank} = 13{,}333 \ fish \ per \ Tank * Weight_{Final}$$

Step 7: Estimate final feed rate per tank

Using the steps previously shown (calculate the weight of the fish the day before harvest), calculate the amount of feed used on the last day for the end

Table 11.3 Initial and final weights and lengths of the three-stage production strategy.

	Initial Wt Size	Final Wt Size	Final Tank Biomass	Feed Rate (% bw/day)	Final Feed Rate
Stage 1:	50 g 13.4 cm	165 g 19.9 cm	2,200 kg	2.81%	62 kg
Stage 2:	165 g 19.9 cm	386 g 26.4 cm	5,146 kg	2.12%	109 kg
Stage 3:	386 g 26.4 cm	750 g 32.9 cm	10,000 kg	1.70%	170 kg

of each stage. Table 11.4 shows the key calculation parameters for the growout stage. The final feeding rates per tank at maximum biomass density for the other stages are summarized in tables 11.3 and 11.4.

Step 8: Determine the controlling flow rate for this design problem (i.e., dissolved oxygen, carbon dioxide stripping, or ammonia-nitrogen removal)

Calculate the required design flow rate for a 100% recirculating flow for the production tank feeding rate of 170 kg feed/day at 38% protein. Calculate the required flow rate for each water quality parameter and then identify the controlling parameter. Compute the required steady-state flow rate for maintaining the following water quality levels: 2 mg/L TAN, 5 mg/L O_2, and 40 mg/L CO_2. (In this example, we are neglecting TSS, which is rarely, if ever, the controlling flow rate.) Assume the following efficiencies for the treatment devices: 35% for TAN, 90% for O_2, and 70% for CO_2. Additionally, the tank water temperature was set to 28°C, and the C_{sat} in the O_2 treatment device is 18.1 mg/L (it uses some pure oxygen).

Start with the General Mass Balance, where C_1 is outflow and C_2 is inflow from the fish culture tank:

$$QC_2 + P = QC_1$$

$$\text{or} \quad Q(C_1 - C_2) = P$$

Table 11.4 The key calculation parameters used for the growout stage.

Growth, cm/month	2.96
Growth, cm/day	0.0970
Length at harvest, cm	32.9
Length, day -1	32.803
Final weight, g	750
Condition factor (K)	2.10
Weight, day -1	741.239
Change in weight, g/fish	8.7613896
Feed Conversion (FC)	1.46
Number of fish	13,300
Total daily feed, kg/day	170

11.20.2   Required design flow rate for dissolved oxygen

Looking initially at Dissolved Oxygen, as this is often the controlling parameter for flow, first calculate the influent dissolved oxygen concentration (C_2) using a Speece cone with 90% oxygen transfer efficiency (TE) and a production tank DO level (target value for minimum) of $C_1 = 5$ mg/L.

$$C_2 = C_1 + TE\ (C_{sat} - C_1)$$

Solve for C_2, and you obtain:

$$C_2 = 5.0 \text{ mg/L} + 0.90^*(18.1 \text{ mg/L} - 5.0 \text{ mg/L})$$
$$C_2 = 16.8 \text{ mg/L}$$

Oxygen production, that is, consumption or (-P), is the sum of fish and bacterial oxygen consumption:

$$P = \frac{0.25 \ kg \ O_2 \ by \ Fish}{kg \ feed} + \frac{0.12 \ kg \ O_2 \ by \ Bacteria}{kg \ feed} = \frac{0.37 \ kg \ O_2}{kg \ feed}$$

$$P = \left(\frac{170 \ kg \ feed}{day}\right)\left(\frac{0.37 \ kg \ O_2}{kg \ feed}\right)\left(\frac{10^6 \ mg}{kg}\right) = \frac{62,900,000 \ mg \ O_2}{day}$$

Note that using 0.37 kg oxygen per kg of feed would be the lowest demand one would use for design purposes. This would represent the case where either a trickling tower or a MBBR is to be used. As previously discussed, using a value of 0.5 to 1.0 for oxygen demand (as opposed to 0.37 in this example) would be a more conservative approach to ensure sufficient flow for all oxygen demands in the system.

Returning to the General Mass Balance for a RAS:

$$Q_1 \ C_2 + P = +Q_1 \ C_1$$
$$Q(16.8 \text{ mg/L}) + (-62,900,000) \text{ mg/day} = Q(5.0 \text{ mg/L})$$

$$Q = \frac{\dfrac{62,900,000 \ mg \ O_2}{day}}{(16.8 - 5.0)\dfrac{mg \ O_2}{liter}} = \frac{5,330,000 \ liter}{day} \quad \text{or } 3,700 \ Lpm$$

Summarizing this calculation, approximately 3,700 Lpm (975 gpm) of influent water at 16.8 mg/L DO is required to satisfy the oxygen demand of the 170 kg of feed per day.

Table 11.5 Summary required flow rates for the production tank.

Water Quality Parameter	Required Flow rate
TSS	3,940 Lpm (1,040 gpm)
TAN*	5,900 Lpm* (1,560 gpm)
Oxygen	3,700 Lpm (975 gpm)
Carbon Dioxide	1,280 Lpm (575 gpm)

Note: *Controlling flow rate.

The other flows for the other water quality parameters are then calculated in a similar manner. The resulting flow rates are shown in table 11.5. The maximum flow rate calculated for any parameter is the *only* flow that guarantees that all the water quality parameters will meet the requirements set. In this case, the controlling flow rate is 5,900 Lpm (1,560 gpm), which was calculated for TAN. At this stage, each of the selected unit processes can be reexamined to determine if more efficient units could be selected that reduce the required flow rate. For example, the biofilter efficiency might be increased by increasing the hydraulic retention time on the bioreactor vessel, thus increasing its efficiency (of course this has capital cost considerations as well). A balance needs to made between the capital cost of equipment and the long-term operational cost of operation and pumping of water.

11.20.3 Calculating tank sizes

The last step in our design process is to determine the tank sizes necessary for each of the three stages for our 500-tonne production farm. The tanks sizes can be determined based on the allowable fish biomass stocking densities. As previously determined (see table 11.3), the final fish biomass per tank was 2,195 kg for the fingerling stage 1; 5,134 kg for stage 2; and 10,000 kg for the growout tank stage 3. Note how the systems support larger biomasses of fish as the fish increase in size. This is because the fish require space (room) by length and not by body mass, and mass is proportional to length cubed.

Using the density equation, the corresponding stocking-harvest densities can be calculated. That is, for the stage 1 system at harvest, the density when moving from stage 1 to stage 2 is:

$$D_{Stage\ 1} = \frac{L}{C_{density}} = \frac{19.9\ cm}{0.24} = 82.9 \frac{kg}{m^3}$$

Correspondingly, the stage 2 final tank density is 110 kg/m³ and the growout tank stage 3 at harvest is 137 kg/m³. The required volume of each tank for each of the three stages is equal to the total biomass for that stage's cohort of fish

Table 11.6 Final tank biomass density and tank dimensions for the three-stage production strategy.

	Fish Density kg/m³ (lb/gal)	Tank Volume m³ (gal)	Tank Depth m (ft)	Tank Dia. m (ft)
Fry production	82.9 kg/m³ (0.69 lb/gal)	26.5 m³ (7000 gal)	1 m (3.2 ft)	5.8 m (19 ft)
Fingerling	110 kg/m³ (0.92 lb/gal)	46.6 m³ (12,300 gal)	1.2 m (3.85 ft)	7.0 m (23 ft)
Growout	137 kg/m³ (1.14 lb/gal)	72.8 m³ (19,200 gal)	1.5 m (3.85 ft)	7.9 m (26 ft)

divided by the stocking density, or for example:

$$V_{Stage\ 1} = \frac{W_{fish}}{D_{fry}} = \frac{2195\ kg}{82.9^{kg}/_{m^3}} = 26.5\ m^3$$

Assume that for management reasons, the tank depth is approximately 1 m in the fry tank (so the area of the tank is 26.5 m²). And from simple geometry, the tank diameters can be determined for stage 1:

$$Dia_{Stage\ 1} = \sqrt{\frac{4\ area}{\pi}} = \sqrt{\frac{4\ 26.5\ m^2}{\pi}} = 5.8\ m$$

Increasing the depth to 1.2 m for the stage 2 tanks and 1.5 m for the growout stage 3 tanks yields diameters of 7 m (22.6 ft) and 7.9 m (26 ft), respectively. Table 11.6 summarizes the tank biomass design and sizing. Note that these calculations are based upon equal time in each stage of the three-stage production strategy. Also, remember that for each stage in our example, there are ten weeks, so this means that ten tanks of the determined dimensions for each of the three stages are required for the production farm, or a total of thirty tanks are required.

11.21   Conclusion

The design and engineering of intensive recirculating aquaculture systems has made significant progress over the past several decades. Systems components have been designed and engineered specially for aquaculture, and are field tested and available from several commercial sources. System integration is just beginning to be investigated to determine optimal configurations of an overall system design based on species, production levels, and water quality requirements. Even with all the successes and improvements, though, it is still challenging to implement and manage recirculation systems that are cost competitive with other less capital-intensive production strategies.

11.22 References

US Environmental Protection Agency (1975) *Process Design Manual for Nitrogen Control.* A design manual prepared for the Office of Technology Transfer of the US EPA.

Timmons, M.B. & Ebeling, J.M. (2010) *Recirculating Aquaculture*, 2nd Edition. Cayuga Aqua Ventures LLC, Ithaca. (Available from www.c-a-v.net.)

Chapter 12

Biofloc-based Aquaculture Systems

Craig L. Browdy, Andrew J. Ray, John W. Leffler, and Yoram Avnimelech

Interest in biofloc-based aquaculture systems has grown over the past twenty years, because these systems can provide comparatively biosecure, more environmentally benign, and financially sustainable aquaculture production. Tilapia and shrimp are well suited to take advantage of natural productivity in aquaculture systems. Each currently represents quickly expanding commercial industries worldwide. Aquaculture production of an important tilapia species, *Oreochromis niloticus,* more than doubled between 2001 and 2007 from 1,030,888 tonnes to 2,121,009 tonnes (FAO 2010). Production of the most commonly cultured marine shrimp, *Litopenaeus vannamei,* more than doubled from 982,663 tonnes in 2003 to 2,296,630 tonnes in 2007 (FAO 2010). An increasing body of research is being devoted to development of biofloc-based applications for these species.

Biofloc technologies can be applied in ponds, tanks, or raceways of various scales. Intensive, biofloc-based production of shrimp is done in lined ponds, from 500 to 20,000 m^2, with a seasonal production of 10 to 20 tons/ha (1 to 2 kg/m^2). These systems are becoming quite common and are expanding quickly. Production of tilapia using biofloc systems in general produces a much higher biomass than that found in intensive shrimp ponds, in the range of 10 to 30 kg/m^2. A third application of the technology targets super-intensive production of shrimp in tanks or raceways. Production levels in these systems can reach close to 10 kg/m^2. The present chapter provides a review of the

Aquaculture Production Systems, First Edition. Edited by James Tidwell.
© 2012 John Wiley & Sons, Inc. Published 2012 by John Wiley & Sons, Inc.

characteristics of these biofloc systems focusing on the commonalities and highlighting management strategies that can differ according to application.

Biofloc systems are based on the concept of cultivating a microbial community within the production unit. This microbial community provides important ecosystem services including the cycling of waste material and provision of supplemental nutrition to the target crop. External inputs include the feed and in many cases supplemental carbon and/or bicarbonate to support target crop growth and to meet the needs of the microbial community. Inputs also include energy for supplemental aeration or oxygenation and mixing to maintain a suspended aerobic microbial consortium. Through proper management of the inputs, target crop density, and cropping or oxidation of organic material during and/or between crops, the grower can achieve a balance, thereby maximizing ecosystem services within the production unit. This can provide for improved cost efficiencies, stable production conditions, and higher overall environmental sustainability of production.

For centuries, fertilized ponds have provided a basis for artisanal production of fish and shrimp. As early as the late 1970s, a group in Israel studied the dynamics of fish culture systems enriched with organic material, developing the concept of a heterotrophic food web (Wohlfarth & Schroeder 1979; Hepher 1985). At the same time, Steven Serfling and Dominick Mendola conceptualized and developed a business based on industrial production of tilapia and shrimp in systems with dense microbial communities with reduced reliance on water exchange. Applying a more holistic approach, the systems focused on maximizing benefits from natural productivity, which developed in the systems. In their early research and commercialization efforts at Solar Aquafarms, some of the concepts that today characterize biofloc technologies were demonstrated, although they never reached the popular or scientific literature (Rosenberry 2007).

In the early 1990s, two groups working independently in Israel at the Technion University and in the United States at the Waddell Mariculture Center (WMC) began to publish a series of papers on the application of reduced and then zero exchange production technologies for tilapia and shrimp, respectively (Avnimelech 1993; Hopkins *et al.* 1993; Avnimelech *et al.* 1994; Hopkins *et al.* 1995b). The research demonstrated the assimilation of excess nitrogen into microbial biomass and its mineralization through nitrification and denitrification processes. At about the same time, a series of studies at the Oceanic Institute in Hawaii reported on the growth enhancement effects of factors in pond water from intensive culture systems (Leber & Pruder 1998; Moss 1995; Moss & Pruder 1995). Commercial farms began to adopt and refine these technologies, and shrimp operations such as Belize Aquaculture confirmed the commercial scale production potential (Boyd & Clay 2002). This increasing body of literature illustrated opportunities for intensification of production while reducing water exchange and improving production efficiencies.

During the late 1990s, there was an increase in awareness of potential negative implications of water exchange and habitat alteration caused by aquaculture. Increasingly limited water resources for a growing human global population and problems with effluent discharge from fresh water and brackishwater aquaculture have been major concerns (Hopkins *et al.* 1995a). These concerns were not only external pressures from regulatory and nongovernmental organizations, but they increasingly arose from within the aquaculture industry itself. Examples of unregulated regional aquaculture overdevelopment demonstrated the potential negative effects of a farm receiving the effluent of neighboring farms. One of the most significant potential consequences is the concentration of pathogens and their spread from one operation to another. Disease epizootics and interest in biosecurity have been the major drivers toward adoption of reduced or zero exchange technologies for all or part of the production cycle, particularly in shrimp farming (Browdy *et al.* 1997; Lotz 1997; Stanley 2000). Adoption of low water exchange, intensive practices for land-based shrimp and tilapia culture, has yielded several distinct advantages for farmers.

- Expenses are reduced as the cost of aeration is typically lower than the cost of exchanging water, and contributions of natural productivity can reduce feed costs. Intensification and reduced water use can minimize land and water costs. For marine shrimp, reduced reliance on water exchange can provide opportunities for production in areas away from sensitive and costly coastal land.
- Operations are environmentally sustainable as water use and effluents are reduced while allowing production to be intensified, thereby reducing ecological footprints per kg product produced.
- Health is enhanced by improving biosecurity and control over introduction of pathogens while cultivating a diverse microbial community, which may improve competitive exclusion of dangerous pathogens.

With growing consumer demands for high-quality tilapia and shrimp products, increasing pressures on producers to reduce production costs, and growing efforts to assure environmental responsibility, interest in biofloc-based production technologies continues to grow.

12.1 Bioflocs

In intensively fed aquaculture systems with reduced water exchange, high inputs of nutrients support the establishment of dense microbial communities. Fish and crustaceans use only a limited portion of the nitrogen and carbon offered in feeds for growth and metabolism. Estimates for average recovery of organic carbon, nitrogen, and phosphorus by shrimp and fish are approximately 13%, 29%, and 16% of feed content, respectively (Avnimelech & Ritvo 2003). The remainder enters the culture system either as uneaten feed or is excreted as metabolic wastes. In some recirculating aquaculture systems, the formation of a microbial community is encouraged on artificial substrates outside the animal culture area in, for example, commercially manufactured bead filters. It is this microbial community that is responsible for cycling excess nutrients. In such systems, particulate matter is often removed by external filtration such as sedimentation, vortex devices, and sand filters. However, in biofloc systems, particles are allowed to form within the culture system and a portion of the microbial community responsible for nutrient cycling is contained within those particles.

Depending upon the intensity of feed inputs, the algal community in a system without water exchange will undergo logarithmic growth, eventually reaching a plateau due to light limitations. During this period, which can last for days in tilapia culture and up to ten weeks in shrimp systems depending on feed inputs, production performance will largely depend upon the composition of the algal community and management of associated pH and dissolved oxygen fluctuations. At this point in the production cycle, pond and tank systems typically undergo a shift from a photoautotrophically dominated community to a more bacterial dominated community. With appropriate mixing and aeration algae, bacteria, zooplankton, feed particles, and fecal matter remain suspended in the aerobic water column and naturally flocculate together, forming the particles that give biofloc culture systems their name (fig. 12.1).

These flocs are held together by physicochemical forces of attraction and a matrix of polymers composed of compounds such as polysaccharides, proteins, and humic complexes (Avnimelech 2009). The floc particles are a diverse mixture

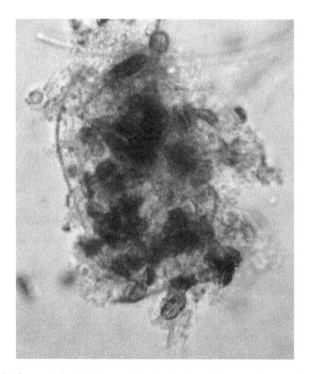

Figure 12.1 A biofloc particle at 100x magnification. This particle contains several algal species and likely contains multiple bacteria and fungi genera. These microbes are grazed on by zooplankton. Shrimp and tilapia may be able to consume these particles and gain nutrition from them, thereby reducing feed costs.

of microorganisms and particulate matter, and they can vary in biochemical composition and physical properties depending upon the type of feed used, target crop, type of aeration, management protocols, physical and environmental factors, and temporal and spatial variables that can change based on site, season, and factors associated with colonization (De Schryver *et al.* 2008; Ray *et al.* 2009). Biofloc communities can have fatty acid profiles distinctly unique from the feed administered to culture systems (Johnson *et al.* 2008), indicating that the microbial community is responsible for some biochemical alterations. Similar particles occur in nature and are often referred to as marine snow (Alldredge & Silver 1988). A considerable amount of research in natural systems has also been conducted on biofilm microbial communities (Costerton *et al.* 1995) and biofloc may also fall under such a description.

Bioflocs themselves represent an interesting ecosystem. The water is microbiologically a limited resources environment, having few nutrients or available organic substrates. Bioflocs are nutrient rich microenvironments embedded within nutrient-poor water. This nutrient density attracts organisms such as protozoa, nematodes, ciliates, and others that graze in or around the biofloc. The biofloc external polysaccharides (EPS) coating adsorbs detritus and free-living microorganisms, adding to its nutritious value. Bioflocs also provide the substrate required by most bacteria and can supply some refuge from predators (De Schryver *et al.* 2008). Biofloc systems contain bacteria, algae, zooplankton, fungi, and viruses. Each group contains taxa that may have positive effects and taxa that can contribute to negative consequences for fish and shrimp culture. Some bacteria, such as several species belonging to the genus *Vibrio,* are known shrimp and fish pathogens and have been historically problematic for aquaculture. Although *Vibrio* spp. occur in biofloc systems, research is still underway to determine whether they pose any risk to target crops. Other bacteria are highly beneficial to biofloc systems. Nitrifying bacteria ultimately transform toxic ammonia to the relatively nontoxic nitrate compound. Many heterotrophic bacteria can directly assimilate ammonia-nitrogen, thereby removing it from the water column. These valuable bacterial processes will be discussed in following sections.

Various forms of micro-, and occasionally macro-algae can be found in biofloc systems. Macro-algae can, at times, be seen growing near the surface or on structures in the water but is quickly grazed by the shrimp or fish if they have access to it. Micro-algae are found in biofloc particles and as free-living cells. Small chlorophytes (green algae), such as *Nanochloropsis* sp., are frequently concentrated within biofloc. It is unclear whether chlorophytes offer any nutritional value, but like most algae, they can assimilate ammonia-nitrogen to make cellular proteins and they photosynthesize in the presence of light.

Diatoms contain relatively high levels of the essential fatty acids eicosapentaenoic acid (EPA) and docosahexaenoic acid (DHA; Volkman *et al.* 1989), providing a potential nutritional advantage for culture animals. Diatoms are found both in and outside of biofloc particles and have been implicated in improving shrimp growth (Ju *et al.* 2009). Other algae such as cryptophytes and chrysophytes are seen occasionally. Algal communities seem to change in abundance and composition over time in biofloc systems and the causes of such changes are often unclear.

Potentially toxic algae can occur in any aquatic system and biofloc systems are no exception. Harmful algal bloom (HAB) taxa such as *Pfiesteria piscicida* have been identified in biofloc systems at WMC, but no negative effects on shrimp, fish, or humans were detected. These small, free-swimming heterotrophic dinoflagellates are uncommon in biofloc systems. A more common group of potentially harmful algae are the cyanobacteria, also known as blue-green algae. There is evidence that this group has hindered shrimp growth in aquaculture systems (Alonso-Rodriguez & Paez-Osuna 2003; Ray *et al.* 2009). Many of the cyanobacteria in biofloc systems such as *Synechococcus* spp. are pico-size (<2 μm) and seem to be contained primarily within biofloc particles. Removing a portion of the biofloc may increase light penetration and reduce cyanobacteria abundance, helping to select against cyanobacteria and favor more beneficial algae (Ray *et al.* 2009).

Zooplankton are an important assemblage in biofloc systems, as they consume both bacteria and algae and may then be consumed by shrimp and fish. Diverse arrays of zooplankton are found in biofloc systems. Free-swimming organisms such as ciliates and micro-flagellates can be seen feeding on biofloc particles. Rotifers and nematodes are commonly observed within the biofloc, consuming the flocculated material.

The size of biofloc particles may be important for animal nutrition, as large particles are likely more accessible to adult fish and shrimp. Most systems contain easily visible bioflocs in the range of mm fractions up to a few mm. Moss and Pruder (1995) demonstrated improved shrimp production when particles were above 5 μm in diameter. The size of biofloc particles can vary between culture systems. It seems that systems with more water pumping activity are prone to having smaller particles due to the severing action of pump impellers. Systems that rely more on airlift mechanisms or standard aeration systems are more likely to have larger particles. Not only might particle size affect whether animals can acquire them, but if particles are to be removed from the system their sizes may dictate what removal technique is appropriate. Settling containers can be used to remove a portion of the biofloc particles if the particles are large enough; however, foam fractionators may be required for smaller particles.

Although biofloc particles are advantageous to fish and shrimp, in the most intensive systems particle concentrations can build to very high levels. Due to intensive feed inputs, over 1,000 mg TSS/L is not uncommon if left unmanaged. It is unclear what the most beneficial concentration of biofloc particles is, but research has shown that managing the concentration is advantageous for intensive shrimp (Ray *et al.* 2010) and tilapia (Rakocy 1989) culture. As described below, high concentrations of biofloc particles imply elevated concentrations of the respiring organisms associated with them. Too much biofloc may increase the oxygen demand, increasing aeration/oxygenation costs and eventually reaching an unsafe level that is stressful to the culture animal (Beveridge *et al.* 1991). An excessive abundance of particles may lead to gill clogging, thereby preventing adequate gas and ion exchange by the culture species (Chapman *et al.* 1987). Having too many particles can shade the water column and promote the occurrence of harmful algae while diminishing the abundance of more beneficial taxa (Brune *et al.* 2003; Hargreaves 2006). Managing biofloc concentration may reduce the age of the microbial community, promoting a younger and more nutritious assortment of organisms (Turker *et al.* 2003). Furthermore, the organisms that assimilate ammonia-nitrogen into their cellular structures must, at some point, be removed from the system or that nitrogen will return to the water and pose risk to the culture animals.

12.2   Oxygen dynamics

Dissolved oxygen (DO) is always one of the most critical water quality parameters to be monitored in the cultivation of aquatic organisms. Culturing organisms in biofloc systems requires that close attention especially be paid to DO. In addition to the oxygen requirements of the shrimp or fish being cultivated, the rich microbial community also consumes DO at a significant rate. The intensity of DO consumption by the microbial community is largely a function of feed inputs required for the particular stocking density (Boyd 2009). A growout study of the Pacific white shrimp, *Litopenaeus vannamei*, conducted at the WMC found that the microbial biofloc contributed significantly to the oxygen demand of the system (fig. 12.2). The shrimp stocking density was approximately 500 shrimp per m^3 in a greenhouse-enclosed raceway measuring 235 m^3. Microbial oxygen demand exceeded shrimp oxygen demand for the first third of the growout period and still required 35 to 40% of the total oxygen demand by the end of the trial (Leffler *et al.* 2010). By the end of the growout period, the total oxygen demand for this biofloc system was approximately 5.6 mg O$_2$/L/hr. It is essential that the system have an emergency backup system to supply oxygen in the event of loss of the primary oxygen source. Other studies at the WMC have found the microbial oxygen demand to approximately equal that of the shrimp throughout the course of the growout period. Tilapia culture in biofloc systems is associated with higher oxygen consumption by the fish, since fish biomass is in the order of 10 to 30 kg/m^3 (Avnimelech 2009).

The simple equation for oxygen required by cultivated organisms is that of aerobic respiration in general:

$$C_6H_{12}O_6 + 6\ O_2 \rightarrow 6\ CO_2 + 6\ H_2O + \text{energy} \tag{1}$$

This is true of the multicellular target species as well as the single-celled bacteria, fungi, algae, and micro-invertebrates that comprise a biofloc community. Biofloc communities contain photoautotrophic, chemoautotrophic, and heterotrophic microbes. Which of these dominates depends upon the amount of organic loading entering the system through the feed inputs, other organic carbon inputs, and removal or cropping of biofloc particles from the system. At low densities,

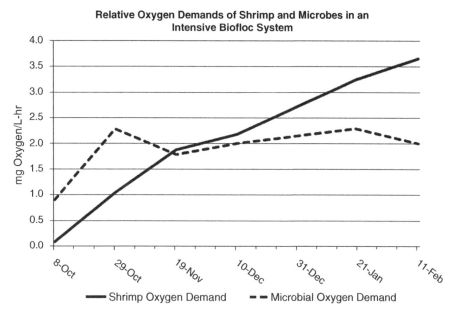

Figure 12.2 Relative oxygen demands of shrimp and microbes in an intensive biofloc system.

relatively little feed input will be required. This means that relatively less un-eaten feed, less feces, and fewer microbes will be available to form biofloc solids. Under such conditions, given sufficient dissolved inorganic nutrients, it is possible for photosynthetic communities dominated by chlorophytes and diatoms to flourish. The abundant photoautotrophic organisms in these situations can supply considerable DO to the system during daylight hours in accordance with the general equation for photosynthesis:

$$6\ CO_2 + 6\ H_2O + \text{solar energy} \rightarrow C_6H_{12}O_6 + 6\ O_2 \qquad (2)$$

The DO supplied by photosynthetic components may be sufficient to fulfill the requirements of the target species and biofloc during daylight hours, potentially remaining above critical levels through the night if stocking densities are low. However, even under these circumstances, and particularly at high stocking densities and feed loadings, DO will have to be mechanically supplied at night and during periods of prolonged cloudiness. As stocking densities and hence nutrient loading rates via the feed increase, biofloc systems become increasingly dominated by bacteria and photoautotrophs decrease in importance. According to Brune (2010), with feeding providing organic-C loading rates below 9 g C/m²-d and full sunlight, a photoautotrophic community is likely to dominate. However as organic-C loading rates exceed 12.5 g C/m²-d, biofloc algal populations decrease and the system becomes increasingly dominated by bacteria.

A great advantage of biofloc systems is the efficiency with which the microbes can process noxious nutrients released by the target species. This permits increasingly high stocking densities resulting in potentially attractive economic results. However, as stocking densities rise and greater quantities of feed are added, the DO demand of both the target species and the biofloc microbes increases. In most pond systems for shrimp and tilapia, aeration can be provided by paddlewheel or aspirator aerators that also provide for mixing and resuspension of organic material as described further on. Aeration rates can be generally estimated based on stocking densities or calculated as a function of feeding rates. It is estimated that yield can be raised by about 375 kg for each hp aeration capacity (Boyd 2009), but reports of higher energy efficiencies up to 1,000 kg/hp are available (Avnimelech 2009).

Most commercial shrimp biofloc ponds have an area in the range of 1,000 to 20,000 m^2 (0.1 to 2 ha). Tilapia ponds are usually smaller, having an area of 100 to 2,000 m^2. Paddle wheels and aspirator aerators are used to both aerate and mix the ponds. Aeration capacity installed is in the range of 20 to 60 hp/ha for shrimp ponds and more than 100 hp/ha in tilapia ponds where fish biomass is in the range of 200,000 kg/ha.

In shrimp tank and raceway systems, shrimp densities of about 300/m^3 can generally be supplied with oxygen by blowers and airstones alone. At a stocking density of 450 shrimp/m^3, Samocha (personal communication) has demonstrated that a shrimp biomass load in a biofloc system can be maintained up to 7.5 kg/m^3 with aeration. This was the result of careful monitoring, management of biofloc particle concentration, and only occasional pure oxygen supplementation for 30 to 60 minutes after each feeding. Work at the WMC in raceway trials stocked with 1 g shrimp at 500 to 900 shrimp/m^3 and reusing biofloc water from previous runs, found it necessary to supply pure oxygen continuously starting as early as two weeks into the sixteen-week growout period.

The physics of oxygen transfer into recirculating aquaculture systems and the engineering approaches for achieving it have been well reviewed by Timmons and Vinci (2007). When it is necessary to introduce pure oxygen into a biofloc system, it can be supplied by an electrical oxygen generator, by compressed oxygen, or by liquid oxygen. Liquid oxygen, if available and reasonably priced, is generally preferable. For a commercial scale, high-density tank or raceway operation, liquid oxygen is more economical than compressed oxygen and can be supplied in controlled amounts as needed (e.g., immediately after feedings). An oxygen generator can be economical but requires an automated backup electrical generator in case of power failure. Oxygen can be introduced into the water through a variety of devices such as aeration cones (downflow bubble contractors), diffusion aerators (airstones), and oxygen injectors (Venturis). A Venturi design is generally economical and efficient, often taking advantage of water already being pumped for another purpose such as circulation or heating. As described below, biofloc systems must be well mixed to maintain particles in suspension. If the system is well mixed, oxygen can be injected in only a few locations and still maintain good dissolved oxygen levels throughout the system.

As shown in equation 1, for every mole of oxygen consumed through aerobic respiration, a mole of carbon dioxide is produced. Thus a super-intensive, biofloc system that may require 5 mg O_2/L/hr is also producing every hour an equal molar quantity of carbon dioxide that can accumulate in the water. Carbon dioxide is toxic to most aquacultured organisms because it reduces the capacity of blood and hemolymph to transport oxygen. While high concentrations may not be lethal to a particular species, they may impact the activity, growth rate, and disease resistance of the cultured organisms. The sensitivity of different species to dissolved CO_2 varies greatly and many are able to acclimate to elevated levels if the increase in concentration is gradual (Timmons & Ebeling 2007). Under low stocking densities when aeration is sufficient for maintaining oxygen, the buildup of CO_2 can be offset by physical degassing into the atmosphere. However, at the high animal biomass and high microbial oxygen demands that are found in super-intensive tank or raceway biofloc systems, oxygen must be injected directly into the water. Under these circumstances CO_2 may not be completely degassed and may accumulate to unhealthy levels. It is possible for DO levels to be sufficiently high within a system while CO_2 concentrations are also high. In a super-intensive shrimp raceway at the WMC, CO_2 levels of 80 mg/L have been measured at the same time that DO levels were maintained consistently between 6 and 7 mg/L. Because of the high metabolic activity attributable to microbes and cultured organisms, and because this demand requires direct oxygen injection, super-intensive biofloc systems may be especially prone to CO_2 accumulation.

Carbon dioxide dissolved in solution also affects the chemical equilibrium:

$$CO_2 + H_2O \leftrightarrow H_2CO_3 \leftrightarrow H^+ + HCO_3^- \leftrightarrow 2H^+ + CO_3^{2-} \tag{3}$$

CO_2 production thus drives the equilibrium to the right, resulting in decreasing pH (i.e., an increase in the H^+ ion concentration). The decreasing pH may negatively impact the growth of the cultured species and should be monitored and chemically adjusted if necessary. Carbon dioxide accumulation both dissolved in the water and in the air above the biofloc system may become a problem especially in enclosed structures. This may cause human health concerns for workers in a poorly ventilated structure with super-intensive biofloc systems. Since CO_2 is heavier than air and will flow downward and away from slightly raised aquaculture containers, ventilation around the base of a greenhouse or building should allow the escape of the CO_2 generated by the system and reduce the buildup of gas in both the water and the air above it.

12.3   Resuspension, mixing, and sludge management

An intrinsic feature of aquatic systems is the sedimentation of particles from the water column to the bottom. Aquatic systems are enriched with respect to organic particles (dead algae, feed residues, etc.) at a rate proportional to the organic loading of those systems. In aquaculture, this enrichment rises with the intensity of the system. The settling of organic particles to the substrate can create an enriched organic sediment layer, accompanied by a high sediment oxygen demand (SOD).

In ponds, diffusion of oxygen from the atmosphere through the water column and to the pond bottom is a slow process. In addition, oxygen is consumed as it diffuses down through the water. As a result of the typically high SOD and slow oxygen supply, anaerobic conditions can develop in the sediment layers of lakes, rivers, and aquaculture systems.

Anaerobic microbial processes are appreciably less efficient than those in aerobic, oxygenated systems, and thus, recycling of the residues in the sediment layer is slow compared to processes in the aerated water column. In addition, anaerobic microbial metabolism leads to the release of toxic reduced metabolites such as sulfides, reduced organic sulfur compounds, organic acids, and ammonia (Avnimelech & Ritvo 2003). The development of an anaerobic bottom layer was found to be the major limitation to increasing production in carbohydrate-fed tilapia ponds equipped with nighttime aeration (Avnimelech *et al.* 1994).

To ensure adequate nutrient cycling and further increase production, the accumulation of anaerobic bottom sludge has to be prevented. This can be achieved, as indicated in a pioneering study by Hopkins *et al.* (1994), by either removing the sludge or resuspending it. In this study, sludge removal remediated 67% of the nitrogen added to the system in feed, resulting in improved water quality. However, removal of this material leaves significant logistical and disposal problems. Discharging sludge with targeted water exchange is restricted or prohibited by environmental and biosecurity regulations and entails high pumping and/or disposal costs.

In typical recirculating aquaculture systems (RAS) the suspended residues are continuously recycled through a series of water treatment components, and the sludge is collected and dumped. An alternative method is to continually resuspend particles and prevent the buildup of anaerobic sediment, keeping the particles within the aerated water body as long as possible. This is the approach taken in biofloc systems where the organic residues are aerobically metabolized, leading to an efficient food chain and recycling, while preventing the production of toxic anaerobic metabolites. Hopkins *et al.* (1994) demonstrated that by leaving sludge in the pond and resuspending it, much of the nitrogen that entered the pond as feed could not be accounted for at the end of the production cycle, indicating volatilization from the system. It is important to note that even the existence of a few anaerobic sludge piles in the pond may affect fish or shrimp growth and water quality (Avnimelech 2009). Under anaerobic conditions hydrogen sulfide (H_2S), which is extremely toxic, can be formed. Even low concentrations of H_2S hinder nitrite oxidation, the second stage of nitrification, thereby resulting in incidental elevated nitrite concentrations when water mixing and resuspension are not efficient.

In the previous section, aeration and oxygenation to meet metabolic demands of target crop and the microbial community were discussed. One approach to reducing the microbial oxygen demand is to aggressively crop the biofloc solids from the system. This can be accomplished in larger ponds by exchanging water to flush solids from a central drain, or in more intensive tank systems by use of settling chambers, filtration, or foam fractionation. However, care must be exercised so as not to remove so much of the biofloc that the processes of waste assimilation and nitrification are decreased. Effective solids management has been shown to increase shrimp growth and productivity in biofloc systems (Ray *et al.* 2010a).

Aeration or supplemental oxygenation are essential to achieve super-intensive animal densities. However, aeration is also employed to achieve several related goals:

- Supply oxygen to the animals to overcome oxygen limitations and thus enable higher stocking densities, growth, and yields
- Distribute the oxygen in the system horizontally and vertically
- Mix water and sediment—water interface
- Control sludge coverage, location, and drainage

It is important that aeration should be designed, deployed, and operated so as to achieve each of these goals, as well as supply oxygen.

In pond systems, achievement of these goals depends upon proper selection and planning of aeration capacity, aerator type, aerator location, and operation mode as well as appropriate pond design. Ponds should be designed to allow for optimal drainage of sludge, between or within growing cycles. In addition, since sludge production is difficult to avoid, the pond should be designed to limit the areal coverage of sludge and sludge depth. Liners are increasingly used in biofloc-based pond systems to improve water movement and control sludge buildup while facilitating sludge oxidation and removal between crops.

The placement of aerators and types of aerators are of prime importance in managing ponds and controlling sludge accumulation, particularly when water exchange is reduced or eliminated. A very common mode of aerator placement is to locate the aerators in parallel to the dikes. This, in turn, creates a peripheral flow near the dikes and an area with no flow, or very limited flow, in the center of the pond. The physical aspects of this radial or elliptical flow were thoroughly analyzed and reported by Peterson *et al.* (2001). Qualitatively, suspended particles settling in regions with fast water movement are resuspended back into the water. However, this is not the case in the central area of radial flow. In this area, water velocity is low, or nonexistent, and settled material is not re-suspended and thus settles and accumulates. A detailed analysis of this situation was conducted by Calle-Delgado *et al.* (2003). It was found that the center area of the pond was stagnant and poorly aerated. Sludge accumulated in this area and living conditions for shrimp were poor. It is important to minimize the size of the sludge pile and to design the pond for efficient sludge drainage. This can be achieved by sloping the pond bottom toward an appropriately sized drainage opening (in the center for a radial flow pattern) and by properly placing and choosing the aerators so as to mix most of the pond water.

For super-intensive tank or raceway-based systems, it is also essential that the biofloc solids remain suspended in the water column. Water within intensive tank or raceway-based biofloc systems is moved by airstones, airlifts, pumps, or a combination of these to create either upwellings or a gentle current to maintain the particles in suspension. At very high fish or shrimp densities, bioturbation by the target crop itself contributes to biofloc resuspension.

Excessive suspended matter and pond bottom sludge must be removed since it consumes large amounts of oxygen and at very high concentrations may clog gills. Settled sludge induces anaerobic processes and can release toxic anaerobic products (Avnimelech 2009). Removal of suspended solids is practiced in super-intensive shrimp culture systems and in some cases in tilapia production tanks (Rakocy 1989; Ray *et al.* 2010b) using sedimentation tanks. In tilapia ponds, the bottom water is drained one to three times daily, stopping when the water changes from turbid to clear. In cases such as ponds in Belize Aquaculture (McIntosh 2001), bottom sludge is drained by opening a valve connecting a sump in the center of the pond to the drainage canals leading to retention ponds.

Issues related to disposal of the sludge, drained pre- or post-harvest, need attention. The sludge contains high concentrations of nutrients, reactive organic matter, and offensive reduced components. Sludge from fresh water systems can be used as a high-value soil amendment. Sludge from marine shrimp ponds on the other hand, contains salt, which precludes most land-based application. However, shrimp biosolids from a WMC pond have been used successfully as fertilizer for more salt tolerant plants like broccoli and bell peppers (Dufault & Korkmaz 2000; Dufault *et al.* 2001). Sludge can be used to produce biogas in anaerobic reactors. In cases where the pond is situated near extensive ponds, the daily drained sludge can be transferred to the extensive pond and can serve as a fish feed supplement, as practiced in Israel. More work is needed to develop methods and regulations to make sludge disposal or reuse more sustainable.

12.4   Nitrogenous waste products

An intrinsic feature of intensive aquaculture systems is the buildup of inorganic nitrogen in the water. Fish and shrimp feeds contain high levels of protein (20 to 45%). Approximately 70 to 75% of the protein nitrogen in feed provided to the cultured organisms is released into the water either as uneaten feed to be broken down by microbes or as metabolic waste products. This nitrogen dissolves into the water as total ammonia nitrogen (TAN). Equilibrium is established in the water between un-ionized NH_3 and ionized NH_4^+ dependent upon pH, salinity, and temperature. Ammonia is lethally toxic to most organisms and even low concentrations can retard growth. Thus, its concentration must be kept as low as possible.

As an example, a fishpond holding 500 g fish/m^2 is fed (2% fish weight per day) with 10 g feed of 30% protein/m^2-day (i.e., 3 g protein or 480 mg N/m^2-day). Excretion of 75% of this amounts to about 360 mg N/m^2-day. For a 1-m-deep pond this amounts to a daily build up of 0.360 mg N/L-day.

The nitrogen accumulation is ten times higher for a pond holding 5 kg fish/m^2. Biofloc systems are ideally suited to deal with ammonia because the rich microbial community can take up excreted TAN rapidly and prevent its concentration from rising to dangerous levels. Three distinct groups of microbes can remove TAN through different processes: (1) assimilation by photoautotrophs (algae and cyanobacteria), (2) assimilation by heterotrophic bacteria, and (3) nitrification by chemoautotrophic bacteria. Typically, all three are present and active to varying degrees in biofloc systems depending upon stocking densities, stage of culture, system designs, and especially management strategy. As production systems are intensified, control of nitrogenous waste becomes increasingly essential.

Usually, algal dominance is found in newly stocked ponds and tanks where the concentration of substrates is too low to support bacterial dominance. Photoautotrophic biofloc systems have been used successfully to manage nitrogen wastes (Brune *et al.* 2003), but they require relatively large spatial areas and low stocking densities.

At higher feed loading rates that accompany higher stocking densities, the heterotrophic assimilation and chemoautotrophic nitrification processes will be favored. Heterotrophy becomes dominant when daily feed addition becomes high (e.g., above 450 kg feed/ha with shrimp; Chamberlain *et al.* 2001). This transition takes place much faster in ponds stocked with tilapia due to the higher biomass and feeding. The intensive growth of bacteria limits algal activity due to reduced light penetration such that only a small part of the water column receives enough light for photosynthesis. It is possible to encourage photoautotrophic activity in biofloc assemblages by controlled removal of particles to increase water clarity. During the early phase of production algae take up ammonium and convert it to protein using energy obtained from photosynthesis. Once succession occurs from an algal dominated community to a more bacterial-based system, and especially if the system is fed with carbonaceous substrates, heterotrophic bacteria have a practically unlimited capacity to assimilate the inorganic nitrogen to build microbial proteins.

Heterotrophic assimilation is an important process in most biofloc systems. Heterotrophic bacteria and other microorganisms use carbohydrates (sugars, starch, and cellulose) as a food to generate energy and to grow:

$$\text{Organic C} \rightarrow CO_2 + \text{Energy} + \text{C assimilated in microbial cells} \qquad (4)$$

The percentage of the assimilated carbon with respect to the metabolized feed carbon is defined as the microbial conversion efficiency (E) and is in the range of 40 to 60%. Nitrogen is required as an important building block of the microbial cell. Thus, microbial utilization of organic carbon is accompanied by the assimilation of inorganic nitrogen. This is a basic microbial process and practically all microbial assemblages perform it.

The amount of carbohydrate supplement (ΔCH) required to reduce the ammonium can be calculated (Avnimelech 1999). Based on equation 4 and the definition of the microbial conversion coefficient, E, the potential amount of microbial carbon assimilation when a given amount of carbohydrate is metabolized (ΔCH) is:

$$\Delta C_{mic} = \Delta CH \times \%C \times E \qquad (5)$$

Where ΔC_{mic} is the amount of carbon assimilated by microorganisms and %C is the carbon contents of the added carbohydrate (roughly 50% for most substrates). The amount of nitrogen needed for the production of new cell material (ΔN) depends on the C/N ratio in the microbial biomass, which is about 4.

$$\Delta N = \Delta C_{mic}/[C/N]_{mic} = \Delta CH \times \%C \times E/[C/N]_{mic} \qquad (6)$$

And using approximate values of %C, E, and $[C/N]_{mic}$ as 0.5, 0.4, and 4, respectively:

$$\Delta CH = \Delta N/(0.5 \times 0.4/4) = \Delta N/0.05 \qquad (7)$$

According to equation 4 (assuming that the added carbohydrate contains 50% C), the CH addition needed to reduce TAN concentration by 1 mg/L N (i.e., 1g N/m^3) is 20 mg (20 g/m^3). This relationship enables a manager finding a high TAN concentration in the pond (following cloudy days, an algae crash, high animal biomass, etc.), to calculate how much carbohydrate substrate must be added to mitigate an otherwise dangerous situation. This mode of action may be considered an emergency, post factum mode. The manager reacts following the excessive rise of TAN or NO$_2$. While heterotrophic bacteria do not efficiently assimilate NO$_2$, the carbohydrate stimulation causes them to take up TAN, leaving less for chemoautotrophic bacteria to convert to NO$_2$.

A different, proactive approach is to add the right amounts of carbohydrate with the feed in order to prevent unwanted TAN increase and to optimize the process. Here, one has to estimate the amount of carbohydrate that has to be added in order to immobilize the ammonia excreted by the fish or the shrimp in real time. As mentioned, fish or shrimp in the pond assimilate only about 25% of the nitrogen added in the feed. The rest is excreted mostly as TAN (some as organic N in feces or feed residue). It can be assumed that the TAN flux into the water, ΔTAN, or generally ΔN, directly by excretion or indirectly by microbial degradation of the organic N residues, is at least 50% of the feed nitrogen flux:

$$\Delta N = \text{Feed} \times \%\text{N feed} \times \%\text{N excretion} \qquad (8)$$

A partial water exchange, sedimentation, or removal of sludge reduces the TAN flux in a manner that can be calculated or estimated. The amount of carbohydrate addition needed to assimilate the TAN flux into microbial proteins is calculated using equations 7 and 8:

$$\Delta CH(g) = [\text{Feed(g)} \times \%\text{N feed} \times \%\text{N excretion}]/0.05 \qquad (9)$$

For example: Assuming 30% protein feed pellets (4.65% N) and assuming that 50% of the feed nitrogen is excreted (%N excretion), we get:

$$\Delta CH(g) = Feed(g) \times 0.0465 \times 0.5/0.05 = 0.465 \times Feed(g) \qquad (10)$$

According to equation 10, the feed having 30% protein should be amended by an additional portion of straight carbohydrates amounting to 46.5% addition to the feed ration.

Another means of controlling TAN is to rely on nitrification. Nitrification is a two-step process in which one group of bacteria oxidizes TAN as its primary energy source, consuming bicarbonate ions as its carbon source, and producing nitrite ions as a by-product. Ammonia-oxidizing bacteria include the genera *Nitrosomonas, Nitrosococcus, Nitrospira, Nitrosolobus,* and *Nitrosovibrio* (Timmons & Ebeling 2007). Nitrite produced by this process can be more toxic than ammonia, altering the hemoglobin molecule to prevent oxygen uptake (Tomasso *et al.* 1979). A second group of bacteria oxidizes nitrite to nitrate to obtain energy, again consuming bicarbonate ions and dissolved carbon dioxide as carbon sources. These bacteria typically belong to the genera *Nitrobacter, Nitrococcus, Nitrospira,* and *Nitrospina* (Timmons & Ebeling 2007). Both groups of these chemosynthetic bacteria are obligate autotrophs and obligate aerobes. Typically, when a new biofloc system is started it will take some time for the nitrifying bacteria to become established. Because nitrite is produced as a product of the first step, there will typically be a lag between establishing the ammonia-oxidizing population and the nitrite-oxidizing population. Nitrate is produced as a product of the second process. It is relatively harmless at concentrations into the hundreds of mg/L and continues to accumulate in the system as nitrification proceeds. A typical example of the dynamics of the nitrogen compounds involved in nitrification is depicted in figure 12.3. As described previously, during the early ammonia and nitrite accumulation periods, dextrose or another simple carbohydrate source can be added to a biofloc system to dampen the spikes of these compounds. Stoichiometry of carbohydrate addition to stimulate the heterotrophic bacterial population has been discussed above. At the WMC, dextrose is used routinely to control ammonia and nitrite spikes in shrimp biofloc systems until the nitrification process becomes fully functional.

Both the ammonia-oxidizing and the nitrite-oxidizing steps of the nitrification process can be represented by the combined equation (Ebeling & Timmons 2007):

$$NH_4^+ + 1.83\, O_2 + 1.97\, HCO_3^- \rightarrow 0.0244\, C_5H_7O_2N$$

$$+ 0.976\, NO_3^- + 2.90\, H_2O + 1.86\, CO_2 \qquad (11)$$

As TAN is removed from the system to provide energy for the bacteria, oxygen and bicarbonate ions are also consumed. As a result, nitrifying bacteria contribute significantly to the oxygen demand on the system and especially to the consumption of alkalinity. In order to maintain the nitrification process as well

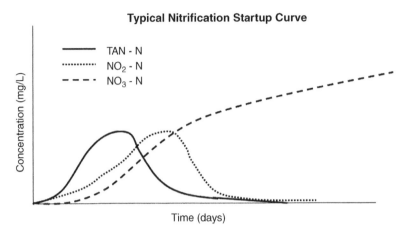

Figure 12.3 Typical nitrification startup curves.

as buffer against decreasing pH, alkalinity, generally in the form of NaHCO$_3$, must be added on a regular basis. Approximately 7.05 g of HCO$_3^-$ is required for every g N removed in order to maintain alkalinity. How much total alkalinity is required is a direct function of the nitrogen loading rate from the feed used to support a given stocking density. The nitrification process also produces nitrate and carbon dioxide. Nitrification is evident in a biofloc system if nitrate steadily increases throughout a production cycle (fig. 12.3). This is distinct from photoautotrophic and heterotrophic assimilation processes in which nitrate levels remain unchanged or decrease. Carbon dioxide production contributes to the CO$_2$ that is also generated by aerobic respiration.

These two management strategies each have advantages and disadvantages and may be more or less suitable to different biofloc production systems depending upon target species, stocking density, and so on. It is important to note that both nitrogen assimilation and nitrification take place in all biofloc systems. The difference among systems is the rate of use of added carbohydrates (or the equivalent use of low protein feed), opposed to use of carbohydrate additions only occasionally when TAN is too high. Table 12.1 compares the implications for water quality and economic considerations that result from the removal of 1 g of ammonia-nitrogen by the mechanisms of nitrification and heterotrophic assimilation.

Heterotrophic assimilation requires the addition of significant amounts of carbohydrate—approximately 15 g of carbohydrate for every gram of TAN removed (table 12.1). Since nitrifying bacteria obtain their energy from oxidizing TAN, no additional organic carbon is required to drive their metabolism. Although simple carbohydrates are relatively inexpensive, the quantities required may become an increasingly significant cost at higher stocking densities. Costs of added carbohydrates are partially offset in some intensive systems by the contribution of the flocs directly to the nutrition of the target crop resulting

Table 12.1 Comparison of stoichiometric balances for the removal of 1 g of ammonia-nitrogen by the mechanisms of nitrification and heterotrophic assimilation (from Ebeling et al. 2006).

per g of NH$_3$-N consumed	Nitrification (g)	Heterotrophic assimilation (g)
Carbohydrate consumed	0	15.17
Alkalinity consumed	7.05	3.57
Oxygen consumed	4.18	4.71
Bacterial Biomass produced	0.20	8.07
CO$_2$ produced	5.85	9.65
NO$_3$-N produced	0.976	0

in improved protein utilization. Nitrification consumes more alkalinity than heterotrophic metabolism, which requires bicarbonate ions to be added to the system, usually in the form of NaHCO$_3$, in order to maintain pH and support continued functioning of both mechanisms. Depending on feeding rates, the use of aerators and high CO$_2$ volatilization can offset the drop of alkalinity in many pond systems. Nitrification can also be slow to establish, resulting in the accumulation of intermediate products and the production of nitrate, which in intensive systems with complete water reuse between cycles, could eventually become problematic. Both processes require approximately equal amounts of oxygen; however, heterotrophic assimilation produces 65% more carbon dioxide per gram of TAN, which can be a significant problem in enclosed systems stocked at high densities.

A substantial difference between these two processes, however, is the quantity of solids produced by heterotrophic assimilation, approximately forty times that of nitrification. Driven by the large input of carbohydrates, assimilation sequesters TAN into bacterial biomass. If left in the system, a portion of the bacteria may be consumed by the culture species; the rest eventually die and release the TAN back into the water. Therefore, as stocking densities rise, the heterotrophic assimilation process must be coupled with a solids removal mechanism. This requires both additional effort and expense, and, once removed, disposal of the solids must be addressed. Nitrification produces relatively minor amounts of solids. Instead, it converts TAN ammonia-nitrogen into nitrate that is harmless to most cultured species even in relatively high concentrations. Where nitrate levels may possibly accumulate to levels of concern would occur in situations where the biofloc water is retained and reused for multiple growout runs that are stocked at very high densities.

In super-intensive shrimp biofloc raceway systems at WMC, nitrification is the dominant mechanism for effectively removing TAN even at stocking densities as high as 900 shrimp/m^3. Carbohydrate in the form of dextrose is only added as necessary at the start of a trial with new water to suppress initial ammonia and nitrite spikes while the nitrifying bacteria are becoming established. Once these populations are established, there are typically no further problems with ammonia and nitrite. When the biofloc water is retained and reused for the next growout run, no significant ammonia or nitrite spikes are usually observed. Nitrate levels increase during the course of each growout, although not as much as would be expected if nitrification was solely responsible for TAN remediation, based on nitrogen entering a system through feed. A nitrogen mass balance study conducted at WMC involved thirty-two outdoor tanks stocked at different densities with different levels of solids removal and different levels of nitrogen inputs. No carbohydrate was added to stimulate assimilation. Shrimp nitrogen conversion efficiencies averaged $27 \pm 2\%$ (mean $\pm$ SE). Cropping of solids reduced nitrate levels 18 to 44% depending on treatment. However, $29 \pm 4\%$ of nitrogen introduced to the tanks left the systems as volatilized ammonia or as nitrogen gas produced through denitrification or other reduction processes.

Under anaerobic conditions nitrate (NO_3^-) can be reduced to nitrogen gas (N_2) through denitrification, resulting in complete removal of nitrogen from the system. Denitrification is conducted by heterotrophic bacteria capable of utilizing NO_3^- as an electron receptor in the absence of O_2 leading first to the production of nitrite (NO_2) and ultimately to the production of N_2 as a byproduct (Wetzel 2001).

$$C_6H_{12}O_6 + 12\ NO_3^- \rightarrow 12\ NO_2^- + 6CO_2 + 6H_2O \qquad (12)$$

$$C_6H_{12}O_6 + 8NO_2^- \rightarrow 4N_2 + 2CO_2 + 4CO_3^{2-} + 6H_2O \qquad (13)$$

Biofloc systems are kept well aerated and mixed to avoid the development of anaerobic accumulations of organic material that might release toxic sulfides and other compounds. However, it has been speculated that denitrification might occur at significant levels in the oxygen-depleted interior of the biofloc particles suspended within the water column. This is a reasonable hypothesis to explain the observation that nitrate generally does not accumulate at the rate predicted by a simple mass balance of nitrogen inputs into a biofloc system. The anammox pathway might also occur in the anaerobic interiors of biofloc particles. This process bypasses a portion of the denitrification reaction by directly converting ammonia and nitrite to N_2 gas. If present, the anammox reaction and the denitrification process are probably complementary and occur simultaneously (Tal *et al.* 2009).

12.5 Temperature

Effects of temperature on bioloc-based systems are similar to those in traditional culture units. The greatest effect of temperature is its relationship with growth rate. In any aquaculture system, temperature is correlated with growth performance, as all cultured animal species are poikilothermic. This limitation constrains growing season in some areas, affecting the number and length of crop cycles in shrimp and the ability to overwinter tilapia. In tropical regions with high temperatures year round, the temperature in biofloc systems is similar to that of traditional systems. However, as one moves to more marginal temperate zones, water exchange can significantly affect temperatures. Thus, ability to culture without exchange has the advantage of allowing for more stable thermal conditions. In shrimp culture, one of the most virulent viral diseases, white spot syndrome virus, has been shown to be highly temperature dependent (Vidal *et al.* 2001). Subsequent to this discovery, management strategies for this disease include maintaining pond temperatures at 29°C and above. These strategies often necessitate avoiding water exchange and even enclosure of production ponds in greenhouses. Use of biofloc technologies allows growers to maintain pond temperatures by eliminating the need for water exchange.

As described above, continuing intensification of shrimp production in biofloc-based systems has resulted in tank and raceway culture systems that have reached production levels exceeding the equivalent of 100,000 kg/ha/crop in experimental systems (Otoshi *et al.* 2007). At these production levels, experimental systems are enclosed in greenhouses, which allow better control of temperature and improved biosecurity (Browdy *et al.* 2009). Li *et al.* (2009) modeled this type of system demonstrating significant gains in heat retention and providing a model to calculate supplemental energy needs to run these types of systems year round in various geographic locations. As discussed below, financial models then can be used to evaluate opportunities for year-round shrimp culture based on biofloc systems in temperate zones close to large markets (Hanson *et al.* 2009).

12.6 Feeds and feeding

Feeds and feeding are critical to any aquaculture operation. Feeds are the most significant part of variable costs of operating a biofloc system (Hanson *et al.* 2009) and feed performance is intimately related with target crop production success both in terms of direct nutrition and in terms of effects on water quality. Recent reviews have been published on current trends in shrimp and tilapia nutrition (Li *et al.* 2006; Lim & Webster 2006; Davis & Sookying 2009; Hardy 2009). Research has focused on efforts to reduce fish meal content in diets, as fish meal represents a limited resource with increasingly volatile price fluctuations. One of the most important trends in aquaculture nutrition research today is to explore the direct and indirect effects of inclusion of alternative protein sources in aquaculture diets. These effects can include changes in nutrient composition and availability, effects on gut health and gut microbial flora, problems with antinutritional factors, etc. When culturing animals in biofloc systems, the natural productivity within the system can make important contributions to nutritional balance. Numerous references indicate the growth-enhancing effect of pond water in general and of culturing fish and shrimp in biofloc systems in particular (Leber & Pruder 1988; Moss *et al.* 1992; Wasielesky *et al.* 2006). Burford *et al.* (2004), Bianchi *et al.* (1990), Avnimelech (2007), and Avnimelech and Kochba (2009) demonstrated for both shrimp and tilapia direct biofloc consumption using stable isotope techniques. Biofloc fatty acid composition has been documented at levels comparable to those found in commercial feeds (Tacon *et al.*

2002). Moss *et al.* (2006) showed that shrimp could be grown in biofloc systems using feeds without supplemental vitamins without significantly reducing growth, and flocs have been shown to contain important minerals and amino acids (Avnimelech 2006; Tacon *et al.* 2002). Thus, biofloc can provide important essential nutrients either to enhance performance when using a complete feed or to allow new directions for reduced cost formulations by taking advantage of floc nutritional contributions. Burford *et al.* (2004) showed that nitrogen conversion efficiencies from feed sources might be increased 18 to 29% in shrimp biofloc-based systems. Avnimelech showed increased protein uptake efficiency in tilapia from 25% to about 50% in biofloc systems. The assimilation of waste nitrogen by the microbial biofloc and reingestion by the target crop can significantly enhance conversion efficiencies, improving environmental sustainability and potentially increasing profitability.

In addition to affecting shrimp or fish growth, variations in protein source or protein levels affect digestibility of the feed and hence feed utilization. In biofloc systems these factors take on increasing importance as waste material is either mineralized or assimilated within the system. Clearly, as stocking densities increase above the most extensive levels, feed inputs become the most important drivers of processes within the system. As discussed above, meeting oxygen demand both by target crops and by the microbial community in biofloc systems is one of the most important aspects of systems management.

Models have been developed demonstrating the direct relationships between feed inputs and aeration requirements in shrimp production systems (Hopkins *et al.* 1991; Boyd 2009). Efficiency of feed utilization has a direct effect on biofloc density, microbial oxygen demand, and sludge production. Two strategies can be found in the literature for managing feed inputs in biofloc systems. One strategy focuses on use of nutrient dense high-protein feeds with highly digestible ingredients offered with feeding strategies, which emphasize control of feed conversion ratios (Kureshy & Davis 2002; Browdy *et al.* 2009). The goal is to provide just enough feed to remain slightly below maximum target crop demand. A second strategy suggests addition of lower protein feeds or mixtures with grain-based supplements to encourage heterotrophic biofloc production and assimilation of waste (Avnimelech 1999; McIntosh 2001; Ebeling *et al.* 2006). Selecting a feed formulation strategy can depend upon the particulars of the biofloc system. For example, when rearing shrimp in ponds at lower densities in low salinity systems, exigencies of ammonia and nitrite toxicity can outweigh dangers from high biofloc densities as long as buildup of anoxic organic material is prevented. In this case, lower protein feeds or higher carbon inputs should be considered. On the other hand, as shrimp rearing densities are increased in higher salinity systems, controlling biofloc densities and managing waste buildup is critically important. In this case, with greater system resiliency in terms of ammonia and nitrite toxicity levels, due to higher salinity, and greater reliance on nitrification to control ammonia and nitrite levels, higher protein feeds and emphasis on feed utilization efficiencies can result in fast target crop growth and minimal waste production. Similarly, culture and feeding of tilapia can be quite different from that of shrimp, as the fish more efficiently consume particles from the biofloc in the water column. For any specific application of the technology, applying an understanding of the dynamic relationships between feeding behavior and feed utilization efficiencies and target crop and microbial community oxygen demand is critical to designing and managing biofloc system efficiencies, particularly at higher densities.

12.7 Economics

Biofloc-based systems can have significant advantages in terms of economic returns when compared to conventional systems, which rely on water exchange. In terms of variable costs, as early as the mid-1990s it was suggested that aerating ponds can be more cost effective than pumping water to maintain dissolved oxygen levels (Hopkins *et al.* 1995b). Use of aeration also allows for increased stocking densities and intensification of production. Perhaps the most important driver of profitability in shrimp culture is maintaining a high survival rate. In sensitivity analyses of shrimp culture in earthen ponds in Texas, it has been estimated that 90.9% of total variation can be assumed by variation in survival (Moss & Leung 2006). Thus it is not surprising that the major driver of conversion to biofloc technologies is disease control. In most shrimp culture systems, management strategies prioritize reducing losses from opportunistic bacterial infections or from excludable pathogens such as viruses. The reduction of water exchange enabled by biofloc technologies and more diverse microbial communities in well-managed biofloc systems, enables biosecurity and reduces opportunities for dominance by harmful microbes, thus driving the system toward more stable survival.

Growth rate and feed conversion efficiency also play important roles in total cost variations in both pond-based and intensive tank or raceway-based shrimp culture systems, representing 23.1 and 19.6%, respectively, for systems analyzed in the United States (Moss & Leung 2006). These authors found that, based on more stable survivability estimates and the harvest of multiple crops per year, a hypothetical super-intensive system for culture of shrimp in biofloc-based raceway systems was found to be a more economically viable alternative than earthen pond culture of penaeid shrimp in the United States. In most major shrimp-producing countries, feed represents the highest operating cost item ranging from a low of 25% to a high of 45% of total operating costs (Tacon *et al.* 2006). Opportunities to better control feed usage and improve feed conversion ratios represent a key component in controlling variable costs. It remains to be seen if feed formulations can be modified to reduce feed costs by leveraging contributions from natural productivity in biofloc systems without affecting long-term survival and growth.

In an analysis of an economic model for super-intensive biofloc tank-based shrimp production Hanson *et al.* (2009) compared the effects of varying biological parameters on net present value (NPV) and internal rate of return (IRR). These authors found that the greatest effects on cost efficiency and returns came from improvements in survival (20% improvement increased IRR by 97%). A

20% increase in stocking densities or growth rate increased NPV by 57 and 45%, respectively. On the other hand, a 20% decrease in feed cost or cost of post-larvae only reduced NPV by 22 and 9%, respectively. Clearly, investments in infrastructure, seed, and feed that can improve survival, growth rate, and stocking densities can have strong, positive impacts on economic returns.

12.8   Sustainability

Sustainable development requires consideration of environmental resource management, social factors, and economic factors. In planning for future sustainable aquaculture development, continued growth, and expansion of the sector must be taken into account. According to FAO projections, there is a need to increase aquaculture production fivefold by the year 2050. This huge expansion must be accomplished in a sustainable way. However, until recently, aquaculture development planning did not include enough sustainability considerations, especially during the period of 1980 to 2000, when shrimp production was considered to be highly profitable and new shrimp ponds were constructed, in some cases with little regard for the environment. The expansion of shrimp ponds in some areas resulted in mangrove destruction; the unregulated release of effluents into rivers, lakes, estuaries, and enclosed marine regions; and the eutrophication of receiving bodies. These developments led to harsh objections from environmental groups and concerned scientists, opposing further unsustainable development and eventually leading to increased environmental regulations and development of certification programs based on best management practices.

A major issue related to the effect of aquaculture on the environment is the release of pollutants with the drained water. Culture systems are fed and often fertilized, leading to relatively high concentrations of nutrients such as nitrogen and phosphorus. Emissions of water into receiving bodies may raise nitrogen and phosphorus concentrations to levels that endanger water quality. Most aquaculture systems also contain high concentrations of organic matter, both soluble and particulate. These compounds are biologically degradable, consuming oxygen as they break down, and often causing anoxic or hypoxic conditions in the water and sediments around the effluent discharge. Concerns have also been raised regarding emissions of pathogenic microorganisms that may impact the natural biota in the receiving water, especially if the pathogens emitted are not normally found in the receiving water. In such a case, the natural biota could have a low resistance toward those pathogens (e.g., Hopkins *et al.* 1995a). Traditional aquaculture systems may exchange a high percentage of their water with the environment and thus potentially endanger the surrounding environmental quality. A production system that does not release large volumes of water to the environment is, in this respect, a potentially more environmentally friendly system.

Another aspect of aquaculture sustainability is the optimal utilization of natural resources, mostly land and water. FAO predicts a fivefold increase in aquaculture production but there is not enough water and land available to raise production fivefold using traditional techniques. A conventional pond, producing 2,000 kg of fish/ha and losing 35,000 m^3 of water per year by evaporation and seepage—and an additional 10,000 m^3 with the water drainage—will consume 45 m^3 of water to produce 1 kg fish, and the production rate per unit area will be 0.2 kg/m^2. In contrast, the water consumption in intensive, zero, or limited water exchange systems is less than 1 m^3 per kg fish production and the productivity of the pond area is in the range of 10 to 100 kg fish/m^2. Using such systems relieves water or land limitations to further production of fish, making more optimal usage of land and water.

An additional environmental issue that can constrain future aquaculture development and reduce economic viability is the dependence of aquaculture on fish products originating in the marine environment. Fish meal and fish oils are common components of fish feed. Marine finfish feeds were traditionally made of about 50% fish meal, marine shrimp feeds contained about 30% fish meal, and tilapia feeds about 15%. According to estimates published by Naylor *et al.* (2000), approximately 5 kg wild fish were needed at that time to produce 1 kg of carnivorous marine fish (Naylor *et al.* 2000). However, much less (<1 kg) is needed to produce omnivorous carps. The dependence upon wild fish for feeds is potentially an important factor endangering marine ecology due to the degradation of fish populations in many marine regions. Furthermore, uncontrolled use of marine fish in aquaculture diets may introduce marine contaminants such as mercury and PCBs to the culture organism. A replacement of this protein source by a more sustainable feeding program is vital for the development of aquaculture. Many efforts have been made in order to develop fish feeds in which plant proteins replace wild fish sources. The improvements in protein utilization demonstrated in biofloc-based systems and potential for further increases in utilization efficiencies represents a major contribution toward the sustainability of aquaculture (Avnimelech 2009). Further improvement of feed quality in bioflocs as previously discussed is a topic worthy of further research.

An environmental issue common to all intensive aquaculture systems is the proper treatment and disposal of accumulated sludge. Sludge is the term for solids removed from the aquaculture system representing a concentrated source of reactive organic matter, rich in nutrients, similar to sludge produced in wastewater treatment systems (without the human pathogen risks). Sludge originating from wastewater treatment plants is usually treated by incineration or by anaerobic processes such as fermentation (producing bio-gas). Sludge is then further treated by composting (with potential agriculture uses depending upon salt content). These treatments add significantly to the cost of wastewater treatment. Presently, there are no well-established methods for treating aquaculture sludge. However, there are some studies and practical experiences of using such sludge as a source of feed for shrimp and fish (Kuhn *et al.* 2008; Schneider *et al.* 2006), as a feed applied to adjacent extensive ponds, or as a soil amendment. An important point that needs further study and quantitative data analysis is the fact that in biofloc systems, residues are suspended over long retention times in aerobic sections of the pond and can be utilized, to a large extent, by culture species like tilapia. These factors probably lead to a better degradation and possibly lower amounts of sludge accumulation compared with typical recirculating aquaculture systems, or possibly lower amounts per kg fish as compared to semi-intensive ponds (Hopkins *et al.* 1993). The problems of sludge minimization, utilization, and handling will likely be the subject of increasing regulation and should be studied further.

12.9   Outlook and research needs

The expanding application of the principles of biofloc technology in many parts of the world have reinforced the potential advantages inherent in this type of system while raising more questions and research needs to overcome problems and enhance efficiencies. Perhaps the most important area of need is an increased understanding of the complex biofloc microbial communities and developing management techniques to direct and optimize their establishment, stability, and control of structure and activity. This is closely related to issues surrounding water use in the system and potential for reuse of water within and among production units. Additional research is needed to better understand factors affecting sludge production and management in biofloc systems so as to maximize efficiency of feed conversion to fish or shrimp flesh while minimizing waste production. Engineering and design of pond and tank systems are important areas of research in this regard, particularly in the context of improving energy efficiencies and reducing carbon footprints. Genetic selection programs have produced improved strains of tilapia and shrimp, demonstrating the potential for increased growth and more robust stocks. Selection of stocks for traits that enhance performance in biofloc-based systems could represent a significant opportunity. As previously mentioned, development of specialty feeds and improvement of feed management are high priorities both because of their impacts on economics and because of effects on water quality and microbial community management. Continued development and application of bioeconomic models and establishment of key production metrics related to energy, water, and other resource uses will also help to focus research efforts, improving system efficiencies and sustainability both from environmental and financial perspectives.

This review and other literature on the topic of biofloc systems tend to focus on the generalities between various system types, levels of intensity, scales of operations, and target species. Clearly, there are important differences between the culture of tilapia and that of shrimp in terms of systems management, biomass density, associated feeding rates, feed composition, and the biology of the species, particularly in terms of their ability to crop biofloc directly from the system. Similarly, system differences between intensive ponds and super-intensive tank and raceway culture systems can lead to important differences in philosophy in terms of feed nutrient densities, strategies for management of nitrogenous wastes, and cropping of excess organic material from the system. When considering the commercial application of these technologies it is important to carefully consider these factors as design and management strategies are developed.

The future outlook for the continued adoption and expansion of biofloc-based production technologies is bright, representing important opportunities to concurrently expand environmental sustainability while opening new options for reducing production costs and improving consistency and profitability.

12.10   Acknowledgment

This is contribution number 668 from the South Carolina Department of Natural Resources Marine Resources Research Institute.

12.11   References

Alldredge, A.L. & Silver, M.W. (1988) Characteristics, dynamics and significance of marine snow. *Progress in Oceanography* 20(1):41–82.
Alonso-Rodriguez, R. & Paez-Osuna, F. (2003) Nutrients, phytoplankton and harmful algal blooms in shrimp ponds: A review with special reference to the situation in the Gulf of California. *Aquaculture* 219:317–36.
Avnimelech, Y. (1993) Control of microbial activity in aquaculture systems: Active suspension ponds. *World Aquaculture* 34:19–21.
Avnimelech, Y. (1999) Carbon/nitrogen ratio as a control element in aquaculture systems. *Aquaculture* 176:227–35.
Avnimelech, Y. (2006) Bio-filters: The need for a new comprehensive approach. *Aquacultural Engineering* 34:172–8.
Avnimelech, Y. (2007) Feeding with microbial flocs by tilapia in minimal discharge biofloc technology ponds. *Aquaculture* 264:140–7.
Avnimelech, Y. (2009) *Biofloc Technology: A Practical Guidebook*. World Aquaculture Society, Baton Rouge.
Avnimelech, Y. & Ritvo, G. (2003) Shrimp and fish pond soils: Processes and management. *Aquaculture* 220:549–67.
Avnimelech, Y., Kochva, M. & Diab, S. (1994) Development of controlled intensive aquaculture systems with a limited water exchange and adjusted carbon to nitrogen ratio. *Israeli Journal of Aquaculture—Bamidgeh* 46:119–31.
Avnimelech, Y. and Kochba, M. (2009) Evaluation of nitrogen uptake and excretion by tilapia in bio floc tanks, using ^{15}N tracing. *Aquaculture* 287:163–8.
Beveridge, M.C.M., Phillips, M.J. & Clarke, R.M. (1991) A quantitative and qualitative assessment of wastes from aquatic animal production. In *Advances in World Aquaculture: Aquaculture and Water Quality*, Volume 3 (Ed. by D.E. Brune, & J.R. Tomasso), pp. 506–33. World Aquaculture Society, Baton Rouge.
Bianchi, M., Bedier, E., Bianchi, A., Domenach, A.M. & Marty, D. (1990) Use of ^{15}N labeled food pellets to estimate the consumption of heterotrophic microbial communities to penaeid prawns diet in closed-system aquaculture. In *Microbiology in Poecilotherms* (Ed. by R. Lésel), pp. 227–30. Elsevier Scientific Publishers, Amsterdam.
Boyd, C.E. (2009) Estimating mechanical aeration requirement in shrimp ponds from the oxygen demand of feed. In *The Rising Tide, Proceedings of the Special Session on Sustainable Shrimp Farming* (Ed. by C.L. Browdy & D.E. Jory), pp. 230–34. World Aquaculture Society, Baton Rouge.

Boyd, C.E. & Clay, J.W. (2002) Evaluation of Belize Aquaculture, Ltd: A superintensive shrimp aquaculture system. *Report prepared under the World Bank, NACA, WWF, and FAO Consortium Program on Shrimp Farming and the Environment.*

Browdy, C., Stokes, A., Hopkins, J. & Sandifer, P. (1997) Improving sustainability of shrimp pond water resource utilization. *Proceedings of the Third Annual Ecuadorian Aquaculture Conference*, CENAIM-ESPOL-CAN, Guayaquil, Ecuador.

Browdy, C.L., Venero, J.A., Stokes, A.D. & Leffler, J. (2009) Superintensive biofloc production systems technologies for marine shrimp *Litopenaeus vannamei*: Technical challenges and opportunities. In *New Technologies in Aquaculture* (Ed. by G. Burnell & G. Allan), pp. 1010–28. Woodhead Publishing, Cambridge.

Brune, D.E. (2010) Algae and aquaculture. In *Abstracts, Aquaculture 2010, Annual Meeting of the World Aquaculture Society*, March 2010, San Diego, California, p. 143. World Aquaculture Society, Baton Rouge.

Brune, D.E., Schwarz, G., Eversole, A.G., Collier, J.A. & Schwedler, T.E. (2003) Intensification of pond aquaculture and high rate photosynthetic systems. *Aquacultural Engineering* 28:65–86.

Burford, M.A., Thompson, P.J., McIntosh, R.P., Bauman, R.H. & Pearson, D.C. (2004) The contribution of flocculated material to shrimp (*Litopenaeus vannamei*) nutrition in a high-intensity, zero-exchange system. *Aquaculture* 232:525–37.

Calle-Delgado, P., Avnimelech, Y., McNeil, R., Bratvold, D., Browdy, C.L., & Sandifer, P. (2003) Physical, chemical and biological characteristics of distinctive regions in paddlewheel aerated shrimp ponds. *Aquaculture* 217:235–48.

Chamberlain, G., Avnimelech, Y., McIntosh, R.P., Velasco, M. (2001) Advantages of aerated microbial reuse systems with balanced C/N. I. Nutrient transformation and water quality benefits. *Global Aquaculture Advocate* 4:53–6.

Chapman, P.M., Popham, J.D., Griffin, J., Leslie, D. & Michaelson, J. (1987) Differentiation of physical from chemical toxicity in solid waste fish bioassay. *Water, Air, and Soil Pollution* 33:295–309.

Costerton, J.W., Lewandowski, Z., Caldwell, D.E., Korber, D.R. & Lappin-Scott, H.M. (1995) Microbial biofilms. *Annual Review of Microbiology* 49:711–45.

Davis, D.A. & Sookying, D. (2009) Strategies for reducing or replacing fish meal in production diets for the Pacific white shrimp, *Litopenaeus vannamei*. In *The Rising Tide, Proceedings of the Special Session on Sustainable Shrimp Farming* (Ed. by C.L. Browdy & D.E. Jory), pp. 108–14. World Aquaculture Society, Baton Rouge.

De Schryver, P., Crab, R., Defoirdt, T., Boon, N. & Verstraete, W. (2008) The basics of bio-flocs technology: The added value for aquaculture. *Aquaculture* 277:125–37.

Dufault, R.J. & Korkmaz, A. (2000) Potential of biosolids from shrimp aquaculture as a fertilizer in bell pepper production. *Compost Science & Utilization* 8(4):310–9.

Dufault, R.J., Korkmaz, A. & Ward, B. (2001) Potential of biosolids from shrimp aquaculture as a fertilizer for broccoli production. *Compost Science & Utilization* 9(2):107–14.

Ebeling, J.M., Timmons, M.B. & Bisogni, J.J. (2006) Engineering analysis of the stoichiometry of photoautotrophic, autotrophic, and heterotrophic removal of ammonia-nitrogen in aquaculture systems. *Aquaculture* 257:346–58.

Ebeling, J.M. & Timmons, M.B. (2007) Stoichiometry of ammonia-nitrogen removal in zero-exchange systems. *World Aquaculture* 38:22–5, 71.

FAO (2010) Fisheries and Aquaculture Online Statistical Database. http://www.fao.org/fishery/statistics/en.

Hanson, T.R., Posadas, B., Samocha, T.M., Stokes, A.D., Losordo, T. & Browdy, C.L. (2009) Economic factors critical to the profitability of super-intensive biofloc recirculating production systems for marine shrimp *Litopenaeus vannamei*. In *The Rising Tide, Proceedings of the Special Session on Sustainable Shrimp Farming* (Ed. by C.L. Browdy & D.E. Jory), pp. 268–83. World Aquaculture Society, Baton Rouge.

Hardy, R. (2009) Protein sources for marine shrimp aquafeeds. In *The Rising Tide, Proceedings of the Special Session on Sustainable Shrimp Farming* (Ed. by C.L. Browdy & D.E. Jory), pp. 115–25. World Aquaculture Society, Baton Rouge.

Hargreaves, J.A. (2006) Photosynthetic suspended-growth systems in aquaculture. *Aquacultural Engineering* 34(6):344–63.

Hepher, B. (1985) Aquaculture intensification under land and water limitations. *Geojournal* 10:253–9.

Hopkins, J.S., Stokes, A.D., Browdy, C.L. & Sandifer, P.A. (1991) The relationship between feeding rate, paddlewheel aeration rate and expected dawn dissolved oxygen in intensive shrimp ponds. *Aquacultural Engineering* 10:281–90.

Hopkins, J.S., Hamilton, R.D. Sandifer, P.A. & Browdy, C.L. (1993) Effect of water exchange rate on production, water quality, effluent characteristics and nitrogen budget of intensive shrimp ponds. *Journal of the World Aquaculture Society* 24: 304–20.

Hopkins, J.S., Sandifer, P.A., Browdy, C.L. & Stokes, A.D. (1994) Sludge management in intensive pond culture of shrimp: effect of management regime on water quality, sludge characteristics, nitrogen extinction and shrimp production. *Aquacultural Engineering* 13:11–30.

Hopkins, J.S., DeVoe, M.R., Sandifer, P.A., Holland, A.F. & Browdy, C.L. (1995a) The environmental impacts of shrimp farming with special reference to the situation in the continental US. *Estuaries* 18:25–42.

Hopkins, J.S., Sandifer, P. & Browdy, C. (1995b) A review of water management regimes which abate the environmental impacts of shrimp farming. In *Swimming Through Troubled Water: Proceedings of the Special Session on Shrimp Farming* (Ed. by C. Browdy & J.S. Hopkins), pp. 157–66. World Aquaculture Society, Baton Rouge.

Johnson, C.N., Barnes, S., Ogle, J., Grimes, D.J., Chang, Y.J., Peacock, A.D. and Kline, L. (2008) Microbial community analysis of water, foregut, and hindgut during growth of Pacific White Shrimp, *Litopenaeus vannamei*, in closed-system aquaculture. *Journal of the World Aquaculture Society* 39(2):251–8.

Ju, Z.Y., Forster, I.P. & Dominy, W.G. (2009) Effects of supplementing two species of marine algae or their fractions to a formulated diet on growth, survival and composition of shrimp (*Litopenaeus vannamei*). *Aquaculture* 292:237–43.

Kuhn, D.D., Boardman, G.D., Craig, S.R., Flick, G.J. & McLean, E. (2008) Use of microbial flocs generated from tilapia effluent as a nutritional supplement for shrimp, *Litopenaeus vannamei*, in recirculating aquaculture systems. *Journal of the World Aquaculture Society* 39:72–82.

Kureshy, N. & Davis, D.A. (2002) Protein requirement for maintenance and maximum weight gain for the Pacific white shrimp, *Litopenaeus vannamei*. *Aquaculture* 204:125–43.

Leber, K.M. & Pruder, G.D. (1988) Using experimental microcosms in shrimp research: The growth-enhancing effect of shrimp pond water. *Journal of the World Aquaculture Society* 19:197–203.

Leffler, J.W., Haveman, J., DuRant, E., Lawson, A. & Weldon, D. (2010) Oxygen demand, ecological energetics and nutrient dynamics in minimal exchange, superintensive, biofloc systems culturing Pacific white shrimp *Litopenaeus vannamei*. In *Abstracts, Aquaculture 2010, Annual Meeting of the World Aquaculture Society*, March 2010, San Diego, California, p. 592. World Aquaculture Society, Baton Rouge.

Li, S., Willits, D.H., Browdy, C.L., Timmons, M.B. & Losordo, T.M. (2009) Thermal modeling of greenhouse aquaculture raceway systems. *Aquacultural Engineering* 41:1–13.

Li, M.H., Lim, C.E. & Webster, C.D. (2006) Feed formulation and manufacture. In *Tilapia: Biology Culture and Nutrition* (Ed. by C. Lim & C.D. Webster), pp. 517–46. Hayworth Press, New York.

Lim, C.E. & Webster, C.D. (2006) Nutrient Requirements. In *Tilapia: Biology Culture and Nutrition* (Ed. By C. Lim & C.D. Webster), pp. 469–502. Hayworth Press, New York.

Lotz, J.M. (1997) Special topic review: Viruses, biosecurity and specific pathogen-free stocks in shrimp aquaculture. *World Journal of Microbiology & Biotechnology* 13:405–13.

McIntosh, R.P. (2001) High rate bacterial systems for culturing shrimp. In *Proceedings from the Aquacultural Engineering Society's 2001 Issues Forum* (Ed. by S.L. Summerfelt, B.J. Watten & M.B. Timmons), pp. 117–29. Aquaculture Engineering Society, Shepherdstown.

Moss, S.M. (1995) Production of growth-enhancing particles in a plastic-lined shrimp pond. *Aquaculture* 132:253–60.

Moss, S.M., Forster, I.P. & Tacon, A.G.J. (2006) Sparing effect of pond water on vitamins in shrimp diets. *Aquaculture* 258:388–95.

Moss, S.M. & Leung, P.S. (2006) Comparative cost of shrimp production: earthen ponds versus recirculating aquaculture systems. In *Shrimp Culture: Economics, Marketing and Trade* (Ed. by P.S. Leung & C.R. Engle), pp. 291–300. Blackwell Publishing, Ames.

Moss, S.M. & Pruder, G.D. (1995) Characterization of organic particles associated with rapid growth in juvenile white shrimp, *Penaeus vannamei* Boone, reared under intensive culture conditions. *Journal of Experimental Marine Biology and Ecology* 187:175–91.

Moss, S.M., Pruder, G.D., Leber, K.M. & Wyban, J.A. (1992) The relative enhancement of *Penaeus vannamei* growth by selected fractions of shrimp pond water. *Aquaculture* 101:229–39.

Naylor, R.L., Goldburg, R.J., Primavera, J., Kautsky, N., Beveridge, M.C.M., Clay, J., Folke, C., Lubchenco, J., Mooney, H., & Max Troell, M. (2000) Effect of aquaculture on world fish supplies. *Nature* 405:1017–24.

Otoshi, C.A., Naguwa, S.S., Falesch, F.C. & Moss, S.M. (2007) Commercial scale RAS trial yields record shrimp production for Oceanic Institute. *Global Aquaculture Advocate* 10(6):74–6.

Peterson, E.L., Wadhwa, L.C., & Harris, J.A. (2001) Arrangement of aerators in an intensive shrimp growout pond having a rectangular shape. *Aquaculture Engineering* 25:51–65.

Rakocy, J.E. (1989) Tank culture of tilapia. *Southern Regional Aquaculture Center Publication 282.* Texas Agricultural Extension Service, Texas A&M University, Texas.

Ray, A.J., Shuler, A.J., Leffler, J.W. & Browdy, C.L. (2009) Microbial ecology and management of biofloc systems. In *The Rising Tide, Proceedings of the Special Session on Sustainable Shrimp Farming* (Ed. by C.L. Browdy & D.E. Jory), pp. 255–66. World Aquaculture Society, Baton Rouge.

Ray, A.J., Lewis, B.L., Browdy, C.L., & Leffler, J.W. (2010a) Suspended solids removal to improve shrimp (*Litopenaeus vannamei*) production and an evaluation of a plant-based feed in minimal-exchange, superintensive culture systems. *Aquaculture* 299:89–98.

Ray, A.J., Seaborn, G., Wilde, S.B., Leffler, J.W., Lawson, A. & Browdy, C.L. (2010b) Characterization of microbial communities in minimal-exchange, intensive aquaculture systems and the effects of suspended solids management. *Aquaculture* 310:130–8.

Rosenberry, B. (2007) http://www.shrimpnews.com/SerflingTribute.html.

Schneider, O., Sereti, V., Eping, Ep.H. & Verreth, J.A.J. (2006) Molasses as a C source for heterotrophic bacteria production on solid fish waste. *Aquaculture* 261:1239–48.

Stanley, D.L. (2000) The economics of the adoption of BMPs: The case of mariculture water management. *Ecological Economics* 35:145–55.

Tacon, A.G., Cody, J.J., Conquest, L.D., Divakaran, S., Forster, I.P. & Decamp, O.E. (2002) Effect of culture system on the nutrition and growth performance of Pacific white shrimp *Litopenaeus vannamei* (Boone) fed different diets. *Aquaculture Nutrition* 8:121–37.

Tacon, A.G., Nates, S.F. & McNeil, R.J. (2006) Overview of farming systems for marine shrimp with particular reference to feeds and feeding. In *Shrimp Culture: Economics, Marketing and Trade* (Ed. by P.S. Leung & C.R. Engle), pp. 301–14. Blackwell Publishing, Ames.

Tal, Y., Schreier, H.J., Sowers, K.R., Stubblefield, J.D., Place, A.R. & Zohar, Y. (2009) Environmentally sustainable land-based marine aquaculture. *Aquaculture* 286:28–35.

Timmons, M.B. & Ebeling, J.M. (2007) *Recirculating Aquaculture*. Cayuga Aqua Ventures, Ithaca.

Timmons, M.B. & Vinci, B. (2007) Gas transfer. In *Recirculating Aquaculture* (Ed. by M.B. Timmons & J.M. Ebeling), pp. 397–438. Cayuga Aqua Ventures, Ithaca.

Tomasso, J.R., Simco, B.A. & Davis, K. (1979) Chloride inhibition of nitrite induced methemoglobinemia in channel catfish (*Ictalurus punctatus*). *Journal of the Fisheries Research Board of Canada* 36:1141–4.

Turker, H., Eversole, A.G. & Brune, D.E. (2003) Filtration of green algae and cyanobacteria by Nile tilapia, *Oreochromis niloticus*, in the partitioned aquaculture system. *Aquaculture* 215:93–101.

Vidal, O.M., Granja, C.B., Aranguren, L.F., Brock, J.A. & Salazar, M. (2001) A profound effect of hyperthermia on survival of *Litopenaeus vannamei* juveniles infected with white spot syndrome virus. *Journal of the World Aquaculture Society* 32:364–72.

Volkman, J.K., Jeffrey, S.W., Nichols, P.D., Rogers, G.I. & Garland, C.D. (1989) Fatty acid and lipid composition of 10 species of microalgae used in mariculture. *Journal of Experimental Marine Biology and Ecology* 128(3):219–40.

Wasielesky, W., Atwood, A., Stokes, A. & Browdy, C.L. (2006) Effect of natural production in a zero exchange suspended microbial floc based super-intensive culture system for white shrimp *Litopenaeus vannamei*. *Aquaculture* 258:396–403.

Wetzel, R.G. (2001) *Limnology*. Academic Press, San Diego.

Wohlfarth, G.W. & Schroeder, G.L. (1979) Use of manure in fish farming a review. *Agricultural Wastes* 1(4):279–99.

Chapter 13

Partitioned Aquaculture Systems

D. E. Brune, Craig Tucker, Mike Massingill, and Jesse Chappell

Competition for land and water resources is driving the need to increase productivity of fish and shellfish aquaculture. At the same time, the public demands reduced environmental impact while expecting high-quality products from aquaculture. Simultaneously, strong international competition continues to force producers to seek ways to reduce production costs. Consequently, pond aquaculture, like other areas of agricultural production, is increasingly pushed to intensify and industrialize.

The conventional aquaculture pond provides a number of ecological services supporting fish and shellfish production. As described in chapter 10, the pond provides confinement space for the aquatic organisms, while algal growth in the pond serves as the base of an aquatic food chain providing some or all of the feed, depending on pond carrying capacity. In addition, algal growth removes potentially toxic CO_2 and ammonia from the pond environment and supplies needed oxygen. Other microbial, as well as physical and chemical, processes assist in treatment of metabolic waste produced by the aquatic animals. In most ponds these functions occur simultaneously in the same space where the animals are cultured as they are allowed to roam freely within the pond. The space required for animal confinement is much less than the area or volume needed to support the waste treatment functions. Consequently, combining animal-confinement and ecosystem support into the same physical space often leads to inefficiencies and management difficulties.

Aquaculture Production Systems, First Edition. Edited by James Tidwell.
© 2012 John Wiley & Sons, Inc. Published 2012 by John Wiley & Sons, Inc.

Intensification of pond aquaculture beyond approximately 1,000 kg/ha requires that in-pond processes be enhanced or supplemented. Applying feeds produced outside the pond environment is the primary technique. Increased feed application rate increases oxygen demand beyond what can be supplied by passive diffusion and pond photosynthesis. Therefore, mechanical aeration is commonly employed to support production intensification. In addition to supplemental feeds and aeration, other techniques to intensify pond culture have been attempted. Crowding of fish into cages or raceways, management of pond sediment, and mixing and treatment of the water column have been attempted with varying degrees of success. In-pond raceways (chapter 15) provide many of the advantages of raceway culture but do not address the increased pond nitrogen loading, and, as a result, fish yields are similar to conventional pond culture. Nitrifying filters like those used to support indoor recirculating systems have been installed, but they are too expensive for large-scale commodity fish production. Investigators have also tried removing and treating pond sediments in attempts to increase fish production by reducing pond ammonia loading and oxygen demand. Water column flocculation has been studied as a technique to reduce the solids content of pond discharge water. Water column mixing was investigated in the late 1970s as a way to better utilize oxygen production capacity of algal growth in fishponds (Busch 1985). Fish polyculture has been shown to increase yields in ponds as a result of the stabilizing effect of algal cropping provided by filter-feeding fish.

Supplementation of ecological services and improved culture practices enabled increased pond catfish production from around 1,100 to 1,700 kg/ha in the early 1960s to typical industry production levels of 3,400 to 5,600 kg/ha by the 1980s (fig. 13.1). In 1959, Swingle recommended a maximum daily feeding rate of 34 kg/ha-d feed to avoid low dissolved oxygen in ponds, limiting pond

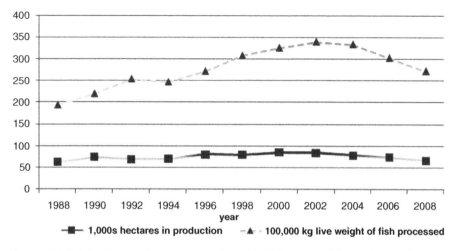

Figure 13.1 Relative increases in pond area and weight of fish processed during the period 1982 to 2002 (from Hargreaves and Tucker 2004).

production to around 1,100 kg/ha (1,000 lb/ac). In 1979, Tucker *et al.* presented data suggesting 3,000 to 4,800 kg/ha catfish production was possible, and by 1984, Busch reported on productivities ranging from 5,000 to 8,000 kg/ha in 1.6 ha (3.5 ac) ponds in Mississippi. In 2004, Hargreaves and Tucker observed that catfish farmers were stocking at rates of 10,000 to 15,000 fish/ha at typical catfish harvest size of 0.68 kg suggesting 6,700 to 10,100 kg/ha (6,000 to 9,000 lb/ac) production. Feed application and supplemental pond aeration are the primary, but not only, techniques enabling these increases. Many other factors interact to define the ceiling of fish yields from ponds, including fish diseases, processing plants' demand for fish, and feed quality and conversion efficiency, to mention a few. In spite of these interacting limitations, as fish-farmers are provided improved techniques to increase pond waste treatment capacity, in particular pond nitrogen removal, the aquaculture industry typically experiences slow but sustained across-the-board increases in fish yield.

By 1995, Clemson researchers had built and operated 0.13 ha (1/3 acre) prototypes of the partitioned aquaculture system (PAS), a technique in which conventional fishpond function is enhanced by installation of slow-moving paddlewheels and fish raceways. This technique represented an adaptation of the "high-rate pond" developed at University of California-Berkeley in the 1960s for treatment of municipal wastewater. The primary advantage provided by the high-rate pond and PAS is to increase algal production within the pond, accelerating ammonia assimilation and oxygen production. The PAS combines the advantages of the high-rate pond with other aquaculture intensification techniques. Of particular importance is the use of tilapia coculture allowing for manipulation of algal population, algal density control, and reductions in zooplankton numbers. As early as 1996 Clemson had successfully demonstrated 11,230 kg/ha (10,000 lb/ac) catfish production, and by 1998, 18,000 kg/ha (16,000 lb/ac) catfish production in six 0.13-ha PAS units. Ultimately, 42,000 kg of catfish and 4,500 kg of tilapia coproduction would be demonstrated in a 0.8-ha PAS prototype at Clemson.

Later, researchers focused on increased mixing and aeration in conventional ponds in an attempt to induce higher rates of waste treatment and fish production capacity in lower-cost conventional pond configurations. Torrans (2005) demonstrated 23,500 kg/ha (20,900 lb/ac) of catfish production in 0.1 ha (1/4 ac) intensively aerated/mixed (at 4 hp/ha) and managed experimental ponds. However, this approach is limited by difficulties encountered when attempting to scale intensive aeration/mixing to larger ponds.

The objective of Clemson University's PAS research program was to stimulate development of a sustainable aquaculture industry, providing a scalable technology enabling increased production with reduced environmental impact, while simultaneously increasing profitability. The PAS technology provides the potential for lower fish production unit-costs and higher productivity as compared to conventional earthen pond production. Subsequent investigators have concentrated on developing lower-cost modifications of the PAS. The Mississippi "split-pond," the Alabama "in-pond raceway," and the California "pondway," represent lower-cost or alternative adaptations of the PAS. The goal of these modifications was to provide a lesser degree of pond ecological enhancement (as compared to the PAS), at minimal additional cost over still water pond culture.

13.1 High rate ponds in aquaculture—the partitioned aquaculture system

By the early 1980s pond-based fish farmers were facing a number of emerging issues including environmental impacts of pond discharge upon public surface waters, a pond productivity "ceiling" of 4,000 to 6,000 kg/ha, maximum pond feed loading of 112 kg of feed/ha-d (100 lb/ac-d), and low dissolved oxygen events driven by zooplankton infestations and cyclic algal blooms and crashes. Other limitations and concerns included labor requirements for fish harvesting and sorting requirements, fish feed conversions of 2.0 to 2.2:1, fish flesh off-flavor events limiting sales, and bird and mammal predation losses, all of which negatively impact fish production and profitability.

Algal biosynthesis is the primary process controlling ammonia concentrations in aquaculture ponds. Algal production in a typical unmixed aquaculture pond is limited to 1 to 3 g-C/m^2-d corresponding to 0.5 mg/L-d of N addition, which is roughly equivalent to a feed loading of 90 to 112 kg/ha-d, limiting pond fish production at the 4,000 to 6,000 kg/ha per season.

In the 1960s, William Oswald and coworkers at the University of California Berkeley developed the use of low RPM paddlewheels to increase algal growth in ponds for use in treating municipal wastewater—a technique called "the high-rate pond." The partitioned aquaculture system (PAS) developed at Clemson University from 1988 to 2008 represents a high-rate pond adapted to expand the carrying capacity of pond aquaculture by increasing the rate of ammonia assimilation through enhancement of algal growth. In addition, the PAS combines a number of previously developed aquaculture intensification techniques into a single integrated system increasing operator control over pond culture processes. Initially in 1988, four 100 m^2 (1/4 ac) pilot-scale PAS units were installed and operated at Clemson University's Calhoun Field Station facility. Beginning in 1997, six 0.13 ha (1/3 ac) PAS were installed, and, later, a single 0.8 ha (2 ac) PAS was installed; these were operated for channel catfish production trials.

The concept of the PAS is to partition fish pond culture into separate physical, chemical, and biological processes linked together by a homogenous water velocity field (fig. 13.2). The physical separation achieves two important goals. First, it allows researchers to precisely quantify the fish culture and pond ecosystem processes. Second, separating the fish culture operation from the "at-large" algal production and water treatment function of the pond allows independent optimization of the processes thereby enhancing and maximizing performance of each component.

The PAS design consists of three physical components: the algal channels with paddlewheel, the fish raceways (with or without a separate paddlewheel) separated from the algal ponds with coarse screens, and a settling sump, to

(a)

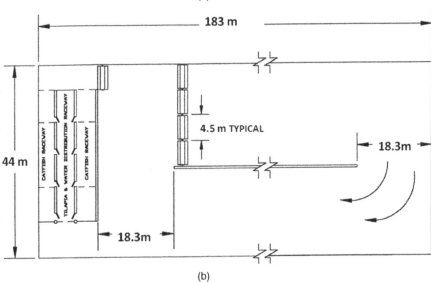

(b)

Figure 13.2 Photograph (a) and schematic (b) of two-acre partitioned aquaculture system showing algal and fish-raceway paddlewheels.

capture and concentrate settleable solids (fish wastes and algal flocs) for removal from the system. In addition, the Clemson PAS included populations of caged or free-roaming tilapia and, in some cases, a separate algal harvesting process for removal of algal biomass production.

The PAS takes advantage of accelerated algal growth potential provided by the high-rate pond observed by Oswald (1988, 1995) and colleagues at the University of California in treatment of municipal wastewater (Benemann 1997; Benemann *et al.* 1980; Green 1994) and is now used by most algal production companies (Benemann & Weissman 1993). The driver of the system is the low-speed paddlewheel that moves water in a racetrack configuration. Paddlewheel-mixed algal growth ponds had not previously been used in aquaculture production. The velocity imparted by the paddle ensures a mixed water column, increasing effective exposure of algal populations to solar radiation thereby increasing pond primary production. This configuration maximizes algal ammonia-N and CO_2 uptake, enhances interfacial gas exchange, minimizes waste solids settling in the pond, and provides an opportunity to harvest excess algal solids. These features improve pond water quality while providing biomass for algal byproducts or for use as a potential bioenergy resource.

13.1.1   Paddlewheels for water movement

The paddlewheel is the central design element transforming the still water fish-pond into a high-rate pond with raceway fish culture. Turning at 1 to 3 rpm, the paddlewheel imparts a water velocity field throughout the algal section and fish raceways. The uniform water velocity field reduces dead zones ensuring that the entire water column is utilized for algal growth and waste treatment. In 0.8-ha units, the paddles were constructed from 16 to 18 gauge mild-steel sheets welded to 0.85 cm (2 inch) schedule 40 steel pipe used as a shaft and supported on standard pillow-block bearings. Bearing failure lead to replacement of pillow-blocks with bearings machined from Teflon. One-piece Teflon bearings provided longer life at lower cost and were easier to replace. Angle-iron struts were welded at the end of the blades to prevent excessive flexure of paddle-blades at the shaft weld joint. Stainless steel should be avoided in this application as flexing of brittle stainless construction materials results in failure of blades and shafts. Typical power requirements at paddlewheel scales examined ranged from 0.75 to 1.13 KW/ha (1.0 to 1.5 hp/ac). Field studies suggest that water velocities ranging from 0.03 to 0.09 m/sec (0.1 to 0.3 ft/sec) are adequate for fish production, while 0.09 to 0.15 m/sec (0.3 to 0.5 ft/sec) are needed if a higher level of algal productivity is desired. At the lower velocities, algal biomass will settle onto the pond bottom increasing benthic oxygen demand and ammonia production. At higher water velocities, (0.15 m/sec or greater) the paddles demand excessive horsepower, and increased water velocity erodes unlined pond berms and bottoms. In field operations paddlewheels were successfully driven using either oil hydraulic motors or variable frequency electrical drives.

13.1.2 Algal production

The most important difference between conventional pond aquaculture and the PAS is the increased algal growth achieved as a result of the increased mixing and uniform water velocity field. The threefold to fourfold increase in pond photosynthesis provides the basis for an increased sustainable feed application rate that supports a threefold to fourfold increase in fish carrying capacity and production.

Stoichiometry of algal biosynthesis in the PAS is similar to that reported by previous investigators (Shelef & Soeder 1980), with minor adjustments in the C:N and C:O_2 ratios observed in the PAS algal biomass production. The adjusted stoichiometry was field determined by Meade as:

$$106 \ CO_2 + 16 \ NH_4 + 52 \ H_2O = C_{106}H_{152}O_{53}N_{16}P + 106 \ O_2 + 16 \ H \qquad (1)$$

The oxygen yield of 2.67 gm-O_2/g-C fixed suggests a 1:1 photosynthetic oxygen to carbon molar ratio. Average algal composition was observed to be approximately 50% carbon (by weight) yielding a C:N weight ratio of 5.7:1 and N:P weight ratio of 7:1.

As suggested by equation 1, harvest and removal of algal biomass from the pond environment using algal sedimentation and removal, or harvest and conversion into filter-feeding animal flesh, yields a net oxygen addition to and net nitrogen removal from the water column. The increased fish carrying capacity of the PAS is provided by the increased rates of nitrogen removal from, and O_2 addition to, the pond environment. This impact was dramatically demonstrated in observed differences in pond ammonia nitrogen concentrations in PAS units fed at similar levels as conventional ponds controls. At feed application rates ranging from 56 to 225 kg/ha-d (50 to 200 lb/ac/d), ammonia levels exceeded 16 mg/L in conventional ponds as opposed to just under 5 mg/L in PAS units (fig. 13.3). PAS field studies demonstrated sustained algal productivities of 6 to 12 g-C/m^2-d throughout the growing season, as opposed to 1 to 3 g-C/m^2-d observed in conventional aerated control ponds (Brune 1991; Brune *et al.* 2001a; Brune *et al.* 2003; Drapcho 1993; Drapcho & Brune 2000). The accelerated photosynthesis provides enhanced fish production per unit of pond area and water volume at significantly reduced aeration energy levels.

13.1.3 Raceway fish culture

The Clemson 0.8-ha (2 ac) PAS prototype was originally designed to include a three-channel raceway with the central channel delivering water into two "side-flow" fish raceways (fig. 13.4). With this configuration water flow enters the raceways as a "distributed plug flow." Adjustable gates are positioned to allow water flow to enter the raceway in proportion to the fish oxygen demand of individual containment cells (D_1 to D_4, as in fig. 13.4). Water velocity in the fish raceways is maintained using a separate paddlewheel. A separate paddle provides

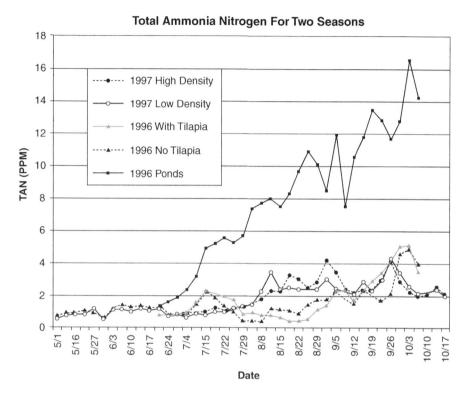

Figure 13.3 Ammonia nitrogen concentrations vs. time in conventional ponds and PAS units at feed application rates ranging from 56 to 225 kg/ha (50 to 200 lb/ac-d).

raceway oxygen and ammonia concentration independent of velocity or water quality in the larger algal basin. High fish density in the individual raceways provides scouring and rapid removal of waste solids. Side-flow raceways were found to be more cost effective than mixed tanks or mixed-loop tanks, the alternative design frequently used to provide uniform water quality to high density fish culture.

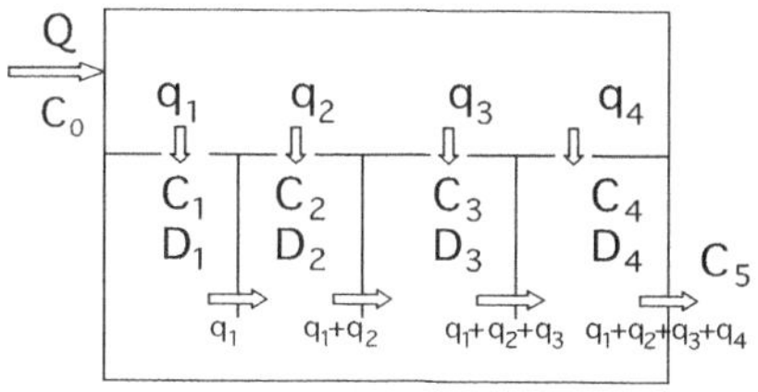

$$\frac{q_n}{D_n} = \frac{Q}{\Sigma D_n}, C_n = C_S = \frac{\Sigma D_n}{\Sigma q_n}$$

Distributed plug flow

Figure 13.4 PAS side-flow raceway water delivery rate in proportion to fish oxygen demand providing uniform oxygen concentrations in fish cages arranged in series.

In 1999, fish culture at variable stocking densities showed that catfish biomass of 128 to 160 kg/m^3 (8 to 10 lb/ft^3) could be sustained within the raceways with no adverse effect on growth. This led to further reductions in system costs by using a single high-density side-flow raceway requiring fewer paddlewheels (Brune *et al.* 2004a). Results from these trials suggest that the lower cost configuration using a single fish raceway and single paddlewheel providing raceway water velocity control would be successful.

In addition to providing greater control of water quality, raceway fish culture was found to offer many other advantages over conventional pond culture. Since the growing fish can be more conveniently sorted and held as similar-sized cohorts in cells or cages, feed application may be fine-tuned to the individual fish weight. As a result, overall feed conversion efficiencies were improved from 2.0 to 2.2:1 (typically observed in pond culture systems) to 1.4 to 1.5:1 in the PAS. Inexpensive netting can be stretched across the top of raceway sections thereby completely eliminating avian fish-predation. Labor requirement for harvesting higher density cells is reduced over conventional ponds seining. And, application of drugs or vaccines is more effective in high-density raceways. In Clemson PAS growth trials, oxygen sensors were added to the fish raceways, making it possible to use the output signal from DO meters to provide feedback control of paddlewheel speed and raceway aerator function, enhancing operator control over fish dissolved oxygen exposure. Computer control of algal basin water velocity (paddlewheel rotation) allows for improved control of pond gas transfer rates providing optimum O_2 and CO_2 transfer throughout the day-night cycle.

13.1.4   Fish carrying capacity and production

From 1995 to 2000, a range of fish stocking rates, sizes, and species combinations were used to test the limits of the PAS carrying capacity (table 13.1). Field data confirmed the PAS capability of threefold to fourfold increases in water treatment capacity over conventional ponds (Brune *et al.* 2001b, 2003, 2004b; Brune 1997; Brune & Wang 1998; Schwartz 1998; Meade 1998). Improvements in water quality dynamics and reduced aeration requirements as a result of tilapia coculture established the technique as standard practice. By 1997, maximum feed application reached 236 kg/ha (210 lb/ac-d) at average seasonal feed application rates of 106 kg/ha-d (94 lb/ac-d). In 1998, average feed application rate was increased to 127 kg/ha (113 lb/ac-d) with maximum feed application exceeding 236 kg/ha (210 lb/ac-d). At this time, fish carrying capacity averaged 18,757 kg/ha (16,694 lb/ac). By 1999, net annual catfish production exceeded maximum carrying capacities as a result of implementation of end-of-year catfish cohort carryover. Loss of carryover fish from proliferative gill disease (PGD) in winter carryover fish limited 1999 production to 12,892 kg/ha (11,474 lb/ac) average production. At this time, the 0.8 ha system yielded 16,682 (14,847 lb/ac) net production. By 2000, annual net catfish production from all units averaged

Table 13.1 PAS catfish and tilapia stocking rates (kg/ha) for the 1995 to 2000 growing seasons.[1]

PAS unit	Area (acres)	1995	1996	1997	1998	1999	2000
1	0.13	477 C	933 C 253 T	1,538 C 933 T	2,458 C 1,389 T	2,178 C 4,056 CC 271 TB	2,376 C 2,245 CC 126 TB
2	0.13	500 C	933 C 353 T	1,538 C 933 T	2,458 C 1,389 T	2,178 C 4,056 CC 271 TB	2,376 C 2,245 CC 126 TB
3	0.13	500 C	933 C 353 T	1,538 C 933 T	2,458 C 1,389 T	2,178 C 4,056 CC 271 TB	2,376 C 2,245 CC 126 TB
4	0.13	500 C	933 C 0 T	1,538 C 1,891 T	2,458 C 1,389 T	2,178 C 4,056 CC 271 TB 1,538 T	2,376 C 2,245 CC 126 TB
5	0.13	500 C	933 C 0 T	1,538 C 1,891 T	2,458 C 1,389 T	2,178 C 4,056 CC 271 TB 1,538 T	2,376 C 2,245 CC 126 TB
6	0.13	500 C	933 C 0 T	1,538 C 1,891 T	2,458 C 1,389 T	2,178 C 4,056 CC 271 TB 1,538 T	2,376 C 2,245 CC 126 TB
Large	0.81					3,012 C 933 T 89 TB	2,926 C 155 TB

[1] C = catfish fingerlings; T = tilapia fingerlings; TB = tilapia breeding pairs; CC = catfish carryover.

19,362 kg/ha (17,232 lb/ac) with a maximum carrying capacity of 15,852 kg/ha (14,108 lb/ac) and a tilapia coproduction of 6,075 kg/ha (5,407 lb/ac).

By 1996, at an average fish carrying capacity of 10,000 kg/ha, the importance of coculture of catfish with tilapia was already evident. During 1997, an optimum end-of-season tilapia/catfish biomass ratio of 1:4 was established at catfish carrying capacities of 15,700 kg/ha. By 1998, 19,000 kg/ha catfish production using multiple stockings was demonstrated. Catfish production increased over the seven years of field trials in both large and small units ultimately peaking at an average of 20,225 kg/ha (18,000 lb/ac; fig. 13.5). Peak feed application rates were demonstrated to be sustainable at 280 kg/ha (250 lb/ac-d) with seasonal averages exceeding 157 kg/ha (140 lb/ac-d).

13.1.5   Stabilizing algal populations with tilapia coculture

Work at Clemson focused primarily on stocking the Nile tilapia, *Oreochromis niloticus,* into the PAS for use as an algal stabilizing technique. Kent SeaTech

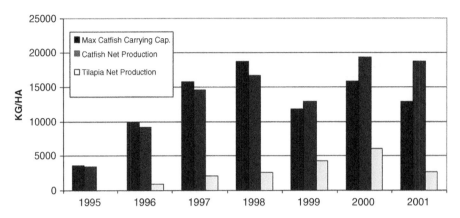

Figure 13.5 Clemson University PAS channel catfish and tilapia net production from 1995 to 2001.

Corporation in southern California used the Mozambique tilapia, *Oreochromis mossambicus,* and its hybrids. Tilapia is a tropical species, which leads to certain disadvantages for year-round use in the southern United States. Although tilapia survive well during summer months, even at water temperatures exceeding 37°C, winter water temperatures of 14 to 18°C, result in reduced tilapia growth and mortality. For this reason, a number of alternative filter-feeding organisms were studied including (in California) the Sacramento blackfish, *Orthodon microlepidotus* (a native California cyprinid), Chinese bighead carp, and a hybrid of common carp and goldfish, *Cyprinus carpio* × *Carassius auratus.* In South Carolina, freshwater mussels and other bivalves were also found to be capable of removing algal cells efficiently. Tilapia coculture was demonstrated to be effective at stabilizing field algal cultures by reducing zooplankton densities and occurrence. Furthermore, tilapia populations were successful in controlling algal standing crop, providing a method to maintain optimal Secchi disk visibilities of 15 to 18 cm. Both tilapia and shellfish populations were demonstrated to be effective in controlling the dominant algal genera in the ponds. Tilapia effectively eliminated algal population dominance by cyanobacteria and associated off-flavor of fish flesh. Tilapia populations were demonstrated to feed more efficiently on cyanobacteria (Turker *et al.* 2002, 2003a–d). Alternatively, shellfish selectively remove green algae, driving the pond algal populations toward cyanobacteria dominance (Stuart *et al.* 2001). In 1998 and in 1999 late-season reductions in catfish off-flavor was observed as 2-acre units were seen to shift from early cyanobacterial dominance to green algae as initial stockings of 100 tilapia breeding-pairs/acres expanded in number and weight as the season progressed.

An additional important function of tilapia grazing in the PAS is to provide a cost-effective technique to maintain young algal cell ages. Maintaining 10 to 25% of cultured fish biomass (channel catfish) as tilapia biomass was seen to reduce average algal cell age from 6 to 10 days to 2 to 3 days (fig. 13.6). The younger rapidly growing algal community yields more net oxygen production at reduced levels of pond respiration, and is less prone to culture instability.

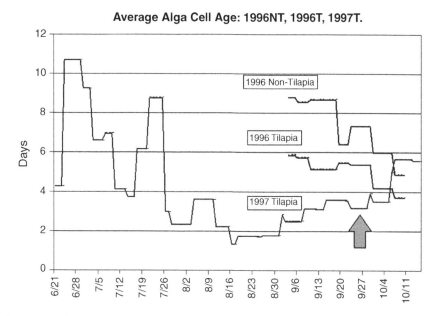

Figure 13.6 Algal cell age reduced with tilapia filtration.

13.1.6 PAS algal harvest

After four years of catfish culture, the data suggested that a significant portion of the algal growth capacity of the PAS was being wasted. Light/dark-bottle determinations of algal photosynthesis revealed that PAS algal production was incorporating nitrogen at rates 2 to 3 times greater than the amount added from fish-feed (table 13.2). In addition, system and sediment oxygen demand was progressively increasing. In the early PAS design, significant quantities of algal biomass were observed to settle in the system at large. If system design does not provide for removal of excess algal production, the biomass settles to bottom and decays, recycling soluble nitrogen and phosphorus back into the water column. At carbon fixation rates of 10 g-C/m^2-d corresponding to 1.8 gm N/m^2-d, calculations indicate that PAS feed assimilation capacity should

Table 13.2 PAS algal nitrogen uptake vs. nitrogen added to system from fish respiration.

Feed Rate kg/ha-d	System	Net Photosynthesis mg O$_2$/L-day	Respiration mg O$_2$/L-day	Nitrogen Addition mg-N/L-day	N-uptake/ N-addition
90	Typical Pond	10	5–8	—	
100	1996 PAS	31	13	0.80	2.2
200	1997 PAS	60	14	1.40	2.4
100	1998 PAS	70	24 ?	1.40	2.8
225	1999 PAS	90	24 ?	1.60	3.2

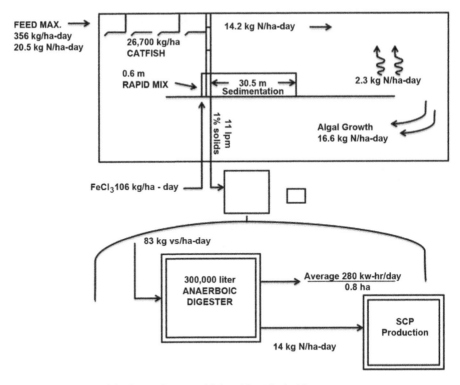

Figure 13.7 Predicted feed assimilation and fish yields with algal harvest.

be in excess of 450 kg/ha (400 lb/ac/d) suggesting potential catfish yields of 33,700 kg/ha (30,000 lb/ac). This indicated that PAS production capacity could likely be expanded (beyond that demonstrated) if nitrogen recycling from pond sediments could be reduced (fig. 13.7). This led to the conclusion that a system was needed to harvest and remove algal biomass from the pond.

The primary limitation standing in the way of widespread use of high-rate algal ponds for biomass production and water treatment is lack of cost-effective methods for harvest, removal, and concentration of algal biomass from ponds. Efficient algal harvest is also the critical cost-limiting step in production of algae for feedstuffs and chemicals. There have been many attempts to develop technology for harvest of single-celled algae from water, mostly based on filtration, centrifugation, or algal settling. Unfortunately, most of these technologies have proved to be inefficient and/or very expensive and, as such, are limited to commercial production of higher-value products (such as pharmaceuticals). Attempts have been made to use iron and aluminum salts to flocculate and settle algae onto a slow moving dewatering belt for removal from the PAS. Although the salts and belt worked well, the chemical costs were prohibitive. However, it was discovered that tilapia holding-cells could be placed over the belts, and the fish would enhance algal bio-sedimentation. This approach proved successful and cost effective in both South Carolina and California. The process is based on the fact that a fraction of the algae consumed by tilapia is converted into fish biomass, but the greater part passes through the fish to be excreted as fecal matter. The algal biomass is now bound into packets that settle rapidly onto the belt. The conveyor belt advances continuously at a critical angle, lifting the thickened algal sludge from the water column (fig. 13.8). Field trials showed that 90 to 95% of algal biomass production might be harvested using tilapia. The harvested algal biomass can be fermented to yield methane gas, which may be used as an energy source reducing dependency of fish production on fossil fuels, or processed into feed for secondary fish production, potentially providing aquaculture protein self-sufficiency.

13.1.7 Reduced environmental impact and water use in the PAS

In typical pond aquaculture, only about 25% of the feed-nitrogen and feed-carbon is incorporated into fish biomass. The remainder of the nitrogen and carbon supplied to the pond is lost through volatilization to the atmosphere, denitrification, methane fermentations, seepage, or water discharges—all of which can result in nutrient enrichment of surface and/or ground waters, and atmospheric enrichment with volatile ammonia and greenhouse gases. The PAS technology allows for improving feed utilization efficiency (in raceway culture), thereby eliminating feed wastage, which results in reductions in atmospheric emissions, water pollutants, and other environmental impacts associated with fish production. The energy requirements of PAS are substantially reduced as a result of reduced aeration requirements due to separation of the algal and fish production processes and increased O_2 production in the algal channels. Paddle-wheel mixing is more energy efficient than airlift or centrifugal pumps, moving large volumes of water at low head (Benemann & Weisman 1993; Brune 1997; Oswald 1995). The lower costs, combined with higher productivity, are key factors encouraging widespread PAS implementation enabling industry wide reductions in aquaculture environmental impacts. The shallow depth of the PAS system 0.5 to 0.6 m (1.5 to 2.0 ft) as opposed to 1.5 m (5 to 6 ft) for conventional ponds requires an initial filling of only 25 to 30% of water volume as compared to conventional pond-based catfish production.

13.1.8 PAS economics

A detailed economic analysis of the PAS was prepared by Goode *et al.* in 2002. In his analysis, 18.2 ha (45 ac) and 73 ha (180 ac) of PAS units producing 413,000 kg and 1,716,000 kg (909,800 and 3,775,670 lb, respectively) of catfish were compared to 65 and 260 ha (160 and 640 ac) of conventional ponds producing 318,000 and 1,293180 kg (700,000 and 2,845,000 lb) of catfish. Goode's analysis suggested that PAS culture could reduce production cost by UD$0.06 to 0.09/kg ($0.13 to 0.19/lb). However this projection was dependent

(a)

(b)

Figure 13.8 Algal sedimentation belts at Clemson (a) and concentrated algal biomass from belt in California (b).

upon successful carryover and growout of the overwintered fish (approximately 20% of the total system harvest). In 1999, PAS growth trails proliferative gill disease produced significant mortality in winter carryover fish. This source of loss (unless controlled) would result in the comparable PAS production costs ranging from US$0.009 to 0.023/kg ($0.02 to 0.05/lb) higher than conventional production costs.

Extending the analysis conducted by Goode *et al.* (2002) to commercial catfish farming is confronted with additional limitations. To date, only the Alabama modification of the PAS (described further on) has been used in a commercial catfish production—and only on a limited scale. Both the Clemson PAS and the Mississippi modification (the "split-pond") have been used under controlled, experimental conditions where catfish populations are predominately stocked in the spring, grown throughout the summer, and harvested in the autumn. While this approach can provide estimates of maximum production and allows optimization of system operation, constraints encountered under commercial conditions will likely reduce production and profitability.

Traditional pond catfish culture constraints include algae-generated fish off-flavor, loss of fish to predators, infectious diseases, and reduction in production potential related to fluctuating demand for fish by processing plants. Incidence of algae-related off-flavor is significantly reduced in the Clemson PAS because of effects of plankton grazing by tilapia cocultured with catfish. Insufficient information exists on the incidence of off-flavors in the Alabama and Mississippi PAS modifications, although it is clear that algal off-flavor is not eliminated in the Mississippi split-pond, which currently does not include tilapia coculture and algal grazing. Loss of fish to predation, particularly avian predators, is reduced to insignificance in all PAS modifications because predation control under fish confinement is relatively inexpensive and simple to implement. However, PAS modifications operated under commercial conditions will face uncertain risks from continuous operation and infectious diseases, and these risks interact significantly.

Catfish processors require fish year-round and some food-sized fish must be stored in ponds through the winter and spring to meet this demand. In effect, pond space that could be used to produce a new crop is used (indirectly) by the processors to hold fully grown fish in inventory. The impact of year-round demand for food-sized fish can be minimized, but not eliminated, by staggering production start dates in different ponds so that multiple populations reach market-size at different times throughout the year. This requires considerable management skill and access to fingerlings of desired sizes for stocking throughout the year, which is currently difficult for the catfish industry. Even if production is staggered so that food-sized fish are available throughout the year, some fish will need to be held in inventory to meet processor demands for fish in winter and spring, because low water temperature limits winter catfish growth in the southeastern United States. Overwintering food-sized catfish to meet winter and early spring processor demand exposes fish to increased risk of infectious disease.

There is no evidence of increased catfish loss to bacterial diseases in any of the PAS modifications. On the contrary, if/when bacterial disease occur, they are easier to manage than in traditional ponds because diseased fish are identified earlier and medicated feed can be delivered more effectively to confined fish. However, there is one common disease of channel catfish that appears to be a significant risk in some modifications of the PAS. Proliferative gill disease (PGD) is a serious disease of channel catfish in traditional earthen ponds. The disease led to loss of fish in both the Clemson PAS and the Mississippi split-pond PAS modification. The disease is caused by the myxozoan parasite *Henneguya ictaluri*, which infects the gills of channel catfish and its blue catfish hybrid, inducing severe hemorrhage of the gill filaments. Changes in gill structure interfere with gas exchange, resulting in hypoxia. The common, sediment-dwelling oligochaete worm *Dero digitata* is the intermediate host for the parasite. In the southeastern United States, the disease occurs in spring and, less commonly, in autumn. Currently, there is no treatment for the disease.

Outbreaks of PGD are not uncommon in PAS modifications, which may be related to the environment provided in the PAS for the oligochaete intermediate host. The worm probably proliferates in the "waste-treatment" side of the system where organic-enriched sediments from algal sedimentation and fish wastes provide food to the benthic invertebrates serving as intermediate host. Because the disease incidence peaks in spring, there is significant interaction between PGD risk and need to overwinter fish both for growout of undersized fish and holding of fish to meet springtime processor demands. These interacting risks could negatively impact economic returns in the PAS and PAS modifications. After the 1999 fish loss in the Clemson PAS, prophylactic lime treatment of PAS sediments was seen to substantially eliminate incidence of PGD is subsequent culture years. However, without automated systems to apply lime to ponds, labor requirements for manual lime application would be excessive.

13.2   PAS fingerling production

In an effort to reduce the need to overwinter catfish, research was directed at use of the PAS to produce larger sized fingerlings for stocking in PAS growout cages. In 2005, raceways in the 0.8 ha (2 ac) Clemson PAS unit were reconfigured into a series of nine, 1.83 m × 2.89 m × 1.22 m deep cinder-block cells. The cells were designed to allow for containment and culture of channel catfish fry to advanced fingerling sizes. The objective was to utilize the improved water treatment capacity and fish culture practices of the PAS to increase size and yield of catfish fingerlings within a single season.

At the Clemson facility, catfish eggs were harvested from broodfish ponds during the second and third weeks of May, after which they were hatched in tanks held at a constant a water temperature of 80°F. Once hatched, the fry were moved into 0.3 cm (1/8 inch) mesh-size rearing containers, and later to 0.44 cm (3/16 inch) mesh nets, then finally into 0.58 cm (1/4 inch) mesh net-pens where they remained until harvest. Each container or net-pen was

Figure 13.9 Continuous application of water and feed to fingerlings held in PAS net-pens.

supplied with water flow delivered with a combination of airlift pumps and submerged aerators (fig. 13.9). At initial stocking the fish were fed a starter feed of 52 to 56% protein, supplied by automated feeders. The fingerlings were harvested after 143 days of culture, yielding an average fingerling weight of 100 to 120 g/fingerling (fig. 13.10). Maximum feed application rates exceeded 135 kg/ha-d of 40% protein feed at a maximum fingerling carrying capacity of 4,200 kg/ha. Fingerling feed uptake rates were pooled from three seasons (2005, 2006, and 2007) yielding a feed application relationship of:

$$\text{Feed rate as percent of body weight per day} = 0.3233(\text{fish weight in grams})^{0.551}$$

$$(2)$$

In 2008, the PAS fingerling growth cells were configured to allow for passive water flow addition to each cell by utilizing the raceway paddlewheel, as opposed to using airlifts or aerators to provide water as was done in the 2005 to 2007 seasons. The intent was to investigate the possibility of using lowering system installation and operational costs. The best configuration, providing controllable high-flow rates through the cells, consisted of combinations of baffles in the water delivery channel and "angled flaps" in the individual bins and net-pens directing water flow into circular paths within the fingerling cells. Water flow rates averaged 1,514 Lpm (400 gpm) in the bins (as opposed to 151 Lpm using

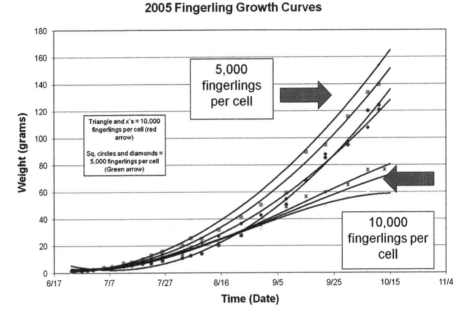

2005 Fingerling Growth Curves

Figure 13.10 Accelerated fingerling growth compared to conventional pond fingerling growth.

airlifts) and 7,570 Lpm (2,000 gpm) in net-pens (as opposed to 1,362 Lpm with aerators). Bin and net-pen hydraulic detention times were reduced to 0.4 to 0.85 minutes using passive flow as opposed to "pumped-flow" detention times of 2 to 5 minutes.

Overall, average fry/fingerling growth rates and yields observed in the 2008 PAS configuration were statistically the same as fingerling growth observed in earlier trials. The experimental trials suggested catfish fingerling growth could be significantly intensified and accelerated in the PAS. Fingerling production trials in the PAS demonstrated that catfish fingerlings in excess of 100 gm size could be produced in a single 140-day season at fingerling densities of 12,350 fish/ha (5,000 to 6,000 fish/acre). Fingerlings of this size may be grown to market-sized 0.68 kg (1.5 lb) fish in a single growout season, potentially reducing the amount of fish biomass that must be overwintered

13.3 Flow-through PAS: the controlled eutrophication process

13.3.1 Eutrophication of the Salton Sea in Southern California

The Salton Sea (SS) is a large inland lake of 945 km^2 (365 square miles) 69 m (227 feet) below sea level, with no outlet, which has been accumulating salt and nutrients from municipal storm water, treated sewage, industrial waste discharges, and agricultural drainage for nearly 100 years. Over geologic time, the Salton Basin and Lake Cahuilla were periodically flooded due to natural diversions of the Colorado River. After each diversion ceased, the area reverted to a dry lake bed. As the region became populated and large-scale irrigation drainage and municipal and industrial waste-streams from the United States and Mexico were dumped into the basin, the SS became a permanent water body, with evaporative losses equaling tributary inflows. Buildup of salts and nutrients from Imperial Valley and Coachella Valley discharges led to the development of an inland hypereutrophic, hypersaline lake. As the SS increased in salt content, it became a stage for a succession of ecosystems ranging from freshwater to brackishwater and ultimately to only the most salt-tolerant organisms. A variety of solutions have been proposed to improve water quality in the SS, ranging from evaporative ponds to concentrate salts to pumping of water for discharge to the Gulf of California. However, reducing the salinity will not resolve the more serious water quality problem associated with high-nutrient loads driving the extreme eutrophication. Large-scale algal blooms, followed by senescence of these blooms, lead to catastrophic low-dissolved oxygen, triggering massive fish mortalities resulting in widespread odor problems. A cost-effective solution to this problem is required.

Recently, the establishment of freshwater impoundment zones within the SS perimeter near the major freshwater river inflows has been proposed. As the SS continues to evolve to a hypersaline environment inhospitable to most freshwater and brackish organisms, this freshwater zone would continue to support fish and the fish-eating birds. Whether the freshwater impoundment zone is implemented or not, lowering the amount of nutrients flowing into the sea is of paramount importance in maintaining acceptable water quality (Setmire *et al.* 1993). The principal tributaries to the SS (the Alamo, New, and Whitewater Rivers) contain 0.5 to 2.00 mg/L of total phosphorus at N:P ratios of 4.6 to 39.6 (Setmire 2000). Phosphorus is considered to be the limiting nutrient in the SS.

13.3.2   PAS aquaculture for eutrophication control

Kent SeaTech Corporation and Clemson University conducted research, developments, and demonstrations involving the application of the PAS for concentrating and removing nutrients from wastewater streams. A modified version of the PAS described as the controlled eutrophication process (CEP), utilizes dense populations of single-celled algae cultivated in high-rate algal ponds and harvested using tilapia driven bio-sedimentation originally developed for aquaculture applications (Brune *et al.* 2007). The CEP relies upon the water treatment function of the PAS. Kent SeaTech and Clemson demonstrated technical and economic feasibility of using CEP technology to reduce nutrient concentrations in aquaculture and wastewater effluent to essentially zero.

The typical PAS installation for aquaculture effluent treatment and reuse has an intensive primary fish production zone in which high-value fish species such as channel catfish or striped bass are cultured. In some applications, a detritivore fish species is placed downstream from the primary fish production zone to consume uneaten food, fecal matter, and particulate waste. Downstream from this area, the treatment zone consists of paddlewheel-mixed high-rate algal growth basins where waste nutrients from the primary fish species are converted into algal biomass. After the treatment zone, a secondary fish-zone houses a filter-feeding fish that consumes algal cells converting a portion into fish biomass. The remaining algal biomass is excreted onto the algal concentration belt for removal from the system.

13.3.3   PAS/CEP for remediation of the Salton Sea

For remediation work, CEP units are configured differently from the PAS used for aquaculture water reuse. In the aquaculture system, no water is discharged to the environment and filter-feeding fish are held in the algal growth basin, resulting in equilibrium algal cell concentrations typically not exceeding 50 to 60 mg/L VS. However, in the CEP application, water flow through the system provides hydraulic retention times of 3 to 4 days. Under these conditions the algal standing crop is typically maintained at densities of 80 to 160 mg/L VS. In aquaculture applications, the PAS process allows the production of fish with minimal impact on the environment.

CEP treatment at the SS operates as a flow-through process treating nutrient-enriched water from agriculture drainage to the Whitewater, New, and Alamo Rivers—the three principal tributaries of the SS. The major components include (1) the nutrient-enriched water, (2) the high-rate pond with paddlewheels, and (3) the algal harvest zone (fig. 13.11). The nutrient-laden water is brought into

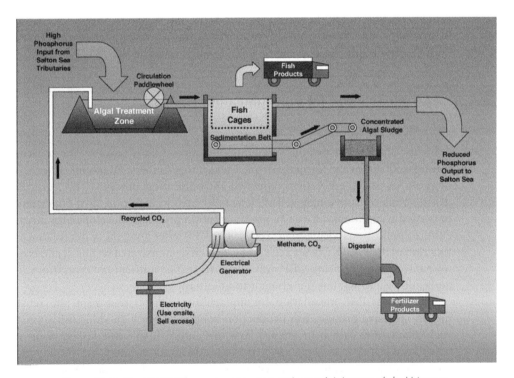

Figure 13.11 The PAS/CEP for waste nutrient control using fish harvest of algal biomass.

Figure 13.12 Solar-drying of CEP produced algal biomass.

the algal treatment zone where paddlewheels continuously mix the water column. Stable algal populations grow at an accelerated rate converting nutrients to algal biomass. The algae-laden water from the treatment zone is passed through high densities of confined tilapia (or other filter-feeder fish) and across an algal sedimentation zone or belt, producing steady-state algal cell density of less than 10 to 15 mg/L. As much as 75 to 90% of the algae-bound N and P can be removed without the need for chemical flocculants. After the algal biomass is harvested, separated, and removed from the water column it can be solar dried (fig. 13.12), and the treated water may be returned to the river for discharge to the sea.

13.4 Photoautotrophic and chemoautotrophic PAS for marine shrimp production

Because of actual and perceived negative environmental impacts of marine shrimp production Clemson efforts where directed (in 2002) to applications of the PAS design in support of marine shrimp production. Four 250 m^2 (0.0625 ac) greenhouse-covered PAS units were designed, installed, and operated at Clemson for marine shrimp (*Litpenaeus vannamei*) production trials.

Initially, in 2003, feed application rates of 280 kg/ha-d were maintained resulting in a final shrimp yield of 16,800 kg/ha. In 2004, feed application rates were pushed to 787 kg/ha-d with a shrimp yield of 25,600 kg/ha. In 2005, feed application rates peaked at 1,662 kg/ha-d with a shrimp yield of 37,339 kg/ha (33,232 lb/ac). In 2003, net photosynthesis in the marine units ranged from 0 to

19 g-C/m²-d with an average of 4.2 g-C/m²-d. In 2004, algal fixation declined to 0 to 15 g-C/m²-d at an average of 3.7 g-C/m²-d. In 2005, photosynthesis further declined to 0 to 6 g-C/m²-d at an average of 1.1 g-C/m²-d at average feed rate of 683 kg/ha-d (608 lb/ac-d).

In contrast, the six 0.13 ha (1/3 ac) freshwater PAS units yielded catfish production approaching 22,000 kg/ha (19,860 lb/ac). Feed application rates to the catfish units ranged between 163 kg/ha-d (145 lb/ac-d) and a maximum of 286 kg/ha-d (255 lb/ac-d). Net photosynthesis in these units ranged from 10 to 120 mg O_2/l-d (1.9 to 22.5 g-C/m²-d).

The decline in photosynthesis in the marine units at the high feed rates was dramatic and very obvious. While peak photosynthetic rates as high as 22.5 g-C/m²-d were observed in the outdoor catfish units, long-term sustained maximum rates of 10 to 12 g-C/m²-d were more typical. However, the shrimp units typically yielded maximum sustained peak photosynthetic rates much lower than 10 g-C/m²-d. By the time external organic carbon addition had reached 6.0 g-C/m²-d, algal carbon fixation declined 5 g-C/m²-d. At 10 g-C/m²-d external carbon application, algal fixation dropped to 3 g-C/m²-d. And when the system was receiving 50 g(organic)-C/m²-d of external carbon (corresponding to 3.5 times the rate of catfish PAS feed-rate), photosynthesis was essentially zero. As a result, the shrimp water column was populated by 200 to 500 mg/L of nitrifying bacterial flocs, which became the dominant nitrogen control mechanism. Shrimp production in excess of 28,000 kg/ha (25,000 lb/ac) is possible using non-photosynthetic systems (table 13.3). However, aeration and mixing energy requirements were seen to increase substantially from 7.7 to 9.2 KW/ha (4 to 5 hp/ac) in an algal-dominated system to an excess of 115 KW/ha (60 hp/ac) within a bacterially dominated system.

Field studies suggest that decline in algal productivity as a result of increased organic loading is likely the result of algal light-limitation arising from heterotrophic and chemoautotrophic bacterial biomass within the water column. This effect is exaggerated in shrimp units where fecal solids and uneaten food

Table 13.3 Field observed and projected algal photosynthetic limits vs. feed loading rate in shrimp PAS production units.

Feed Rate kg/ha-day	Production[1] kg/ha	Nitrogen Loading[2]		Photosynthesis[3]	
		gm-N/m²	mg-N/L	g-C/m²-d	mg-O_2/L
112	4,210/5614	0.5	0.9	2.6	14.1
224	8,420/11,227	0.9	1.9	5.3	28.2
560	21,050/28,070	2.4	4.7	13.4	70.3
1,120	42,100/56,135	4.7	9.4	26.7	140.6[4]
1,684	63,150/112,270	7.1	14.1	40.0	210.9[4]

[1] Growing season = 120 days (catfish) or 200 days (shrimp); CF = 2/1; average feed = 50% of peak feed rate.
[2] 35% protein, 75% N-release to water, water column = 0.5 meter deep.
[3] Algal C/N = 5.6/1, C/O_2 = 1/1 molar.
[4] Photosynthetic rate required to match nitrogen loading rate; these rates are not observed in practice.

add to the accumulation in the water column. Aggressive solids management, using a combination of settling and aeration basins, coupled to tilapia polishing basins may be used to reduce this light limitation. Advantages of maintaining enhanced algal production at high organic load within the PAS include reduced system oxygen demand, reduced aeration and mixing energy requirements, reduced alkalinity destruction/consumption, and improved pH management—all leading to increased feed application and fish/shrimp carrying capacity.

13.5   Alabama in-pond raceway system

Work in west Alabama (supported by Alabama Cooperative Extension Service, Alabama Experiment Station, and the Alabama Catfish Producers Association) has focused on development and demonstration of an "in-pond raceway" fish production system applying technologies and management approaches to utilize existing earthen ponds to increase production yield and efficiency (fig. 13.13). Fish production costs on most southern catfish farms currently exceed US$1.65/kg ($0.75/lb). The objective of this research was to develop and demonstrate in-pond systems, strategies, and technologies providing improvements supporting sustainable and profitable US aquaculture enterprises.

Commercial scale in-pond raceways were installed and operated on farms in west Alabama in 2007 to 2008. The design represents a rearrangement of partitioned aquaculture systems (PAS) pioneered by Clemson University that employ confined fish production using filter-feeding fish to harvest manure waste

Figure 13.13 Alabama in-pond raceway system.

and excess algae productivity. The Alabama system was installed in a 2.4 ha (6 ac) earthen pond of average depth of 1.7 m (5.5 ft). Six fish-production cells 4.9 m (16 ft) wide and 11.6 m (38 ft) long were constructed from concrete blocks on a reinforced concrete pad (fig. 13.13). The cells were arranged side by side sharing common walls. Each cell was equipped with a 0.38 KW (0.5 hp) water-mover (paddlewheel) at the upstream end rotating at 0.7 to 1.5 RPM, providing a raceway water exchange every 1 to 2 minutes. For additional life support, an aeration grid was installed downstream from the water-mover. It was supplied with low-pressure air from a 1.1 KW (1.5 hp) Sweetwater blower. Fish were confined in cells using two PVC coated steel mesh panels extending across the width of each cell, one placed upstream adjacent to the aeration grid and the second was placed downstream 2.4 m (8 ft) from the end of the raceway. An automatic, timer-controlled feeding system was installed in each cell and was supplied with bulk feed via an overhead tube originating at a bulk-feed storage tank.

At the discharge of the raceway cell a V-shaped manure pit was installed at a "quiescent zone," designed to allow settling of waste or other particles as they passed from the cells. The V- shaped trap extends across width of the raceway. A baffle was installed on the long axis of the pond forcing water discharging from the production cells to circulate around the open-pond, preventing short-circuiting to the intake channel.

A different version of the in-pond system was installed at a farm in Greene County, Alabama. This system consisted of two floating cells of dimensions similar to the fixed system described previously (fig. 13.14). The main difference

Figure 13.14 Floating raceways.

is that the water is moved through the floating raceway by an airlift pump attached directly to the inlet to each cell. The airlift was used to aerate, pump, and mix the inlet water. The airlift devices were powered by 1.1 KW (1.5 hp) Sweetwater blowers. Water exchange rates were adjusted to be the same as in the paddlewheel driven fixed-cell system.

In the fixed-cell systems, the units were stocked with 9,000 to 10,000 juveniles averaging 80 to 200 grams in weight. Dissolved oxygen levels remained satisfactory and stable (>1.5 mg/L and <15.0 mg/L) in all cells. Ammonia and nitrite levels remained well below stressful levels. Results during 2008 have demonstrated mean catfish survival of 91 to 95% with one cell at 80%. Feed conversion efficiency (FCR) averaged 1.3:1 (ranging from 1.1:1 to 1.6:1) at a production average of 28,000 kg/ha (25,000 pounds/ac).

Aeration energy use was reduced by approximately 50% compared to conventionally aerated catfish production ponds in the area. In both configurations, observations suggest the in-pond production strategy significantly and reliably increased fish yield/acre to 33,700 kg/ha (30,000 lb/ac) at survival rates >90% and at feed efficiencies routinely below 1.5:1. A fish production management guide and economic model is currently under development. We anticipate that improvements in survivorship, feed efficiency, management of disease, and overall increased production will significantly reduce the cost of production, and substantially improve enterprise profitability.

13.6 Mississippi split-pond aquaculture system

A modification of the PAS has been constructed at Mississippi State University's National Warmwater Aquaculture Center that takes advantage of existing pond infrastructure in the major catfish-growing area of the southern United States. These systems are referred to as "split-ponds" to differentiate them from the PAS. Split-ponds are constructed by retrofitting existing earthen ponds, as opposed to a complete redesign as is done with construction of the PAS. As originally conceived, the goal of the split-pond was to take advantage of the fish-confinement benefits of the PAS, such as facilitation of feeding, inventory, harvest, health management, and protection from predators. During development of the system, it became evident that loading rates and fish production were significantly greater in the split-pond than in traditional ponds. The split-pond thus represents an intermediate level of intensification between traditional earthen ponds and the PAS.

The split-pond has a relatively smaller algal basin (about 70 to 80% of the total area) and a larger fish-holding area than the PAS. Fish are held at only 5 to 10 times the density of traditional ponds. The split-pond is constructed by dividing an existing earthen pond into two unequal sections with an earthen levee. The levee is then breached with two sluiceways. One sluiceway is fitted with a large, slow-turning paddlewheel (fig. 13.15), which moves water out of the fish-confinement area. The second sluiceway provides return flow from the algal basin into the fish-confinement area (fig. 13.16). Metal screens are installed

Figure 13.15 Paddlewheel at Mississippi split-pond levee.

Figure 13.16 Return-flow sluiceway.

in each sluiceway to prevent fish escape. Aerators in the fish-confinement area provide supplemental dissolved oxygen supply at night. The system is currently operated as a catfish monoculture, although it is easily adapted to coculture with tilapia or other fish.

The key design parameters for the split-pond are water flow rate and aerator capacity. During daylight hours, water flowing from the algal basin provides oxygen for fish in the confinement area. The required flow rate is calculated from estimates of the oxygen consumption rate of the confined fish at the highest expected biomass loading and most extreme water temperature. The paddlewheel size required to supply the desired flow rate is determined empirically. At night, dissolved oxygen cannot be supplied by photosynthesis in the algal basin, so the large paddlewheel is turned off to stop water exchange between the two sections. At this time dissolved oxygen is provided by mechanical aeration within the fish confinement zone. Aeration requirements are predicted by matching aerator oxygen transfer rate to maximum projected fish oxygen consumption rate. Water exchange and mechanical aeration never occur at the same time in the split-pond.

As an example, a 2-ha earthen pond at Mississippi State University was reconfigured into a 1.42-ha algal basin and 0.40-ha fish-confinement area. Based on previous studies in prototype split-pond units, the 1.82-ha system was designed to hold a maximum fish biomass of 40,000 kg, all of which were contained in the 0.40-ha confinement area. Required water flow from the algal basin was estimated at 50 m^3/minute at an inflow dissolved oxygen concentration of 5 mg/L. A six-blade 3.66-m-long, 2.4-m-diameter paddlewheel operated at 2.5 rpm at a wetted paddlewheel depth of 1.1 m, producing a water flow of 60 m^3/minute, more than adequate to supplying sufficient oxygen supply to fish during daylight hours. Nightly aeration in the fish confinement area was provided by two 7.5-kW paddlewheel aerators.

After seven years of study, net annual catfish production in the split-pond ranged from 17,000 to nearly 20,000 kg/ha—2 to 4 times that achieved in traditional ponds and marginally less than in the PAS system. At stocking rates of 25,000 fish/ha, fish were grown from an initial weight of 50 to 70 g/fish to 0.80 to 0.90 kg/fish in a seven-month growing season. Feed conversion ratios (weight of feed fed divided by net fish weight gain) were approximately 1.8:1.

Although daily feeding rates exceeded 250 kg/ha for extended periods, total ammonia-nitrogen concentrations seldom exceeded 1 mg/L, except when phytoplankton communities crashed and in late autumn when cooler water temperatures lead to slower rates of nitrogen assimilation by the pond microbial community. When phytoplankton populations crash, rates of dissolved oxygen production and nitrogen assimilation in the algal basin decline markedly and water exchange between the algal basin and the fish-confinement area becomes a liability rather than an asset. During those periods, which have occurred at a frequency of less than once per year, water exchange is stopped and the fish confinement area is aerated for a few days until the phytoplankton community recovers. The system is then returned to normal operation.

Although the split-pond may not achieve the fish production levels obtained in the PAS, it offers several advantages over traditional ponds. Aerating the small fish-confinement area is more effective at maintaining adequate levels of dissolved oxygen than in traditional ponds. Fish in the confinement area also are easier to feed and harvest. These attributes, combined with greater fish production, render the split-pond an attractive alternative for commercial catfish culture.

13.7   California pondway system

From 1983 to 2009, the Kent SeaTech (KST) fish production facility located in the agriculturally productive Coachella Valley of southern California was used for culture of hybrid bass, tilapia (*T. mossambica*), channel catfish (*Ictalurus punctatus*), and hybrid carp (*Cyprinus carpio* × *Carassius auratus*). This desert region provides very warm summers and mild winters that offer a farmer-friendly environment with high-quality geothermal groundwater aquifers supporting striped bass production, and 7.5 to 8.5 months of open-pond growing season.

During this time, KST conducted research on the use of algal-based systems for fish production and cost-effective wastewater treatment; the majority of this work was done in cooperation with Clemson University in South Carolina. In 2005, KST designed, installed, and operated commercial algae-based fish production systems utilizing Clemson's partitioned aquaculture system (PAS) concept and KST's pondway design.

Fifteen existing earthen ponds, originally used for the production of hybrid striped bass fingerlings, were modified into a single Kent SeaTech pondway/PAS system. The individual ponds were combined into a single pondway unit by removing sections of levees separating the ponds (fig. 13.17), allowing water flow between the ponds that was powered by two 7.5 hp steel paddlewheels, 3 m (10 ft) in diameter and 6 m (20 ft) long (fig. 13.18). The paddlewheels provided 113,000 to 151,000 Lpm (30,000 to 44,000 gpm) water flow through the pond's 2,600 m (8500 ft) lineal distance at a total head loss of 6.3 to 7.6 cm (2.5 to 3 inches). Total volume of the recirculating ponds at 0.9 m (3 ft) depth (area of 9.6 ha or 24 acres) was 89 million liters (23.5 million gallons) with an average water velocity of 0.05 to 0.07 m/sec (0.18 to 0.26 ft/sec).

Approximately 3,785 to 7,570 Lpm (1,000 to 2,000 gpm) of wastewater effluent from KST's striped bass production facility was fed continuously into the pondway after holding in two non-recirculating ponds. This water discharge averaged 3 mg/L ammonia, with a pH of 6.3 to 7.0, 20 to 40 mg/L of TSS (fish feces and uneaten feed), and at dissolved oxygen levels of 8 to 15 mg O_2/L. At 3,785 to 7,570 Lpm water addition, the pondway hydraulic retention time ranged between 8 and 16 days. Although the discharge water added ammonia and organic particulate load to the pondway system, the water addition acted indirectly as an algal harvest mechanism reducing algal cell age. Discharge water from the pondway system was sold to adjacent vegetable farms and seasonal

Figure 13.17 Overview of California pondway systems.

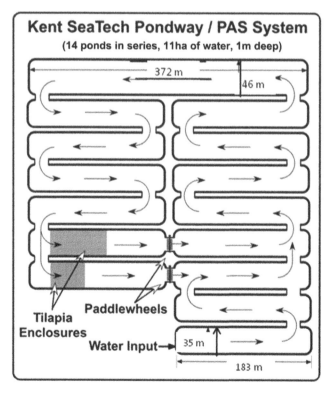

Figure 13.18 Schematic showing water flow in the California pondway system.

duck clubs. Excess water discharge beyond these needs was discharged to the Coachella Valley Stormwater Channel provided for by an NPDES discharge permit.

To limit construction costs, the original 0.9 m (3 ft) water depth of the preexisting ponds was maintained. The added water depth provided thermal buffering capacity against the extreme daytime water temperatures and hot air temperatures (38 to 50°C) of this Southern California region. During summer operations, the pondway water conditions ranged between a temperature of 25 to 38°C, with predawn dissolved oxygen concentration of 1 to 3 mg with afternoon concentrations of 10 to 22 mg O_2/L with a pH of 6.6 to 8.5, total ammonia-nitrogen concentration ranging from 0.3 to 3 mg/L (but usually below 1) and total suspended solids concentration ranging from 30 to 90 mg/L.

Two of the ponds were fitted with concrete floored and fenced fish enclosures. The first growout enclosure became operational in 2005, the second coming on line in 2007. The enclosures contained the production fish, both tilapia and carp/goldfish hybrids, partitioning them from the pondway algal growth basins. Water inflow to the fish production enclosures was split between two variable speed paddlewheels, which could be adjusted to independently control flow-rate across the two separate fish populations

The enclosures contained 1.5 KW (2.0 hp) high-speed paddlewheel aerators to supply nighttime aeration needs. In addition, each section contained a pure-oxygen diffuser hose or "aerotube" for emergency aeration needs. Each enclosure routinely held 22,700 kg (50,000 lb) of fish, with a maximum carrying capacity of 45,000 kg (100,000 lb) and 68,000 kg (150,000 lb), respectively. During periods of predawn low oxygen concentrations, the two pondway paddlewheels were shut down to temporarily maximize aerator oxygenation efficiency within the fish enclosures. Fish-feeding was restricted until sunrise or later, so that photosynthesis increased dissolved oxygen levels. Feeding was discontinued several hours before darkness so fish metabolism would decline before the nighttime algal respiration depressed oxygen levels.

At the beginning of each season, young tilapia and hybrid carp were moved from KST's geothermal well-water tanks where they had been overwintered. The young fish were stocked into the pondway at sizes ranging from 0.04 to 0.18 kg/fish (0.1 to 0.4 lb). Young tilapia became abundant throughout the pondway system due to natural spawning. These fish were very effective at consuming incoming particulates matter from the striped bass wastewater. The young tilapia could be collected throughout the summer at feeding station traps or trapped at warm-water inlets in late fall. KST produced a significant proportion of tilapia stock in this manner during the first two years of pondway use. Because of *Tilapia zilli* contamination, the pondway units were emptied in winter (slower growth rendered *T. zilli* undesirable).

Each enclosure contained multiple fish-holding sections—four in unit 1 and six in unit 2—providing for separation of tilapia and carp populations and, more importantly, allowing grading operations and fish-cohort separation. Continuous grading allows market ready fish to be removed before their larger size negatively impacts system feed conversion efficiency. The primary market for tilapia is as live fish; hence, fish size is very important to obtain best prices. The layout of the fish enclosures was designed to minimize harvest labor requirement and reduce fish handling stress. The enclosure concrete floor, shallow water depth, and inclusion of fish concentration zones allow for efficient, rapid, and safe daily loading of live fish onto waiting trucks.

The algal-based production system proved to be reasonably stable. Algae species dominance was observed to change during the season, shifting primarily from diatoms to green algae as water temperature increased. In spite of these changes there were only limited occasions when short periods of depressed algal productivity or algae die-offs impacted system performance during the three years of operation. At no time did algal events result in significant fish mortality. However, fish feeding was reduced for periods lasting as long as 1 to 2 weeks until normal algae densities could be reestablished.

After springtime stocking, pondway feeding rates began at 682 kg/d (1,500 lb). As water temperature increased, feeding was ramped up to 2,272 to 2,727 kg/d (5,000 to 6,000 lb) of total feed for both enclosures. Feeding rates were lower than added to tank-based systems and was spread out over more feedings per day to reduce oxygen demand and avoid low dissolved oxygen concentrations. The KST pondway/PAS systems produced tilapia yields of 17,727 kg (39,000 lb) in 2005; 60,454 kg (133,000 lb) in 2006; 69,900 kg (134,000 lb) in 2007; and 143,182 kg (315,000 lb) in 2008.

There are several advantages to the pondway approach, all resulting in more efficient fish production at lower cost. Many of the advantages are derived from confinement of the fish in a more manageable area so that the culturist can control the environment, accelerate growth rates, supplement oxygen levels, manage water quality, maximize feeding, improve feed conversion efficiency, treat disease, control parasites, eliminate bird predation, and manage fish size with grading, transfer, and harvest operations. The system also results in lower costs for utilities, chemical use, and water treatment to regulate pH, and less labor is needed for fish transfers and harvests. Likewise, there are cost advantages in treating the recycled water in pond systems as opposed to using more expensive filtration systems, including reduced costs of installation, operation, and maintenance. The pondway method is suitable for a wide variety of warm and temperate water fish species, including striped bass, catfish, tilapia, walleye, and many other candidate species.

During twenty-nine years of operation, KST utilized 26.5°C geothermal well water to grow tilapia, carp, catfish, and hybrid striped bass in approximately 100 concrete tanks on a year-round basis relying on a unique biological water treatment system recirculating 57,000 Lpm (15,000 gpm), or 80% of water needs. KST's staff of more than fifty employees produced and marketed over 50 million pounds of hybrid striped bass worldwide, and they sold several million pounds of tilapia and hybrid carp sold primarily to local markets. In 2006, as a result of increasing production costs and competition from imported frozen striped bass at US$3.30/kg ($1.50/lb), Kent SeaTech reduced hybrid striped bass production from 1.4 million kg/yr to 0.45 million kg/yr. At this time Kent began to increase pondway tilapia production. Unfortunately, at the same time, imported frozen tilapia at US\$1.32/kg began to appear in high volumes. Simultaneously, energy and feed costs rose dramatically.

In spite of improved economics of the pondway production system, in 2008, managers at KST decided that positive margins for fish sales in the United States were no longer possible. Consequently, the decision was made to alter Kent's business direction. In 2008, Kent SeaTech reemerged as Kent BioEnergy, transitioning to algae/aquaculture production for wastewater treatment and production of algae-based products including biofuels and biopower generation, animal and human food, pharmaceuticals, and other higher-value algal products.

13.8   References

Benemann, J.R., Koopman, B.L., Weissman, J.C., Eisenberg, D.E. & Goebel, R.P. (1980) Development of microalgae harvesting and high-rate pond technology. In *Algal Biomass* (Ed. by G. Shelef & C.J. Soeder), pp. 457–99. Elsevier, Amsterdam.

Benemann, J.R. (1997) CO_2 mitigation with microalgae systems. *Energy Conversion Managment* 38(Suppl.1):S475–79.

Benemann, J.R. & Weissman J.C. (1993) Food, fuel and feed production with microalgae. *Proceedings of the First Biomass Conference of the Americas*, NREL/CP-200-5768, pp. 1427–40. National Renewable Energy Laboratory, Golden.

Brune, D.E. (1991) Pond aquaculture. *Engineering Aspects of Intensive Aquaculture, Publication No. NRAES-49*. Northeast Regional Agriculutural Engineering Service, Ithaca.

Brune, D.E. (1997) Water quality dynamics as a basis for aquaculture system design. In *Striped Bass Culture* (Ed. by R. Harrel), pp. 99–126. Elsevier, Amsterdam.

Brune, D.E., Collier, J.A. & Schwedler, T.E. (2001a) *Partitioned Aquaculture System*. United States Patent No. 6,192,833 B1.

Brune, D.E., Reed, S., Schwartz, G., Collier, J.A., Eversole, A.G. & Schwedler, T.E. (2001b) High rate algal systems for aquaculture. In *Proceedings of the Aquacultural Engineering Society Issues Forum*, Natural Resources, Agriculture and Engineering Service Publication 157. Cooperative Extension, Ithaca.

Brune, D.E., Schwartz, G., Eversole, A.G., Collier, J.A. & Schwedler, T.E. (2003) Intensification of pond aquaculture and high rate photosynthetic systems. *Aquacultural Engineering* 28:65–86.

Brune, D.E., Schwartz, G., Collier, J.A., Eversole, A.G. & Schwedler, T.E. (2004a) Partitioned Aquaculture Systems. In *Biology and Culture of Channel Catfish* (Ed. by C.S. Tucker & J.A. Hargreaves), pp. 561–84. Elsevier, Amsterdam.

Brune, D.E., Schwartz, G., Eversole, A.G., Collier, J.A. & Schwedler, T.E. (2004b) Partitioned Aquaculture Systems. *Southern Regional Aquaculture Center*. Publication No. 4500. Southern Regional Aquaculture Center, Stoneville.

Brune, D.E., Eversole, A.G., Collier, J.A. & Schwedler, T.E. (2007) *Controlled Eutrophication Process*. United States Patent No. 7,258,790.

Brune, D.E. & Wang, J.K. (1998) Recirculation in photosynthetic aquaculture systems. *Aquaculture Magazine* 24(3):63–71.

Busch, R.L. (1985) Channel catfish culture in ponds. In *Channel Catfish Culture* (Ed. by C.S. Tucker), pp. 13–78. Elsevier, Amsterdam.

Drapcho, C.M. & Brune, D.E. (2000) Modeling of oxygen dynamics in the Partitioned Aquaculture System. *Aquacultural Engineering* 21(3):151–62.

Drapcho, C.M. (1993) *Modeling Algal Productivity and Diel Oxygen Profiles in the Partitioned Aquaculture System*. Doctoral Dissertation, Department of Agriculture & Biological Engineering Department, Clemson University.

Goode, T., Hammig, M. & Brune, D.E. (2002) Profitability comparison of the Partitioned Aquaculture System with a traditional catfish farm. *Aquaculture Economics & Management* 6(1–2):19–38.

Green, F.B. (1994) *Energetics of Advanced Integrated Wastewater Pond System*. Doctoral Dissertation, Energy & Resources Program, UC Berkeley.

Hargreaves, J.A. & Tucker, C.S. (2004) Industry development. In *Biology and Culture of Channel Catfish* (Ed. by C.S. Tucker & J.A. Hargreaves), pp. 1–14. Elsevier, Amsterdam.

Meade, J.L. (1998) *Algal and nitrogen dynamics in the Partitioned Aquaculture System*. Doctoral Dissertation, Biosystems Engineering, Clemson University.

Oswald, W.J. (1988) Microalgae and wastewater treatment. In *Micro-algal Biotechnology* (Ed. by M.A. Borowitzka & L.J Borowitzka), pp. 357–394. Cambridge University Press, Cambridge.

Oswald, W.J. (1995) Ponds in the Twenty-First Century. *Water Science Technology* 31(12):1–8.

Schwartz, G. (1998) *Photosynthesis and oxygen dynamics of the Partitioned Aquaculture System*. Master's Thesis, Biosystems Engineering, Clemson University.

Setmire, J.G., Schroeder, R.A. & Densmore, J.N. (1993) Detailed study of water quality, bottom sediment, and biota associated with irrigation drainage in the Salton Sea Area, California, 1988–1990. *US Geological Survey Water-Resources Investigations Report 93-4014*, Denver.

Setmire, J.G. (Ed.) (2000) *Eutrophic Conditions at the Salton Sea*. A topical paper from the Eutrophication Workshop convened at the University of California at Riverside, September 7–8, 2000. US Bureau of Reclamation.

Shelef, G. & C. J. Soeder (Eds.) (1980) *Algal Biomass Production and Use*. Elsevier, Amsterdam.

Stuart, K.R., Eversole, A.G. & Brune, D.E. (2001) Filtration of green algae and cyanobacteria by freshwater mussels in the Partitioned Aquaculture System. *Journal of the World Aquaculture Society* 32:105–11.

Swingle, H.S. (1959) Experiments on growing fingerling channel catfish to marketable size in ponds. *Proceedings of the Conference of the Southeast Association of Game and Fish Commission* 12:63–72.

Torrans, E.L. (2005) Effect of oxygen management on culture performance of channel catfish in earthen ponds. *North American Journal of Aquaculture* 67:275–88.

Tucker, L., Boyd, C.E. & McCoy, E.W. (1979) Effects of feeding rate on water quality, production of water quality and economic returns. *Transactions of the American Fisheries Society* 108:389–96.

Turker, H., Eversole, A.G. and Brune, D.E. (2002) Partitioned Aquaculture Systems: Tilapia filter green algae and cyanobacteria. *Global Aquaculture Advocate* 5(1):68–9.

Turker, H., Eversole, A.G. & Brune, D.E. (2003a) Comparative Nile tilapia and silver carp filtration rates of partitioned aquaculture system phytoplankton. *Aquaculture* 220:449–57.

Turker, H., Eversole, A.G. & Brune, D.E. (2003b) Effect of Nile tilapia, *Oreochromis niloticus*, size on phytoplankton filtration rate. *Aquaculture Research* **34**:1087–91.

Turker, H., Eversole, A.G. & Brune, D.E. (2003c) Filtration of green and cyanobacteria by Nile tilapia, *Oreochromis niloticus,* in the Partitioned Aquaculture System. *Aquaculture* **215**:93–101.

Turker, H., Eversole, A.G. & Brune, D.E. (2003d) Effects of temperature and phytoplankton concentration on Nile Tilapia, *Oreochromis niloticus,* filtration rates. *Aquaculture Research* **34**:453–9.

Chapter 14

Aquaponics—Integrating Fish and Plant Culture

James E. Rakocy

Aquaponic systems are recirculating aquaculture systems that incorporate the production of plants without soil. Recirculating systems are designed to raise large quantities of fish in relatively small volumes of water by treating the water to remove toxic waste products and then reusing it. In the process of reusing the water many times, nontoxic nutrients and organic matter accumulate. These metabolic byproducts need not be wasted if they are channeled into secondary crops that have economic value or in some way benefit the primary fish production system. Systems that grow additional crops by utilizing byproducts from the production of the primary species are referred to as integrated systems. If the secondary crops are aquatic, or if terrestrial plants grown in conjunction with fish, this integrated system is referred to as an aquaponic system.

Plants grow rapidly in response to dissolved nutrients that are excreted directly by fish or generated from the microbial breakdown of fish wastes. In closed recirculating systems with very little daily water exchange (less than 2%), dissolved nutrients accumulate and approach concentrations that are found in hydroponic nutrient solutions. Dissolved nitrogen, in particular, can occur at very high levels in recirculating systems. Fish excrete waste nitrogen directly into the water through their gills in the form of ammonia. Bacteria convert ammonia to nitrite and then to nitrate. Ammonia and nitrite are toxic to fish, but nitrate is relatively harmless and is the preferred form of nitrogen for growth of higher plants, such as fruiting vegetables. It is the symbiotic relationship between fish

Aquaculture Production Systems, First Edition. Edited by James Tidwell.
© 2012 John Wiley & Sons, Inc. Published 2012 by John Wiley & Sons, Inc.

and plants that makes the consideration of an aquaponic system a reasonable system design.

Aquaponic systems offer several advantages. In RAS, the disposal of accumulated waste is always a major concern. Recirculating systems are promoted as a means of reducing the volume of waste discharged to the environment. Certainly the volume is reduced, but the pollution load (organic matter, dissolved nutrients) per unit of discharge is correspondingly higher. This more concentrated discharge may pose a threat to the environment in some situations or generate an additional expense if the wastewater is discharged to a municipal sewer system for further treatment. Effluent is discharged from the system to eliminate organic sediment and prevent nutrient buildup.

In aquaponic systems, the plants recover a substantial percentage of these nutrients, thereby reducing the need to discharge water to the environment and therefore extending water use (i.e., by removing dissolved nutrients through plant uptake, the water exchange rate can be reduced). Minimizing water exchange reduces operating costs of aquaponic systems in arid climates and heated greenhouses where water or heated water represents a significant expense. Lennard (2006) demonstrated that nitrate accumulation in culture waters was reduced by up to 97% (table 14.1) in the aquaponic system when compared with the fish-only system.

Profitability is always a major concern when considering a recirculating system. Recirculating systems are expensive to construct and operate, and profitability often depends on serving niche markets for live fish such as tilapia, whole fresh fish on ice, or other high-value products. A secondary plant crop, which receives most of its required nutrients at no additional cost, improves system profit potential. The daily feeding of fish provides a steady supply of nutrients to plants, which reduces or eliminates the need to discharge and replace depleted nutrient solutions or adjust nutrient solutions as is required in hydroponics. The carbon dioxide vented from fish culture water can increase plant yields in enclosed environments. The plants purify the culture water and can, in a properly sized and designed facility, eliminate the need for separate and expensive biofilters. Biofiltration represents a major capital expense and a minor operational expense. In well-designed aquaponic systems, the hydroponic component can provide sufficient biofiltration for the fish, and therefore the cost of purchasing and operating a separate biofilter is avoided. These costs are

Table 14.1 Fish growth, lettuce yield, and nitrate removal for fish-only systems and aquaponic systems (Lennard 2006).

Parameter	Fish-only	Aquaponic
Fish FCR	0.87 ± 0.01	0.88 ± 0.0
Lettuce yield (kg/m^2)	NA	5.77 ± 0.19
NO$_3$ accumulation (mg/l)	52.20 ± 5.28	1.43 ± 1.09
NO$_3$ removal (%)	0	97

charged to the hydroponic subsystem, which, in the case of lettuce, generates approximately two-thirds of the system's income. The profitability of recirculating systems can thus be improved substantially with aquaponics, if there is a good market for the vegetable crop.

The expense of water quality monitoring is reduced in aquaponic systems as waste nutrients are generated daily at uniform levels and there is generally excess wastewater treatment capacity. An aquaponics system also generates savings in several areas of construction and operation by sharing operational and infrastructural costs for pumps, blowers, reservoirs, heaters, and alarm systems. Initial capital investment is reduced, in that an aquaponics system can be erected with a modest increase in acreage over that required for a hydroponic facility. Aquaponic systems do require high capital investment, moderate energy inputs, and skilled management. The premium prices available in niche markets may be required for an aquaponic facility to be profitable.

There are, of course, disadvantages to aquaponic systems. The most obvious of these is the large ratio of plant growing area in comparison to the fish rearing surface area. A large ratio of plant surface to fish surface is needed to achieve a balanced system in which nutrient levels stay relatively constant. For example, in the UVI raft system the ratio of plant growing area to fish surface area ratio is 7:3. Larger ratios are needed as solids removal efficiency decreases. In essence, aquaponic systems emphasize plant culture, which is an advantage if viewed by a horticulturist. Most of the labor expended in the facility is devoted to seeding, transplanting, maintaining, harvesting, and packing plants. Additionally, a new set of skills is required for the plant component, so a commercial operation would do better with both an aquaculturist and horticulturist on staff. Another disadvantage is that the horticulturist must rely on biological control methods rather than pesticides to protect the plants from pests and diseases. However, this restriction can be viewed as an advantage in that the plant products can be niche marketed as "pesticide free."

14.1 System design

The design of aquaponic systems closely mirrors that of recirculating systems in general, with the addition of a hydroponic component and the possible elimination of a separate biofilter and devices (foam fractionators) for fine and dissolved solids removal. Fine solids and dissolved organic matter generally do not reach levels that require foam fractionation in aquaponic systems at the recommended design ratio. The essential elements of an aquaponic system consist of a fish rearing tank, a settleable and suspended solids removal component, a biofilter, a hydroponic component and a sump (fig. 14.1; Rakocy & Hargreaves 1993).

Effluent from the fish-rearing tank is treated first to reduce organic matter concentration in the form of settleable and suspended solids. Next, the culture water is treated to remove ammonia and nitrite by fixed-film nitrification, which often occurs in the hydroponic component. As water flows through the

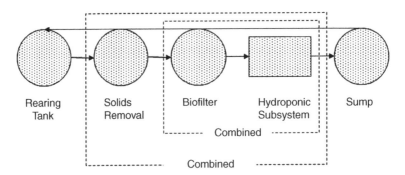

Figure 14.1 Optimum arrangement of aquaponic system components.

hydroponic unit, some dissolved nutrients are recovered by plant uptake. Finally, water collects in a reservoir (sump) where it is returned to the rearing tank. The location of the sump may vary. For example, if elevated hydroponic troughs are used, the sump can be located after the biofilter and water would be pumped up to the troughs and returned by gravity to the fish-rearing tank.

The system can be configured such that a portion of the flow is diverted to a particular treatment unit (Naegel 1977; Wren 1984). For example, a small side-stream flow may go to a hydroponic component after solids removal, while most of the water passes through a biofilter and returns to the rearing tank.

The biofiltration and hydroponic components can be combined by using a plant support media, such as gravel (Lewis *et al.* 1978; Sutton & Lewis 1982; Rakocy 1984; Watten & Busch 1984) or sand (McMurtry *et al.* 1990), which also functions as biofilter media. Raft hydroponics, which consist of floating sheets of polystyrene and net pots for plant support, can also provide sufficient biofiltration if the plant production area is sufficiently large (Rakocy 1995). Combining biofiltration with hydroponics is a desirable goal because eliminating the expense of a separate biofilter is one of the main advantages of aquaponics. An alternative design combines solids removal, biofiltration, and hydroponics in one unit. The hydroponic support media (e.g., pea gravel) captures solids and provides surface area for fixed-film nitrification, although with this design, it is important not to overload the unit with suspended solids. An overload of suspended solids is always a threat due to variations in fish feeding activities and efficiency of the solid removal component. For these reasons, gravel or sand beds *should be avoided* for large commercial-scale operations.

14.1.1 Aquaponics research at the University of the Virgin Islands (UVI)

Aquaponics research at the University of the Virgin Islands (UVI) has focused on the culture of tilapia in outdoor tanks equipped with raft hydroponics. As

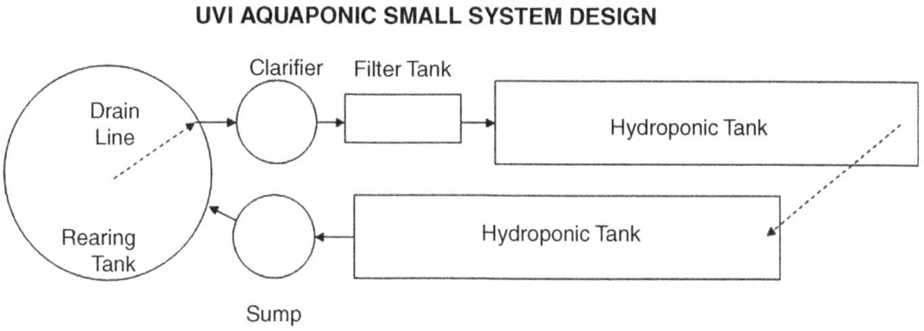

UVI AQUAPONIC SMALL SYSTEM DESIGN

Figure 14.2 Design of UVI experimental aquaponic system.

the UVI system developed, there were many design evolutions. Most of the experimental work was conducted in six replicated systems that consisted of a rearing tank (12.8 m^3), clarifier (1.9 m^3), two hydroponic tanks (13.8 m^2), and a sump (1.4 m^3; fig. 14.2). The hydroponic tanks (28-cm deep) were initially filled with gravel supported by wire mesh above a false bottom (7.6 cm). The gravel bed, which served as a biofilter, was alternately flooded with culture water and drained. Coarse gravel proved to be difficult to work with, so it was removed in favor of a raft system, consisting of floating sheets of polystyrene 1.22 m × 2.44 m × 3.8 cm (4 ft × 8 ft × 1.5 in). A rotating biological contactor (RBC) was then used for nitrification. Effluent from the clarifier was split into two flows, one going to the hydroponic tanks and the other to the RBC. These flows merged in the sump, from which the treated water was pumped back to the fish-rearing tank. A 0.7 m^3 filter tank was added later to remove suspended solids.

The fish-rearing tank was situated under an opaque canopy, and the clarifier and filter tank sump were covered with plywood. Shading inhibits algae growth, lowers daytime water temperature, and creates more natural light conditions for the fish. The rearing tank in this particular design proved to be too large relative to the plant growing surface area of the hydroponic tanks, or, conversely, the hydroponic tanks were too small relative to the size of the rearing tank. When the rearing tank was stocked with tilapia at commercial densities (107 fish/m^3), the daily feed ration to the system was so high that nutrients rapidly accumulated to levels that exceeded the recommended upper limits for hydroponic nutrient solutions (2,000 mg/L as total dissolved solids or TDS; Rakocy *et al.* 1993). The optimum ratio between the fish feeding rate and plant growing area was determined, using Bibb lettuce as the baseline plant, to be 57 g of feed/day/m^2 of plant growing area (Rakocy 1989a). At this ratio, the nutrient accumulation rate decreased, and the hydroponic tanks were able to provide sufficient nitrification. This success enabled the RBCs to be removed when the fish stocking rates were reduced to levels that allowed feed to be administered near the optimum rate for good plant growth. The optimum ratio will vary depending on plant species, production method (i.e., staggered vs. batch culture) and water exchange rate.

The experimental system was scaled up two times. In the first scale-up, the length of each hydroponic tank was increased from 6.1 m to 29.6 m. The optimum design ratio of 57 g feed/day/m² of plant growing area allowed the rearing tank to be stocked with tilapia at commercial levels (for a diffused aeration system) without excessive nutrient accumulation.

> **Rule of Thumb**
>
> 57 g of feed/day per square meter of plant growing
> area for staggered production of Bibb lettuce with minimum water exchange
> (<1%/day)

In the second scale-up, the number of hydroponic tanks (29.6 m in length) was increased to six and the number of fish-rearing tanks was increased to four (fig. 14.3). This production unit design represents a realistic commercial scale, although there are many possible size options and tank configurations. For example, the number of hydroponic tanks could be reduced from six to two, as in the experimental unit, by increasing the width or length as long as the total area remained the same. Likewise, the rearing tanks (7.8 m³) could be increased in size and/or number, provided there was a corresponding increase in the surface

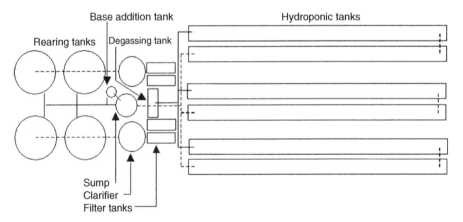

Tank Dimensions

Rearing tanks: Diameter: 3 m, Height: 1.2 m, Water volume: 7,800 L
Clarifiers: Diameter: 1.8, Height of cylinder: 1.2 m, Depth of cone: 1.1 m, Slope: 45°, Water volume: 3,785 L
Filter and degassing tanks: Length: 1.8 m, Width: 0.76 m, Depth: 0.61 m, Water volume: 700 L
Hydroponic tanks: Length: 30.5 m, Width: 1.2 m, Depth: 41 cm, Water volume: 11,356 L
Sump: Diameter: 1.2 m, Height: 0.9 m, Water volume: 606 L
Base addition tank: Diameter: 0.6 m, Height: 0.9 m, Water volume: 189 L
Total system water volume: 111,196 L
Flow rate: 387 Lpm, Pump: ½ hp, Blowers: 1½ hp (fish) and 1 hp (plants)
Total land area: 0.05 ha.

Pipe Sizes

Pump to rearing tanks: 7.6 cm
Rearing tanks to clarifier: 10 cm
Clarifier to filter tanks: 10 cm
Between filter tanks: 15 cm
Filter tank to degassing tank: 10 cm
Degassing to hydroponic tanks: 15 cm
Between hydroponic tanks: 15 cm
Hydroponic tanks to sump: 15 cm
Sump to pump: 7.6 cm
Pipe to base addition tank: 1.9 cm
Base addition tank to sump: 3.2 cm

Figure 14.3 Layout of UVI aquaponic system with tank dimensions and pipe sizes.

area of the hydroponic tanks. There is more flexibility in sizing and configuring outdoor systems, which are restricted to tropical or subtropical climates for year-round production of fish, than is available for indoor systems. In temperate indoor systems, the components must be configured tightly to conserve space, and the unit cannot exceed the width of a greenhouse bay, which ranges from 6.7 to 9.45 m. UVI's commercial-scale unit (second scale-up) could be configured to occupy as little as 0.05 ha of land.

14.2 Fish production

Tilapia is the most common fish cultured in aquaponic systems. Although some aquaponic systems have used channel catfish, largemouth bass, crappies, rainbow trout, pacu, common carp, koi carp, goldfish, Asian sea bass (barramundi), and Murray cod, most commercial systems are used for raising tilapia. The majority of freshwater species, which can tolerate crowding, including ornamental fish, will do well in aquaponic systems. One species reported to perform poorly is hybrid striped bass because they cannot tolerate high levels of potassium, which is often supplemented to promote plant growth.

To recover the high capital cost and operating expenses of aquaponic systems and to earn a profit, the fish-rearing component and the hydroponic vegetable component must be operated near maximum production capacity on a continuous basis. Three fish-stock management methods can be used to maintain fish biomass close to the system's maximum carrying capacity, the maximum amount of fish a system can support without restricting fish growth. Operating a system near its carrying capacity utilizes space efficiently, maximizes production, and reduces variation in the daily feed input to the system, an important factor in sizing the hydroponic component. The basic methods of fish management are sequential rearing, stock splitting, and multiple rearing units. It is important to determine the best method for varying circumstances prior to designing a commercial system, as each method has different requirements. Changing rearing methodology in a production mode is costly and interruptive to steady fish production.

14.2.1 Sequential rearing

Sequential rearing involves the culture of several age groups (multiple cohorts) of fish in the same rearing tank. When one age group reaches marketable size, it is selectively harvested with nets and a grading system, and an equal number of fingerlings are immediately restocked in the same tank. There are three problems with this system: (1) the periodic harvests stress the remaining fish and could trigger disease outbreaks; (2) stunted fish avoid capture and accumulate in the system, wasting space and feed; and (3) it is difficult to maintain accurate stock records over time, which leads to a high degree of management uncertainty and unpredictable harvests. Nevertheless, sequential rearing has been used successfully on a commercial scale with tilapia, a very hardy species, despite its drawbacks. Sequential rearing may be risky for species less tolerant to the repeated stress of partial harvest.

14.2.2   Stock splitting

Stock splitting involves stocking very high densities of fingerlings and periodically splitting the population in half as the carrying capacity of the rearing tank is reached (Van Gorder 1991). A typical stock splitting system would divide the initial population three times so that the fish would go from one tank to two tanks and then from two tanks to four tanks and finally from four tanks to eight tanks, from which marketable fish are harvested. Alternatively, single cohorts can be moved to successively larger tanks, which reduces stress on the fish and generally takes less time than trying to physically divide a single cohort into two equal populations, based on weight or number, to be placed in tanks of the original size.

If the splitting cohort technique is chosen, a total of fifteen rearing tanks are required to be able to harvest eight tanks at the end of each growth interval, a period of five or six weeks for tilapia. This method of fish management avoids the carryover problem of stunted fish and improves the accuracy of stock inventory assessment. However, the moves can be very stressful on the fish unless a swimway is installed, which connects all the rearing tanks to a narrow channel and the fish can be herded into it through a hatch in the wall of the rearing tanks and maneuvered into another rearing tank by movable screens. With swimways, the division of the populations in half involves some guesswork because the fish cannot be weighed or counted. An alternative method involves crowding the fish with screens and pumping them into a hauling tank or directly into another tank using a pescalator.

14.2.3   Multiple rearing units

With multiple rearing units, the entire population is moved to larger rearing tanks when the carrying capacity of the initial rearing tank is reached. The fish are either herded through a hatch between adjoining tanks or into swimways connecting distant tanks. Multiple rearing units usually come in modules of two to four tanks and are connected to a common filtration system. After the largest tank is harvested, all of the remaining groups of fish are moved to the next largest tank, and the smallest tank is restocked with fingerlings.

A variation of the multiple-rearing-unit concept is the division of a long raceway into compartments with movable screens. As the fish grow, their compartment is increased in size and moved closer to one end of the raceway where they will eventually be harvested. These should be cross-flow raceways or mixed-cell raceways to ensure uniform water quality throughout the length of the tank. In a cross-flow raceway influent water enters the raceway through a series of ports down one side of the raceway while effluent water leaves the raceway through a series of drains down the other side. This system ensures that water is uniformly good throughout the length of the raceway.

Another variation is the use of several tanks of the same size. Each rearing tank contains a different age group of fish, but they are not moved during the production cycle. This system does not utilize space efficiently in the early stages of growth, but the fish are never disturbed and the labor involved in moving the fish is eliminated.

UVI's current commercial-scale, aquaponic system uses multiple rearing tanks to simplify stock management, as the fish are not moved during their twenty-four-week growout cycle. The system consists of four fish rearing tanks (7.8 m³ each, water volume), two cylindro-conical clarifiers (3.8 m³ each) with a 45° slope, four filter tanks (0.7 m³ each), one degassing tank (0.7 m³), six hydroponic tanks (11.3 m³ each, 29.6 m × 1.2 m × 0.4 m), one sump (0.6 m³) and one base addition tank (0.2 m³). The total hydroponic surface area is 214 m², and the total system water volume is 110 m³. Tilapia production is staggered in the four rearing tanks so that one rearing tank is harvested every six weeks. At harvest the rearing tank is drained and all of the fish are removed. The rearing tank is then refilled with the same water and immediately restocked with fingerlings for a twenty-four-week production cycle

The system is used to culture Nile tilapia (*Oreochromis niloticus*) and red tilapia. Fry are sex-reversed with 17α-methyltestosterone according to the INAD (Investigations in New Animal Drugs) protocol to obtain a consistently high percentage (~99%) of male fingerlings. Nile tilapia fingerlings are stocked at a rate of 77 fish/m³ to obtain a harvest size of 800 to 900 g. These large fish are processed for the fillet market. Red tilapia fingerlings are stocked at 154 fish/m³ to achieve an average weight near 500 g for the whole fish West Indian market. Every six weeks approximately 500 to 600 kg of fish are harvested. Annual production has been 9,152 lb (4.16 mt) for Nile tilapia and 10,516 lb (4.78 mt) for red tilapia (table 14.2). However, production can be increased to 11,000 lb (5 mt) with close observation of the ad libitum feeding response.

Tilapia grow well at high densities if good water quality is maintained. In the commercial-scale system, dissolved oxygen (DO) levels in the rearing tanks are maintained at 5 to 6 mg/L by high DO in the incoming water and by diffused aeration with air delivered through twenty-two air stones around the perimeter

Table 14.2 Average production values for male mono-sex nile and red tilapia in the UVI aquaponic system. Nile tilapia are stocked at 77 fish/m³ (0.29 fish/gallon) and red tilapia are stocked at (154 fish/m³; 0.58 fish/gallon).

Tilapia	Harvest Weight per Tank (kg)	Harvest Weight per Unit Volume (kg/m³)	Initial Weight (g/fish)	Final Weight (g/fish)	Growth Rate (g/day)	Survival (%)	FCR
Nile	480 kg (1,056 lb)	61.5 (0.51 lb/gal)	79.2	813.8	4.4	98.3	1.7
Red	551 kg (1,212 lb)	70.7 (0.59 lb/gal)	58.8	512.5	2.7	89.9	1.8

of the tank. A 1.5-hp (1.1 kW) blower provides air to the rearing and degassing tanks. Vigorous aeration vents carbon dioxide gas into the atmosphere and prevents its buildup. A high water exchange rate quickly removes suspended solids and toxic waste metabolites (ammonia and nitrite) from the rearing tank. A 1/2-hp in-line pump produces a flow of 380 Lpm (100 gpm) and an average retention time of 1.4 hours/rearing tank. However, flows to the individual rearing tanks are adjusted so that the tank with the highest biomass receives the highest flow rate, which exceeds 130 Lpm (35 gpm) for a retention time of less than 1 hour. The other rearing tanks receive proportionately lower flow rates relative to their biomass. Values of ammonia-nitrogen and nitrite-nitrogen in the rearing tanks are approximately 1 to 2 mg/L and <1 mg/L, respectively.

Through careful attention to management of the water quality parameters of DO, ammonia-nitrogen, and nitrite-nitrogen, it has been possible to grow tilapia at high densities. Other water quality variables of importance to the system are water temperature, pH, and alkalinity. Water temperature ranges from a low of 23°C in the winter to a high of 29°C in the summer. The average water temperature has been 27°C, which is lower than the optimum temperature (30°C) for tilapia and higher than the optimum temperature (20 to 22°C) for many vegetables. The system water temperature is lower than that in nearby ponds because none of the system's surface area is exposed to direct sunlight. The pH is generally maintained at 7.0 by adding equal amounts calcium hydroxide and potassium hydroxide. Total alkalinity averages approximately 100 mg/L as calcium carbonate ($CaCO_3$).

In general, it is recommended that the carrying capacity in aquaponic systems should not exceed 60 kg/m^3 (0.50 lb/gallon). This density will promote fast growth and efficient feed conversion and reduce crowding stress that may lead to disease outbreaks. Pure oxygen is generally not needed to maintain this density.

The logistics of working with both fish and plants are challenging. In the UVI system one rearing tank is stocked every six weeks. Therefore, eighteen weeks are required before a system is fully stocked. If multiple units are used, then fish may be stocked as frequently as once weekly and harvested and sold once weekly. Similarly, staggered crop production requires frequent seeding, transplanting, harvesting, and marketing. Therefore, the overarching goal in the design process is to reduce labor requirements wherever possible and make operations as simple as possible. For example, the purchase of four fish-rearing tanks adds extra expense. One larger tank could be purchased instead and partially harvested and partially restocked every six weeks. However, this operation requires additional labor, which is a recurring cost, and makes management more complex. In the long run, having smaller, multiple tanks, in which the fish are not disturbed until harvest (hence, less mortality and better growth) will be more cost effective.

14.3   Solids

Fish generate fecal waste, most of which should be removed from the waste stream before it enters the hydroponic tanks. Other sources of particulate waste are uneaten feed and organisms (e.g., bacteria, fungi, and algae) that grow in the system. If this organic matter accumulates in the system, it will depress dissolved oxygen (DO) levels as it decays and produce carbon dioxide and ammonia. If deep deposits of sludge form, they will decompose anaerobically (without oxygen) and produce methane and hydrogen sulfide, which are very toxic to fish.

Suspended solids have special significance in aquaponic systems. Suspended solids entering the hydroponic component may accumulate on plant roots and produce a deleterious effect by creating anaerobic zones, which kills roots, and blocking the flow of nutrients to the plant. However, some accumulation of solids may be beneficial. As solids undergo decomposition by microorganisms, inorganic nutrients essential to plant growth are released to the water, a process known as mineralization. Mineralization supplies several essential nutrients. Without sufficient solids for mineralization, more nutrient supplementation is required, thereby increasing the operating expense and management complexity of the system. However, it may be possible to minimize nutrient supplementation if fish stocking and feeding rates are increased relative to plants. Another benefit of solids is brought about by the action of decomposing microorganisms. Microbes associated with decomposing solids are antagonistic to plant root pathogens and help maintain healthy root growth. Therefore, it appears that a delicate balance must be reached between excessive accumulation of suspended solids and insufficient accumulation.

Some common devices used for removing solids from recirculating systems include settling basins, tube or plate separators, the combination particle trap and sludge separator, centrifugal separators, microscreen filters, and bead filters. Sedimentation devices (e.g., settling basins, tube or plate separators) primarily remove settleable solids (>100 microns) while filtration devices (e.g., microscreen filters, bead filters) remove settleable and suspended solids. Solids removal devices vary in regards to efficiency, solids retention time, effluent characteristics (both solid waste and treated water), and water consumption rate. While many devices may be appropriate for aquaponic systems, there is no research on the relationship between techniques for solids removal and the performance of hydroponic vegetables.

Sand and gravel hydroponic substrates are sometimes used to remove solid waste from the water flow stream. The solids remain in the system to provide nutrients to the vegetables through mineralization. As solids accumulate in the media, there is an increase in the cation exchange capacity (CEC), i.e., the ability of the media to adsorb and retain cations, positively charged nutrients, which are available for plant growth. Since cation concentrations are often high in aquaponic systems, CEC is generally not an important factor to plant growth. The use of sand is becoming less common, but one popular aquaponic system uses small beds 2.4 m by 1.2 m (8 ft by 4 ft) containing pea gravel ranging from 3 to 6 mm (1/8 to 1/4 inch) in diameter. The hydroponic beds are flooded several times daily with system water and then allowed to drain completely, and the water is returned to the rearing tank. During the draining phase, air is brought into the gravel. The high oxygen content of air (compared to water) speeds the decomposition of organic matter in the gravel. The beds are inoculated with worms (*Eisenia foetida*), which improve bed aeration and assimilate organic matter.

The organic waste produced in aquaponic systems does not break down completely. The fraction of organic waste that resists microbial decomposition is referred to as being refractory. Particulate refractory compounds will slowly accumulate in substrates such as pea gravel or on the bottom or raft system troughs. Dissolved refractory compounds give the culture water a brown or tea color, which contains tannic acid, humic acid, and other humic substances. These compounds have mild antibiotic characteristics and are beneficial to the system's fish and plants. Humic compounds form metalo-organic complexes with Fe, Zn, and Mn and thereby increase the availability of these micronutrients to plants.

14.3.1   Solids removal

The most appropriate device for solids removal in a particular system may depend primarily on the organic loading rate (daily feed input and feces production) and secondarily on the plant growing area. For example, if large amounts of fish (high organic loading) are raised relative to the plant growing area, then a highly efficient solids removal device such as a microscreen drum filter is desirable. Microscreen drum filters capture fine organic particles (e.g., 60 micron and larger), which are retained by the screen for only a few minutes prior to backwashing and removal from the system. In this system, the dissolved nutrients are excreted directly by the fish or produced by mineralization of very fine particles, and dissolved organic matter may be sufficient for the size of the plant growing area. At the other extreme, if small amounts of fish (low organic loading) are raised relative to the plant growing area, then solids removal may be unnecessary, as more mineralization is needed to produce sufficient nutrients for the relatively large plant growing area. However, un-stabilized solids, solids that have not undergone microbial decomposition, should not be allowed to accumulate on the tank bottom and form anaerobic zones. A reciprocating pea gravel filter (subject to flood and drain cycles), in which incoming water is spread evenly over the entire bed surface, may be the most appropriate device in this situation because solids are evenly distributed in the gravel and exposed to high oxygen levels (21% in air as compared to 0.0005 to 0.0007% in fish culture water) on the drain cycle, thereby enhancing microbial activity and increasing the mineralization rate.

UVI's initial commercial-scale aquaponic system on two cylindro-conical clarifiers to remove settleable solids (Rakocy 1984). The fiberglass clarifiers each were 1.9 m^3 (500 gallons in volume). The cylindrical portion of the clarifier is situated above ground and has a central baffle that is perpendicular to the incoming water flow (fig. 14.4). The lower conical portion with a 60° slope was buried below ground. A drainpipe is connected to the apex of the cone. The

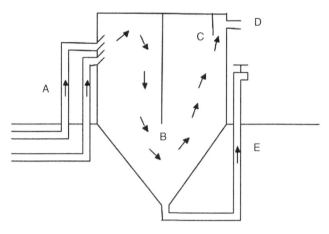

Figure 14.4 Cross sectional view (not to scale) of UVI clarifier showing drain lines from two fish rearing tanks (A), central baffle (B) and discharge baffle (C), outlet to filter tanks (D), sludge drain line (E) and direction of water flow (arrows).

drainpipe rises vertically out of the ground to the middle of the cylinder and is fitted with a ball valve. Rearing tank effluent enters the clarifier just below the water surface. The incoming water is deflected upward by a 45° pipe elbow to dissipate the current. As water flows under the baffle, turbulence diminishes and solids settle on the sides of the cone. The solids accumulate there and form a thick mat that eventually rises to the surface of the clarifier. To prevent this, approximately twenty male fingerlings are required to graze on the clarifier walls and consolidate solids at the base of the cone. Solids are removed from the clarifier three times daily. Hydrostatic pressure forces solids through the drain line when the ball valve is opened. A second, smaller baffle keeps floating solids from being discharged to the filter tanks.

The fingerlings serve another purpose. They swim into and through the drain lines and keep them clean. Without tilapia, the 10.2-cm (4-in) drain lines would require manually cleaning nearly every day due to bacterial growth in the drain lines, which constricts water flow. A cylindrical screen attached to the rearing tank drain prevents fingerlings from entering the rearing tank.

In UVI's current commercial-scale system, the clarifier volumes have been increased to 3.8 m^3 (1,000 gal) to achieve a twenty-minute retention time, which resulted in much more effective removal of settleable solids. The clarifiers were constructed with standard fiberglass material. The slope of the conical portion was reduced to 45°. Although a 60° slope is more efficient for solids removal, installation of a heavy fiberglass tank with such a large cone is not feasible. Each clarifier is stocked with thirty male fingerlings to enhance clarifier performance (e.g., for grazing on surfaces and resuspending solids from upper reaches of the clarifier).

Although the cylindro-conical clarifier can remove 21% of the dry weight of feed added to the system during a production cycle (Rakocy *et al.* 1991), large quantities of solids are not removed. Twarowska *et al.* (1997) determined that 35.3% of feed input to a tilapia culture system can be captured as settleable and non-settleable solids (based on volatile solids analysis) by using a particle trap and particle separator (17.6% removal) in combination with a microscreen drum filter (17.7% removal). These results indicate that the clarifier removes approximately 59% of the total removable solid waste. Although fingerlings are needed for effective clarifier performance, their grazing and swimming activities are also counterproductive in that they resuspend some solids that exit through the clarifier outlet. The fish in the clarifier grow rapidly and should be replaced with small (~50 g) fingerlings every 12 weeks.

With clarification as the sole method of solids removal, large quantities of solids were discharged into the hydroponic tanks where they settled out and formed sludge deposits more than 5-cm deep at the influent end. This was a very undesirable condition that adversely affected the plant component, as the sludge would float to the surface, engulf the plant roots, and either kill the plants or greatly reduce their growth. A series of experiments resulted in arriving at a design solution that incorporated another water treatment stage consisting of additional tanks filled with orchard netting for the removal of fine solids. Two rectangular filter tanks $0.7 \, \text{m}^3$ (185 gal) were installed after each clarifier. Effluent from the clarifier flows through these tanks in series. The netting is washed once or twice a week with a high-pressure water sprayer, and all the water in the filter tanks is discharged, and the sludge is discharged to lined holding ponds. Prior to cleaning, a small sump pump is use to carefully return the filter tank water to the rearing tanks without dislodging the solids. This process conserves water and nutrients but increases dissolved organic matter and biofouling.

Effluent from the UVI rearing tanks is highly enriched with dissolved organic matter, which stimulates the growth of filamentous bacteria in the drain lines, clarifiers, and filter tanks. The bacteria appear as translucent, gelatinous, light-tan filaments. Tilapia consume the bacteria and control its growth in the drain lines and clarifiers, but bacteria do accumulate in the filter tanks. Without the filter tanks, the bacteria would overgrow plant roots. The bacteria do not appear to be pathogenic, but they do interfere with the uptake of dissolved oxygen, water, and nutrients, thereby affecting plant growth. The feeding rate to the system and the flow rate from the rearing tank determine the extent to which filamentous bacteria manifest, but it can be contained by providing a sufficient area of orchard netting, either by adjusting screen tank size or using multiple screen tanks. In systems with lower organic loading rates (i.e., feeding rates) or lower water temperature (hence, less biological activity), filamentous bacteria diminish and are not a problem.

The organic matter that accumulates on the orchard netting between cleanings forms a thick sludge. Anaerobic conditions develop in the sludge, which leads to formation of gases such as hydrogen sulfide, methane, and nitrogen. Therefore, a degassing tank is used in the UVI system to receive the effluent from the filter tanks. A number of air diffusers vent the gasses into the atmosphere before the culture water reaches the hydroponic plants. The degassing tank has an internal standpipe well that splits the water flow into three sets of hydroponic tanks.

In UVI's commercial-scale system, the combination clarifier and filter tanks have maintained very low levels of total suspended solids. In one trial with leaf lettuce, suspended solids values averaged 4.2 mg/L in the filter tank effluent (Rakocy *et al.* 1997). The hydroponic tanks also contribute to suspended solids removal. There is no phytoplankton in the water, which is clear but darkly colored with humic substances.

Solids that are discharged from aquaponic systems must be disposed in an environmentally acceptable manner. There are several methods for effluent treatment and disposal. Effluent can be stored in aerated ponds and applied as relatively dilute sludge to land after organic matter has stabilized. This method is advantageous in dry areas where sludge can be used to irrigate and fertilize field crops. The solid fraction of the sludge can be separated from the water, using geotextile membranes and polymers, and combined with other waste products from the system (vegetable matter) to form compost. As an example, solids from the UVI commercial-scale aquaponic system are discharged through drain lines into two lined 16-m^3 ponds, which are continuously aerated with diffused air. As one pond is being filled over a two- to four-week period, water from the other pond is used to irrigate and fertilize field crops. Urban area facilities might have to discharge solid waste into sewer lines for disposal at the municipal sewage treatment plant.

14.4 Biofiltration

A major concern in aquaponic systems is the removal of ammonia, a metabolic waste product excreted through the gills of fish. Ammonia will accumulate and reach toxic levels unless it is removed by the process of nitrification (referred to more generally as biofiltration), in which ammonia is oxidized first to nitrite, which is toxic. Then nitrite is oxidized to nitrate, which is relatively nontoxic. Two groups of naturally occurring bacteria (*Nitrosomonas* and *Nitrobacter*) mediate this two-step process. Nitrifying bacteria grow as a film (referred to as biofilm) on the surface of inert material or they adhere to organic particles. Biofilters contain media with large surface areas for the growth of nitrifying bacteria. Aquaponic systems have used biofilters with sand, gravel, shells, or various plastic media as substrate (Rakocy & Hargreaves 1993). Biofilters perform optimally at a temperature range of 25 to 30°C, a pH range of 6.0 to 9.0, saturated DO, low BODs <20 mg/L, and total alkalinity of 100 mg/L or greater. Nitrification is an acid-producing process that destroys alkalinity. Therefore, an alkaline base must be added frequently depending on feeding rate to maintain relatively stable pH values. Some means of removing dead biofilm is necessary to prevent media clogging, short circuiting of water flow, decreasing DO values, and declining biofilter performance.

In aquaponic systems that use hydroponic subsystems with insufficient surface area for biofiltration or culture species that demand excellent water quality a separate biofilter is required. A moving bed biofilter is the the best biofilter in these situations because it is self cleaning and requires virtually no maintenance.

If a separate biofilter is required or if a combined biofilter (biofiltration and hydroponic substrate) is used, the standard equations used to size biofilters may not apply to aquaponic systems, as additional surface area is provided by plant roots and a considerable amount of ammonia removal is due to direct uptake by plants. However, the contribution of various hydroponic subsystem designs and plant species to water treatment in aquaponics systems has not been studied. Therefore, aquaponic system biofilters should be sized fairly close to the recommendations for recirculating systems.

Nitrification efficiency is affected by pH. The optimum pH range for nitrification is 7.0 to 9.0, although most studies indicate that nitrification efficiency is greater at the higher end of this range. Most hydroponic plants grow best at pH in the range of 5.8 to 6.2. The acceptable range for hydroponic systems is 5.5 to 6.5. The pH of a solution affects nutrient solubility—especially trace metals. Essential nutrients such as iron, manganese, copper, zinc, and boron are less available to plants at pH above 7.0 while the solubility of phosphorus, calcium, magnesium, and molybdenum sharply decreases at pH below 6.0. Compromise between nitrification and nutrient availability is reached in aquaponic systems by maintaining pH close to 7.0.

Nitrification is most efficient when water is saturated with DO. The UVI commercial-scale system maintains DO levels near 80% saturation (6 to 7 mg/L) by aerating the hydroponic tanks with numerous small air diffusers: one every 1.2 m (4 ft) distributed along the long axis of the tanks. Reciprocating (ebb and flow) gravel systems expose nitrifying bacteria to high atmospheric oxygen levels during the dewatering phase. The thin film of water that flows through NFT channels absorbs oxygen by diffusion, but dense plant roots and associated organic matter can block water flow and create anaerobic zones, which precludes the growth of nitrifying bacteria and further necessitates the installation of a separate biofilter.

Ideally, aquaponic systems should be designed so that the hydroponic subsystem also serves as the biofilter, which eliminates the capital cost and operational expense of a separate biofilter. Granular hydroponic media such as gravel, lava cinders provide sufficient substrate for nitrifying bacteria and generally serve as the sole biofilter in some aquaponic systems, although they have a tendency to clog, as stated previously. If serious clogging occurs due to organic matter overloading, gravel and sand filters can actually produce ammonia as organic matter decays, rather than remove it. If this occurs, the media must be washed, and the system must be redesigned by installing a solids removal device prior to the granular biofilter, or the organic loading rate must be decreased by reducing fish stocking and feeding rates. McMurtry *et al.* (1990) relied on sand for hydroponic substrate and biofiltration while the early work at UVI (Rakocy 1984) utilized gravel media for both functions. In some aquaponic systems, nitrification in the hydroponic component has supplemented that in the biofilter. Lewis *et al.* (1978) used pea gravel for hydroponic substrate in conjunction with an RBC for biofiltration. Watten and Busch (1984) used crushed-gravel hydroponic substrate and a trickling biofilter. When gravel was eliminated from UVI's experimental units, an RBC was installed for biofiltration. When the experimental units were scaled up the first time, two RBCs were installed.

A series of three stock management experiments were conducted in the scaled-up unit, in which the surface area of each RBC was 92.9 m² (1000 ft²) and the total growing area for hydroponic lettuce was 71.4 m² (768 ft²). In the first two experiments the feeding to the system ranged from 3 to 6 kg/day (6.6 to 13.2 lb/day), and good water quality was maintained. A feeding rate of 4 kg/day (8.8 lb/day) was nearly equivalent to the optimum design ratio (57 g of feed/day/m²) for staggered lettuce production. In the third experiment the RBCs were eliminated to obtain a preliminary determination of the biofiltration capacity of the hydroponic component. The maximum feeding rate to the system was increased to 8 kg/day (17.6 lb/day), and water quality continued to be good. Average levels of ammonia-nitrogen and nitrite-nitrogen in the rearing tank were 1.3 and 0.7 mg/L, respectively. This experiment showed that a combination of direct ammonia uptake by lettuce plants and nitrification on the hydroponic tank floor, walls, and the underside of the floating polystyrene sheets can provide sufficient nitrification to maintain good water quality at a feeding rate that is two times greater than the optimum design ratio. The total surface area in each hydroponic tank was 92.9 m² (1000 ft²), not including the additional surface area provided by the plant roots. This is approximately the same surface area that would be available in an RBC.

An experiment was conducted in UVI's replicated systems to determine the waste treatment capacity of raft hydroponics (Gloger *et al.* 1995). Three identical systems were stocked with large *Oreochromis niloticus* (480 g/fish) so that feed could be incremented weekly until the waste treatment capacity was reached. The fish were initially underfed. Romaine lettuce (Parris Island) at a density of 29.6 plants/m² (2.75 plants/ft²) was produced continuously on a four-week, staggered cycle. The results were based on the hydroponic growing area, expressed as g/m²/day. The mean values and ranges were as follows: feeding rate, 159 g/m²/day (77 to 230); wet weight plant production, 338 g/m²/day; dry weight plant production, 16.9 g/m²/day; ammonia-nitrogen removal, 0.56 g/m²/day (−1.23 to 2.29); nitrite-nitrogen removal, 0.62 g/m²/day (−2.68 to 2.28); BOD removal, 4.78 g/m²/day (0.0 to 4.92); COD removal, 30.29 g/m²/day (−10.58 to 69.72); total nitrogen uptake by lettuce, 0.83 g/m²/day; total phosphorus uptake by lettuce, 0.17 g/m²/day. If the ammonia-nitrogen removal rate is divided by two, to account for the surface area of the tank floor and the underside of the polystyrene sheet, the resultant value (0.28 g/m²/day) falls within the range of removal rates reported for recirculating-system biofilters (Losordo 1997). The maximum sustainable feeding rate was equivalent to 180 g/m²/day—about three times greater than the optimum design ratio for the staggered production of Bibb lettuce in aquaponic systems.

Raft hydroponics not only provides adequate waste treatment for correctly sized aquaponic systems and eliminates the need and expense of separate biofiltration units, but its excess treatment capacity ensures safe and stable water quality. After an initial acclimation period of about one month, experience with the UVI systems has shown that it is not necessary to monitor ammonia and nitrite values on a frequent basis, but weekly checks would be a prudent management practice.

Aquaponic systems using nutrient film technique (NFT) as the hydroponic component may require a separate biofilter. NFT consists of narrow plastic channels for plant support with a film of nutrient solution flowing through them. Compared to raft culture, the water volume and surface area of NFT are considerably smaller because there is just a thin film of water and no substantial sidewall area and no raft underside surface area for colonization by nitrifying bacteria.

14.5   Hydroponic subsystems

A number of hydroponic subsystems have been used in aquaponics (Rakocy & Hargreaves 1993). Gravel hydroponic subsystems are common in small operations. To ensure adequate aeration of plant roots, gravel beds have been operated in a reciprocating (ebb and flow) mode, where the beds are alternately flooded and drained (Lewis *et al.* 1978; Lewis *et al.* 1980; Sutton & Lewis 1982; Wren 1984; Rakocy 1984), or in a dewatered state, in which culture water is applied continuously to the base of the individual plants through small-diameter plastic tubing (Watten & Busch 1984). Depending on its composition, gravel can provide some nutrients for plant growth (e.g., calcium is slowly released as the gravel reacts with acid produced during nitrification; Rakocy & Nair 1987).

Gravel has several negative aspects. The weight of gravel requires strong support structures. It is subject to clogging with suspended solids, microbial growth, and the roots that remain after harvest. The resulting reduction in water circulation together with decomposition of organic matter leads to the formation of anaerobic zones, which impairs or kills plant roots. The small plastic tubes used to irrigate gravel are also subject to clogging with biological growth. Moving and cleaning gravel substrate is difficult due to its weight. Planting in gravel is also difficult and plant stems can be damaged by abrasion in outdoor systems exposed to wind. Having high porosity, gravel retains very little water if drained. Disruption in flow will lead to the rapid onset of water stress (wilting). The sturdy infrastructure required to support gravel and the potential of clogging impose a size limitation on gravel beds.

One popular gravel-based aquaponic system uses pea gravel in 1.2 m × 2.4 m (4 ft × 8 ft) beds that are irrigated through a distribution system of PVC pipes over the gravel surface. Numerous small holes in the pipes distribute culture water on the flood cycles. The beds are allowed to drain completely between flood cycles. Solids are not removed from the culture water and organic matter does accumulate, but the beds are tilled between planting cycles, allowing some organic matter to be dislodged and discharged.

Sand has been used as hydroponic media in aquaponic systems (Ferguson 1982; McMurtry *et al.* 1990). Ferguson (1982) raised twenty-eight types of vegetables in trays of sand that were stacked three-tiers high in a commercial greenhouse. The greenhouse produced an average of (19,140 lbs) of vegetables during ten-week growth cycles in trays (2 ft to 3 ft wide) with a total length of 1,155 m. The trays were made of parachute cloth so that water applied to the upper tray slowly percolated through to irrigate the trays beneath. McMurtry *et al.* (1990) constructed hydroponic sand beds (24.6 ft × 4.9 ft × 1.6 ft) on sloped ground that was covered by polyethylene sheets. The beds were adjacent to in-ground rearing tanks with their floors sloping to one side. A pump in the deep end of the rearing tank was activated for thirty minutes five times daily to furrow irrigate the adjacent sand bed. The culture water percolated through the sand and returned to the rearing tank. The potential of sand substrates becoming clogged with solids can be reduced by regulating the solids loading rate. A coarse grade of sand is needed to reduce the potential for clogging over time, and some solids should be removed prior to irrigation.

Perlite is another media that has been used in aquaponic systems. Perlite is placed in shallow aluminum trays 8-cm (3 in) deep with a baked enamel finish. The trays vary from 20 cm to 1.2 m (8 in to 4 ft) in width and can be fabricated to any length, but 6 m (20 ft) is the maximum recommended length. At intervals of 6 m (20 ft), adjoining trays should be separated by 8 cm (3 in) or more in elevation so that effluent drops to the lower tray and becomes re-aerated. A slope ratio of 1 in 144 (or a 1-in drop in 12 ft of run or a 0.7% slope) is needed for water flow. A small trickle of water enters at the top of the tray, flows through the perlite, keeping it moist, and discharges into a trough at the lower end. Solids must be removed from the water before it enters the perlite tray. Full solids loading will clog the perlite, form short-circuiting channels, create anaerobic zones, and lead to nonuniform plant growth. Shallow perlite trays provide minimal area for root growth and are better for smaller plants such as lettuce and herbs.

Nutrient film technique (NFT) has been successfully incorporated into a number of aquaponic systems (Burgoon & Baum 1984; Head 1984). NFT consists of many narrow plastic troughs 10 to 15 cm wide (4 to 6 in) in which plant roots are exposed to a thin film of water that flows down the troughs delivering water, nutrients, and oxygen to the roots of the plants. The troughs are lightweight, inexpensive, and versatile. Troughs can be mounted over rearing tanks to efficiently utilize vertical greenhouse space. However, this practice is discouraged if it interferes with fish and plant operations such as harvesting. High plant density can be maintained by adjusting the distance between troughs to provide optimum plant spacing during the growing cycle. Aquaponic systems utilizing NFT require effective solids removal to prevent excess solids accumulation on roots, which can lead to root death and poor plant growth. With NFT, a disruption in water flow can lead quickly to wilting and death. Water is delivered at one end of the troughs by a PVC manifold with discharge holes above each trough and collected at the opposite, down-slope end in an open channel or large PVC pipe. The use of microtubes, used in commercial hydroponics, is not recommended because they will clog. The holes should be as large as practical to reduce cleaning frequency.

A floating or raft hydroponic subsystem is ideal for the cultivation of leafy green and other types of vegetables (Zweig 1986; Rakocy *et al.* 1989b). Long channels with closed-cell polystyrene sheets support vegetables at the water surface with roots suspended in the culture water (Jensen & Collins 1985). The system provides maximum exposure of roots to the culture water and avoids clogging, although suspended solids captured by the roots can cause root death if concentrations are high (Zweig 1986). The sheets shield the water from direct sunlight and maintain lower than ambient water temperatures. A disruption in pumping does not affect the plant's water supply as in gravel, sand, and NFT subsystems. The sheets are easily moved along the channel to a harvesting point where they can be lifted out of the water and placed on supports at an elevation that is comfortable for the workers.

The UVI system uses three sets of two raft hydroponic tanks that are 30.5-m (100-ft) long by 1.22-m (4-ft) wide by 40.6-cm (16-in) deep and contain 30.5 cm (12 in) of water. The channels are lined with a low-density polyethylenes liner 20-mils thick (a mil equals 1/40 of a millimeter or 1/1000 of an inch) and covered by expanded polystyrene sheets (rafts), which are 2.44-m (8-ft) long by 1.22-m (4-ft) wide by 3.8-cm (1.5-in) thick. Net pots are placed in holes in the raft and just touch the water surface. (5-cm) net pots are generally used for leafy green plants while 7.62-cm (3-in) net pots are used for larger plants such as tomatoes or okra. Holes of the same size are cut into the polystyrene sheet. A lip at the top of the net pot secures the net pot and keeps it from falling through the hole into the water. Seedlings are nursed in a greenhouse and then placed into net pots, and their roots grow into the culture water while their canopy grows above the raft surface.

A disadvantage of rafts in an aquaponic system is that roots are exposed to harmful organisms associated with aquaculture systems. For example, if tilapia fry gain access to the raft tanks, they consume plant roots or thereby severely stunt growth, although it is relatively easy to prevent the entry of tilapia by using a fine mesh screen. Similarly, blooms of zooplankton, especially ostracods, will consume root hairs and fine roots, retarding plant growth. Other pests are tadpoles, snails, and leeches that consume roots and nitrifying bacteria. These problems are surmounted by stocking some carnivorous fish that prey on pests in the hydroponic tanks. At UVI, snails are controlled with shellcracker sunfish (*Lepomis microlophus*), and zooplankton are controlled with black tetra (*Gymnocorymbus ternetzi*).

14.6 Sump

Water flows by gravity from gravel, sand, and raft hydroponic subsystems to a sump, which is the lowest point in the system. The sump contains a pump or pump inlet that returns the treated culture water to the rearing tanks. If NFT troughs or perlite trays are located above the rearing tanks, the sump would be positioned upstream of them so that water could be pumped up to the hydroponic component for gravity return to the rearing tanks. There should be only one pump to circulate water in an aquaponic system.

The sump should be the only tank in the system where the water level decreases as a result of overall water loss from evaporation, transpiration, sludge removal, and splashing. A mechanical valve is used for the automatic addition of replacement water from a storage reservoir or well. Municipal water should not be used unless it is de-chlorinated, and surface water should not be used because it may contain disease organisms. A water meter should be used to record additions. Unusually high water consumption indicates a leak.

The sump is a good location for the addition of base to the system. Soluble base such as potassium hydroxide causes high and toxic pH levels in the sump. However, as water is pumped into the rearing tank, it is diluted and pH decreases to acceptable levels. The UVI system uses a separate base addition tank located next to the sump. As water is pumped from the sump to the fish rearing tanks, a small pipe, tapped into the main water distribution line, delivers a small flow of water to the base addition tank, which is well aerated with one large air diffuser. Base is added to this tank as needed to maintain a pH of 7.0 in the system. The base dissolves, gradually enters the sump, and is pumped to the rearing tanks where it is quickly diluted in large volumes of turbulent water. Gradual addition of base avoids spikes in pH values, which are harmful to both fish and plants.

14.7   Construction materials

A wide range of materials is used to construct aquaponic systems. Budget limitations often play a major role in selecting inexpensive and questionable materials such as vinyl-lined, steel-walled swimming pools. Plasticizers used in vinyl manufacture are toxic to fish. The liners must be washed thoroughly or aged with water for several weeks before fish can be added safely to a tank of clean water. After a few growing periods, vinyl liners shrink upon drying, become brittle, and crack, while the steel walls gradually rust. Nylon-reinforced, neoprene-rubber liners are not recommended, either. Tilapia eat holes in rubber liners at the folds by grazing on microorganisms. Moreover, neoprene-rubber liners are not impervious to chemicals. If herbicides and soil sterilants are applied under or near rubber liners, these chemicals can diffuse into culture water, accumulate in fish tissue, and kill hydroponic vegetables.

Wood is not considered to be a good construction material for aquaponic systems because it prone to rotting in the high-humidity environment. If wood is used, it should be untreated, as treated lumber contains toxic compounds, such as arsenic, to inhibit bacterial growth. If these compounds leach into the water, they could affect the beneficial bacteria that the system depends on and contaminate the fish and vegetables. Untreated wood must be waterproofed with fiberglass matt and resin on the inside and epoxy paint on the outside. Wooden tanks must not be in contact with soil to prevent the entry of termites. In general, wooden tanks have a short life span.

Fiberglass is the best construction material for the rearing tanks, sump, and filter tanks. Fiberglass tanks are sturdy, durable, nontoxic, movable, and easy to plumb. An alternative to fiberglass is concrete, which is cheaper in many countries, although it lacks the flexibility of fiberglass construction. Commercially available NFT troughs, made from extruded polyethylene, are specifically designed to prevent puddling and water stagnation leading to root death and are preferable to makeshift structures (rain gutters, PVC pipes, etc.). Plastic troughs are commercially available for floating hydroponic subsystems, but they are expensive. A suitable alternative is to use of polyethylene liners and concrete block, poured concrete or wooden plank walls. Four types of liners have been tested in UVI's commercial-scale system. They are high-density polyethylene [1.5 mm (60 mil) and 0.5 mm (20 mil)], low-density polyethylene [0.5 mm (20 mil)], and a thick grade of nylon-reinforced vinyl with an under layer of high-density polyethylene. All of these liners are performing well after five to ten years, but it appears that 0.5-mm high-density polyethylene liners (HDPE) are best. They are easy to install, relatively inexpensive and durable, and have an expected service life of twelve to fifteen years. Initially, HDPE liners were black, but recently UV-resistant white liners have been introduced. White liners are preferable in that they reflect light and do not become as hot as black liners. This is an important characteristic in the tropics and during summers in temperate climates where the goal is to avoid high water temperatures.

14.8 Component ratios

Aquaponic systems are generally designed to meet the size requirements for solids removal (for those systems requiring solids removal) and biofiltration (if a separate biofilter is used) for the amount of fish being raised. After the size requirements are calculated, it is prudent to add excess capacity as a safety margin. However, if a separate biofilter is used, the hydroponic component is the safety factor because a significant amount of ammonia uptake and nitrification will occur regardless of hydroponic technique.

Another key design criterion is the ratio between the fish rearing and hydroponic components. The key aspect of the criterion is the ratio of daily feed input to plant growing area. If the ratio of daily feeding rate to plant growing area is too high, nutrient salts will accumulate rapidly and may reach phytotoxic levels. Higher water exchange rates will be required to prevent excessive nutrient buildup. If the ratio of daily feeding rate to plants is too low, plants will develop nutrient deficiencies and more nutrient supplementation will be required. Fortunately, hydroponic plants grow well over a wide range of nutrient concentrations.

The optimum ratio of daily fish feed input to plant growing area will maximize plant production while maintaining relatively stable levels of dissolved nutrients. A volume ratio of 1 m^3 of fish-rearing tank to 2 m^3 of pea gravel 3 to 6 cm (1/8 to 1/4 in) in diameter as hydroponic media is recommended for reciprocating (flood and drain) gravel aquaponic systems. This ratio requires that tilapia are raised to a final density of 60 kg/m^3 (0.5 lb/gallon) and fed appropriately. With the recommended ratio no solids are removed from the system. The hydroponic beds should be cultivated (stirred up) between crops and inoculated with red worms to help break down and assimilate the organic matter. With this system nutrient supplementation may not be necessary.

Rule of Thumb

Pea Gravel Hydroponic Media
1 m^3 of fish tank volume to 2 m^3 of hydroponic media

As a general guide for raft aquaponics, a ratio in the range of 60 to 100 g of fish feed/m^2 of plant growing area per day should be used. Ratios within this range have been used successfully in the UVI system for the production of tilapia, lettuce, basil, and several other plants. In the UVI system all solids are removed, with a residence time of <1 day for settleable solids (>100 microns) removed by a clarifier, and 3 to 4 days for suspended solids removed by an orchard netting filter. The system uses rainwater, and supplementation is required for potassium, calcium, and iron.

Rule of Thumb

60 to 100 g of fish feed per day per square meter of plant growing area for the staggered production of leaf lettuce, herbs and many other crops

---

Another factor to consider when determining the optimum feeding rate ratio is the total water volume of the system, which affects nutrient concentrations. In raft hydroponics, approximately 75% of the system water volume is in the hydroponic component whereas gravel beds and NFT troughs contain minor amounts of system water. Theoretically, in systems producing the same quantity of fish and plants, a daily feeding rate of 100 g/m^2, for example, would produce total nutrient concentrations nearly four times higher in gravel and NFT systems (e.g., 1,600 mg/L) as compared to raft systems (e.g., 400 mg/L), but total nutrient mass would be equal among systems. Nutrient concentrations outside acceptable ranges affect plant growth. Therefore, the optimum design ratio varies depending on the type of hydroponic component. Gravel and NFT systems should have a feeding rate ratio that is approximately 25% of the recommended ratio for raft hydroponics.

Other factors involved in determining the optimum feeding rate ratio are the water exchange rate, nutrient levels in the source water, degree, and speed of solids removal, and type of plant being raised. Lower rates of water exchange, higher source-water nutrient levels, incomplete or slow solids removal resulting in the release of more dissolved nutrients through mineralization, and slower-growing plants would allow a lower feeding rate ratio. Conversely, higher water exchange rates, low source-water nutrient levels, rapid and complete solids removal, and fast-growing plants would allow a higher feeding rate ratio.

The optimum feeding rate ratio is influenced by the plant culture method. With batch culture all plants in the system are planted and harvested at the same time. During their maximum growth phase, there is a large uptake of nutrients, which requires a higher feeding rate ratio during that period. In practice, however, a high feeding rate ratio is used throughout the production cycle. With a staggered production system, plants are in different stages of growth, which levels out nutrient uptake rates and allows good production with slightly lower feeding rate ratios.

In properly designed aquaponic systems, the surface area of the hydroponic component is quite large compared to the surface area of the fish rearing tanks if it is stocked at commercial levels. The commercial-scale unit at UVI has a ratio of 7.3:1. The total plant growing area is 214 m^2 (2,304 ft^2) compared to total fish-rearing surface area of 29.2 m^2 (314 ft^2). With diffused aeration, final tilapia densities have reached a mean of 76 kg/m^3 (0.63 lb/gal) With pure oxygen, final densities can be increased up to 120 kg/m^3 (1 lb/gal) or greater. Therefore, smaller rearing tanks (18.7 m^2 or 201 ft^2) could be used to produce the same amount of fish and a ratio of 11.5:1 would be more appropriate.

> **Rule of Thumb**
>
> 7.3 to 1 ratio of plant beds to fish tank surface area for a final fish harvest density of 76 kg/m^3 (0.63 lb/gal)

14.9   Plant growth requirements

Maximum plant growth in aquaponic systems requires sixteen essential elements for proper nutrition. These nutrients are referred to below, in the order of their concentrations (mg/L) in plant tissue with carbon and oxygen being the highest. The essential elements are arbitrarily divided into macronutrients, those required in relatively large quantities, and micronutrients, those required in considerably smaller amounts. Three of the macronutrients, carbon (C), oxygen (O), and hydrogen (H), are supplied by water (H_2O) and carbon dioxide gas (CO_2). All of the remaining nutrients are absorbed from the culture water. Other macronutrients include nitrogen (N), potassium (K), calcium (Ca), magnesium (Mg), phosphorus (P), and sulfur (S). The seven micronutrients include chlorine (Cl), iron (Fe), manganese (Mn), boron (B), zinc (Zn), copper (Cu), and molybdenum (Mo). All of these nutrients must be in proper balance for optimum plant growth. High levels of one nutrient can influence the bioavailability of others. For example, excessive amounts of potassium may interfere with uptake of magnesium or calcium while excessive amounts of either of the latter nutrients may interfere with the uptake of the other two nutrients (Gerber 1985).

An elevated CO_2 level in the atmospheric environment of unventilated hydroponic structures has given dramatic increases in crop yield in northern latitudes (Jensen & Collins 1985). Kimball (1982) summarized the data on CO_2

enrichment from more than 360 observations on 24 crops in a series of 50 reports. The study showed that doubling atmospheric CO_2 increased agricultural yields by an average of 30%. The high cost of energy to generate CO_2 has discouraged its use in conventional hydroponic systems. However, an enclosed aquaponic system is ideal for generating CO_2 due to the huge amounts that are constantly vented from the culture water.

Luther (1990, 1991a–c, 1993) cites a growing body of evidence that healthy plant development relies on a wide range of organic compounds in the root environment. These compounds, generated by complex biological processes involving microbial decomposition of organic matter, include vitamins, auxins, gibberellins, antibiotics, enzymes, coenzymes, amino acids, organic acids, hormones, and other metabolites. These compounds are directly absorbed and assimilated by the plants and stimulate growth, enhance yields, increase vitamin and mineral content, improve fruit flavor, and hinder the development of pathogens. Various fractions of dissolved organic matter (e.g., humic acid) form organo-metalic complexes with Fe, Zn, and Mn and thereby increase the availability of these micronutrients to plants (Chen & Solvitch 1988). Luther states that although inorganic nutrients are necessary for plant survival, plants need organic metabolites from the environment to reach full hereditary potential.

Maintaining high DO levels in the culture water is extremely important for optimal plant growth, especially in aquaponic systems with their high organic loads. Hydroponic plants are subject to intense root respiration and they draw large amounts of oxygen from the surrounding water. If DO is deficient, root respiration decreases, resulting in reduced water absorption, decreased nutrient absorption, loss of cell tissue from roots, and a reduction in plant growth (De Wit 1978). Low DO levels correspond with high concentrations of carbon dioxide, conditions that promote the development of plant root pathogens. Chun and Takakura (1993) tested four nutrient-solution DO levels for hydroponic lettuce and found that root respiration, root growth, and transpiration were greatest at saturated DO levels.

Climatic factors also influence hydroponic vegetable production. Production is generally best in regions with maximum intensity and duration of light. Jensen and Collins (1985) reported that growth rates of lettuce plants in an Arizona greenhouse correlated positively with levels of available light up to the highest levels measured, although radiation levels in the Arizona desert are two to three times that of more temperate climates. When 30% shade cloth was used to cover lettuce plants in the UVI system, also in a region of intensive solar radiation, the plants elongated, the leaves twisted around the stem, and production declined. Growth slows substantially in temperate greenhouses during the winter due to low solar radiation. Supplemental illumination can improve wintertime production, but it is not generally cost effective but it is costly.

Water temperature is far more important than air temperature for hydroponic plant production. The best water temperature for most hydroponic crops is around is 22 to 24°C (68 to 75°F). However, water temperature can go as low as the mid-60s for most common garden crops and slightly lower for winter crops such as cabbage, brussels sprouts, and broccoli. Maintaining the best water temperature requires heating during the winter in temperate greenhouses and year-round cooling in tropical greenhouses. In addition to evaporative cooling of tropical greenhouses, chillers are often used to cool the nutrient solution. In tropical outdoor systems, complete shading of the fish rearing and filtration components lowers system water temperature. In raft hydroponics, the polystyrene sheets shield water from direct sunlight and maintain temperatures that are several degrees lower than those in open water bodies. Seasonal adjustment in selection of plant crop varieties may be necessary for both temperate and tropical aquaponic production. Plants cultured in outdoor aquaponic systems must be protected from strong winds, especially following transplanting when seedlings are fragile and most vulnerable to damage.

14.10   Nutrient dynamics

Collectively dissolved nutrients are measured as total dissolved solids (TDS), expressed as mg/L, or as the capacity of the nutrient solution to conduct an electrical current (EC), expressed as millimhos per centimeter (mmho/cm). In a hydroponic solution the recommended range for TDS is 1,000 to 2,200 (1.5 to 3.5 mmho/cm). In an aquaponic system considerably lower levels of TDS (200 to 400 mg/L) or EC (0.3 to 0.6 mmho/cm) will produce good results because nutrients are generated continuously. A concern with aquaponic systems is nutrient accumulation. High feeding rates, low water exchange, and insufficient plant growing areas can lead to the rapid buildup of dissolved nutrients to potentially phytotoxic levels. Phytotoxicity is encountered at TDS concentrations above 2,200 mg/L or EC above 3.5 mmho/cm. Because aquaponic systems are characterized by variable environmental conditions such as daily feed input, solids retention, mineralization, water exchange, nutrient input from source water or supplementation, and variable nutrient uptake by different plant species, it is difficult to predict the exact level of TDS or EC and how it is changing. Therefore, the culturist should purchase an inexpensive conductivity meter and periodically measure TDS or EC. If dissolved nutrients are steadily increasing and approaching 2,200 mg/L as TDS or 3.5 mmho/cm as EC, increasing the water exchange rate or reducing the fish-stocking rate and feed input will quickly reduce nutrient accumulation. Since these methods either increase costs (i.e., more water consumed) or lower output (i.e., less fish produced), they are not good long-term solutions. Better but more costly solutions involve increased solids removal (i.e., upgrade the solids removal component) or enlarged plant growing areas.

Early work with UVI's experimental systems showed that conductivity measurements of TDS increased steadily as increasing quantities of feed were added to the system (Rakocy *et al.* 1993). Phytotoxic levels were reached after the addition of approximately 10 kg feed/m^3 of system volume. In an experiment to determine the optimum ratio of daily feed input to leaf lettuce growing area, the concentration of TDS increased by 147.5 g/kg of dry weight of feed at the optimum ratio of 57 g/day/m^2. However, during the first 8 months of operation of an early model of UVI's commercial-scale system, 26.9 kg of feed/m^3 of system

Table 14.3 Accumulation rate (g/kg dry weight feed) total dissolved solids (TDS), major cations and anions from two experimental aquaponic systems and a commercial-scale aquaponic system using raft hydroponics for lettuce production

Nutrient	Exp. System 1[a]	Exp. System 2[b]	Commercial System[c]
TDS	215.2	147.5	26.2
NO_3-N	35.6	14.9	3.7
PO_4-P	3.0	-	0.2
SO_4-S	1.9	1.8	0.6
K	66.0	-	4.2
Ca	-	7.3	2.3
Mg	1.2	1.8	0.4

[a] Supplementation with K but not Ca. Minor, one-time supplementation with P. From Rakocy *et al.* 1993.
[b] Optimum feed to growing area ratio (57 g/day/m^2) from ratio study (Rakocy *et al.* 1993). K and Ca supplementation.
[c] From the first eight months of operation of an early model of the commercial-scale system. K and Ca supplementation.

volume was applied and the highest total nutrient concentration was only 890 mg/L as TDS. The accumulation rate for TDS was 26 g/kg of dry weight of feed (table 14.3). Three factors contributed to the lower nutrient-accumulation rate. The actual mean ratio of daily feed input to plant growing area (49.5 g/day/m^2) turned out to be less than the optimum design ratio. Lettuce productivity was greater in the commercial-scale system. Romaine and leaf lettuce plants grew to a size of 250 to 650 g in four weeks (8.9 to 23.2 g/day) in the commercial-scale system compared to an average Bibb lettuce size of 131 g in three weeks (6.2 g/day) in the experimental system. The average reduction of nutrients on passage through the hydroponic component during the first eight months was 4.2 mg/L as TDS. Substantial amounts of solids were removed by the filter tanks, and, consequently, less mineralization may have occurred than in the experimental systems.

The major ions that contribute to increased conductivity are nitrate (NO_3^-), phosphate (PO_4^{-2}), sulfate (SO_4^{-2}), K^+, Ca^{+2}, and Mg^{+2}. Levels of NO_3^-, PO_4^{-2}, and SO_4^{-2} are usually sufficient for good plant growth while levels of K^+ and Ca^{+2} are generally insufficient for maximum plant growth. Potassium is added to the system in the form of potassium hydroxide (KOH) while Ca is added as calcium hydroxide [$Ca(OH)_2$]. In some systems Mg may be limiting. In the UVI commercial-scale system, KOH and $Ca(OH)_2$ are added in equal amounts, usually in the range of 500 to 1,000 g in the UVI system. The bases are added alternately several times weekly to maintain pH near 7.0. Adding basic compounds of K and Ca serves the dual purpose of supplementing essential nutrients and neutralizing acid. Magnesium can be supplemented by using dolomite [$CaMg(CO_3)_2$] as the base to adjust pH. The addition of too much Ca can lead to the precipitation of phosphorous from culture water in the form of dicalcium phosphate ($CaHPO_4$). All of the macronutrients with the exception of orthophosphate have a strong correlation with feed input (Rakocy *et al.* 1993).

The accumulation of nitrate ions is a concern with aquaponic systems. The discharge from one experimental system at UVI contained 180 mg/L as NO_3-N

(Rakocy 1994). The installation of the filter tanks in the UVI commercial-scale system provided a mechanism for controlling nitrate levels through the process of denitrification, the reduction of nitrate ions to nitrogen gas by anaerobic bacteria. Large quantities of organic matter accumulate on the orchard netting between cleanings. Denitrification occurs in anaerobic pockets that develop in the sludge. The entire water column moves through the accumulated sludge, which provides good contact between nitrate ions and denitrifying bacteria. The frequency of cleaning the netting regulates the degree of denitrification. When the netting is cleaned frequently (e.g., twice per week) sludge accumulation and denitrification are minimized, which leads to an increase in nitrate concentrations. When the netting is cleaned less frequently (e.g., once per week) sludge accumulation and denitrification are maximized, which leads to a decrease in nitrate levels. Nitrate-nitrogen levels can be regulated within a range of 1 to 100 mg/L or higher. High nitrate concentrations promote the growth of leafy green vegetables while low nitrate concentrations promote fruit development in vegetables such as tomatoes.

Denitrification recovers the alkalinity that is lost in the nitrification process. Sometimes an aquaponic system will go for long periods of time with no change in pH and no need to add bases such as calcium hydroxide or potassium hydroxide. Stable pH in an aquaponic system indicates that too much denitrification is occurring. If there is no need to add base, the plants could develop calcium and potassium deficiencies. When pH is stable, the frequency of cleaning the filter tanks should be increased and any anaerobic zones that have been created due to the accumulation of solids in the system should be removed. Solids on the bottom of the hydroponic tanks should be cleaned once annually.

In a study using raft hydroponics for lettuce production, Seawright (1995) obtained similar results for macronutrient accumulation with two important exceptions: P accumulated in relation to feed input, but there was no significant relationship between feed input and N. Sodium bicarbonate ($NaHCO_3$) was added for pH control. The addition of Ca bases in the UVI system may have contributed to the precipitation of P from the culture water in the form of calcium phosphate [$Ca_3(PO_4)_2$]. Seawright removed solid waste from the system once per week and therefore denitrification may have been greater than in the UVI system where solid waste was removed three times per day from the clarifier and up to twice per week from the filter tanks. Although Seawright did not supplement with K, it accumulated with respect to feed input. However, plants require high levels of K and supplementation is needed in aquaponic systems. Seawright's finding that Ca was negatively correlated with feed input agreed with earlier work at UVI (Rakocy & Nair 1987). Some investigators have found that Mg is limiting (Pierce 1980; Head 1984; Zweig 1986). Magnesium was supplemented by using dolomite [$CaMg(CO_3)_2$] as the base to adjust pH.

Sodium bicarbonate ($NaHCO_3$) should *never* be added to an aquaponic system for pH control. The accumulation of Na is a concern in aquaponic systems because high Na levels in the presence of chloride are toxic to plants (Resh 1995). The maximum Na concentration in hydroponic nutrient solutions should not exceed 50 mg/L (Verwer & Wellman 1980). Higher Na levels will interfere with

Table 14.4 Mean concentration (mg/L) of micronutrients from two experimental aquaponic systems and a commercial-scale aquaponic system using raft hydroponics for lettuce production

System	Micronutrient					
	Fe	Mn	Cu	Zn	B	Mo
Exp. 1[a]	1.79	0.04	0.12	0.78	0.16	-
Exp. 2[b]	0.48	0.13	0.07	0.68	-	-
Commercial[c]	0.57	0.05	0.05	0.44	0.06	0.006
HNF[d]	5.0	0.5	0.03	0.05	0.5	0.02
HNF[e]	5.0	0.5	0.1	0.1	0.5	0.05

[a] From Rakocy et al. (1993).
[b] Optimum feed to growing area ratio (57 g/day/m^2) from ratio study (Rakocy et al. 1993).
[c] From the first eight months of operation of a commercial-scale system.
[d] Hydroponic nutrient formulation for lettuce grown in the tropics (Resh 1995).
[e] Hydroponic nutrient formulation for lettuce grown in Florida and California (Resh 1995).

the uptake of K and Ca (Douglas 1985). In lettuce, reduced Ca uptake leads to tip burn, resulting in an unmarketable plant (Collier & Tibbitts 1982). In UVI's systems, tip burn has occurred during the warmer months. Soluble salt (NaCl) levels in fish feed are relatively high. In the initial commercial-scale system, Na reached 51.0 mg/L in the sixth month and then declined to 37.8 mg/L in the eighth month, possibly due to rainfall dilution. The Na accumulation rate through the sixth month was 2.56 g/kg of dry weight feed. If Na exceeds 50 mg/L and the plants appear to be affected, a partial water exchange (dilution) may be necessary. Rainwater is used in UVI's systems because the groundwater of semiarid islands generally contains too much salt for aquaponics.

With the exception of Zn^{+2}, the micronutrients Fe^{+2}, Mn^{+2}, Cu^{+2}, B^{+3}, and Mo^{+6} do not accumulate significantly in aquaponic systems with respect to cumulative feed input (table 14.4; Rakocy et al. 1993). The Fe^{+2} derived from fish feed is insufficient for hydroponic vegetable production and must be supplemented (Lewis et al. 1980; MacKay & Van Toever 1981; Zweig 1986). Chelated Fe^{+2} should be applied at a rate to achieve a Fe^{+2} concentration of 2.0 mg/L. Chelated Fe^{+2} has an organic compound attached to the metal ion to prevent it from precipitating out of solution and making it unavailable for plant nutrition. The best chelate is Fe-DTPA because it remains soluble at pH 7.0. Fe-EDTA is commonly used in the hydroponics industry, but it is less stable at pH 7.0 and needs to be replenished frequently. Fe^{+2} may also be applied in a foliar spray from which Fe^{+2} is absorbed directly through plant leaves. A comparison of Mn^{+2}, B^{+3}, and Mo^{+6} levels with standard nutrient formulations for lettuce shows that their concentrations in aquaponic systems are several times lower than their initial levels in hydroponic formulations. Deficiency symptoms for Mn^{+2}, B^{+3}, and Mo^{+6} are not detected in aquaponic systems, and so their concentrations appear to be adequate for normal plant growth. Concentrations of Cu^{+2} are similar in aquaponic systems and hydroponic formulations, while Zn^{+2} accumulates in aquaponic systems to levels that are four to sixteen times higher than initial levels in hydroponic formulations. Nevertheless, Zn^{+2} concentrations usually remain within the upper limit for fish safety, which is 1 mg/L in hydroponic solutions (Douglas 1985).

Seawright (1995) worked on the development of a "designer diet" for aquaponic systems that would generate nutrients in proportion to their requirements for hydroponic plant nutrition, thereby creating stable and balanced nutrient concentrations over prolonged periods. Data on the change in nutrient concentrations in relation to dietary nutrient input were collected for the co-culture of *O. niloticus* and romaine lettuce (Jericho). Data were used to develop a mass balance model theoretically capable of predicting the nutrient inclusion rates required in fish diets to maintain stable dissolved nutrient concentrations in aquaponic systems. The model was validated by applying a specially formulated "designer diet" to an aquaponic system and maintaining near-equilibrium concentrations of Ca, K, Mg, N, and P; suitable concentrations of Mn and Cu; and acceptable accumulation rates of Na and Zn. The results showed that Cu, Fe, and Mn are not good candidates for dietary manipulation because of low bioavailability. Phosphorus is a good candidate at sub-neutral pH, but it precipitates from solution at basic pH. Sulfur, B, and Mo were not tested. The fish grew well on the diet, but the development of bacterial diseases indicated that elevated nutrient levels might have lowered their disease resistance.

14.11   Vegetable selection

Many types of vegetables have been grown in aquaponic systems (tables 14.5–14.7). However, the goal is to select a vegetable for culture that will generate the highest level of income per unit area per unit time. Using this criterion, culinary herbs are the best choice. They grow very rapidly and command high market prices. The income from herbs such as basil, cilantro, chives, parsley, portulaca, and mint are many times higher than that from fruiting crops such as tomatoes, cucumbers, eggplant, and okra. For example, in experiments in UVI's commercial-scale system, basil production was 11,000 lb annually at a value of US$110,000 compared to annual production of 6,400 lb of okra at a value of US$6,400.

Fruiting crops also require longer culture periods (ninety days or more) and are subject to more pest problems and diseases. Lettuce is another good crop for aquaponic systems because it can be produced in a short period (three to four weeks in the system), and, as a consequence, pest pressure is relatively low. Unlike fruiting crops, a high proportion of the harvested biomass is edible. Other suitable crops include Swiss chard, pak choi, Chinese cabbage, collard and watercress. The cultivation of flowers has potential in aquaponic systems. Good results have been obtained with marigold and zinnia in UVI's aquaponic system. Traditional medicinal plants and plants used for the extraction of modern pharmaceuticals have not been cultivated in aquaponic systems, but there may be potential in growing some of these plants.

Table 14.5 Varieties and yields of leafy greens (lettuce, pak choi, Chinese cabbage, spinach) evaluated in aquaponic systems[a]. Environment codes: TEMP = Temperate Zone, TROP = Tropical Zone, O = Outside, GH = Greenhouse.

Variety	Yield (g/plant)	Environment	Reference
Ostinata	162	TEMP/GH	Baum 1981
Reskia	236	TEMP/GH	Burgoon & Baum 1984
All Year Round	236	TEMP/GH	Burgoon & Baum 1984
Karma	98	TEMP/GH	Burgoon & Baum 1984
Ravel	45–50	TEMP/GH	Burgoon & Baum 1984
Salina	44	TEMP/GH	Burgoon & Baum 1984
El Captain	43	TEMP/GH	Burgoon & Baum 1984
Bruinsma Columbus	0	TEMP/GH	Burgoon & Baum 1984
Salad Bowl	-	TEMP/GH	Head 1984
Pak Choi	-	TEMP/GH	Head 1984
Winter Bloomsdale (Spinach)	-	TEMP/GH	Head 1984
Buttercrunch	-	TEMP/GH	Zweig 1986
Buttercrunch	193	TROP/O	Rakocy 1989b
Summer Bibb	180	TROP/O	Rakocy 1989b
Le Choi	508	TROP/O	Rakocy 1989b
Pak Choi	422	TROP/O	Rakocy 1989b
50-Day Hybrid (Chinese Cabbage)	638	TROP/O	Rakocy 1989b
Tropical Delight (Chinese Cabbage)	589	TROP/O	Rakocy 1989b
Summer Bibb	107–116	TEMP/GH	Parker *et al.* 1990
Sierra	182–340	TROP/O	Rakocy *et al.* 1997
Nevada	149–360	TROP/O	Rakocy *et al.* 1997
Jerhico	267–344	TROP/O	Rakocy *et al.* 1997
Parris Island	181–446	TROP/O	Rakocy *et al.* 1997

[a] From Rakocy and Hargreaves (1993) and some more recent data.

14.12   Crop production systems

There are three strategies for producing vegetable crops in the hydroponic component. These are staggered cropping, batch cropping, and intercropping. A staggered crop production system is one in which groups of plants in different stages of growth are cultivated simultaneously in the hydroponic subsystem. This production system allows regular harvest of produce and relatively constant uptake of nutrients from the culture water. This system is most effectively implemented where crops can be grown continuously, as in the tropics, subtropics, or temperate greenhouses with environmental control (Zweig 1986). At UVI, the production of leaf lettuce is staggered so that a crop can be harvested weekly on the same day, which facilitates marketing arrangements. Bibb lettuce reaches market size in three weeks from transplanting. Therefore, three growth stages of Bibb lettuce are cultivated simultaneously, and one-third of the crop is harvested weekly. Red leaf lettuce and green leaf lettuce require four weeks to reach marketable size. The cultivation of four growth stages of these lettuce

Table 14.6 Varieties and yields of tomatoes evaluated in aquaponic systems[a]. Environmental Codes: TEMP = Temperate Zone, TROP = Tropical Zone, O = Outside, GH = Greenhouse.

Variety	Yield (kg/plant)	Environment	Reference
Tropic	0	TEMP/GH	Landesman 1977
Floradel	5.4	TEMP/O	Lewis *et al.* 1978
Campbells 1327	4.6	TEMP/O	Lewis *et al.* 1978
San Marzano	4.6	TEMP/O	Lewis *et al.* 1978
Sweet 100	6.2	TEMP/O	Lewis *et al.* 1980
Better Boy	4.5	TEMP/O	Lewis *et al.* 1980
Rampo	3.9	TEMP/O	Lewis *et al.* 1980
Campbells 1327	2.3	TEMP/O	Lewis *et al.* 1980
Sweet 100	1.9	TEMP/GH	Baum 1981
Spring Set	1.6	TEMP/GH	Baum 1981
Sweet 100	0	TEMP/GH	MacKay *et al.* 1981
Jumbo	0	TEMP/GH	MacKay *et al.* 1981
Michigan Ohio Forcing	0	TEMP/GH	MacKay *et al.* 1981
Burpee Big Boy	0.2	TEMP/O	Markin 1982
Floradel	8.9, 9.1	TEMP/O	Sutton & Lewis 1982
Korala #127	6.8[b]	TEMP/GH	Burgoon & Baum 1984
Vendor	4.1	TEMP/GH	Head 1984
Tropic	0.5, 3.2	TROP/O	Watten & Busch 1984
Homestead	2.7	TROP/O	Watten & Busch 1984
Red Cherry	1.5	TROP/O	Watten & Busch 1984
Prime Beefsteak	1.4	TROP/O	Watten & Busch 1984
Vendor	0	TROP/O	Nair *et al.* 1985
Tropic	0	TROP/O	Nair *et al.* 1985
Jumbo	0	TROP/O	Nair *et al.* 1985
Perfecta	0	TROP/O	Nair *et al.* 1985
Laura	2.3–3.4	TEMP/GH	McMurtry 1989
Kewalo	2.5–5.0	TEMP/GH	McMurtry 1989
Sunny	10.1	TROP/O	Rakocy 1989
Floradade	9.0	TROP/O	Rakocy 1989
Vendor	3.7	TROP/O	Rakocy 1989
Cherry Challenger	2.9	TROP/O	Rakocy 1989
Champion	4.6[b]	TEMP/GH	McMurtry *et al.* 1990

[a] From Rakocy and Hargreaves 1993
[b] kg/m^2

Table 14.7 Varieties and yields of cucumbers evaluated in aquaponic systems[a]. Environmental Codes: TEMP = Temperate Zone, TROP = Tropical Zone, O = Outside, GH = Greenhouse.

Variety	Yield (kg/plant)	Environment	Reference
Triumph	4.1	TEMP/O	Lewis *et al.* 1980
Patio Pik	1.6	TEMP/O	Lewis *et al.* 1980
Corona (Stokes)	28.6[b]	TEMP/GH	Burgoon & Baum 1984
Bruinsma Vetomil	8.2[b]	TEMP/GH	Burgoon & Baum 1984
Superator	4.1	TEMP/GH	Head 1984
Sprint 4405	0.7	TEMP/GH	Wren 1984
Burpee Hybrid II	7.3[b]	TEMP/GH	McMurtry 1990

[a] From Rakocy and Hargreaves (1993).
[b] kg/m^2.

varieties allows one-fourth of the crop to be harvested weekly. In 3 years of continuous operation of UVI's commercial-scale system, 148 crops of lettuce were harvested, which demonstrates the system's sustainability. Leafy green vegetables, herbs, and other crops with short production periods are well suited for continuous, staggered production systems.

A batch cropping system is more appropriate for crops that are grown seasonally or have long growing periods (>3 months), such as tomatoes and cucumbers. Various intercropping systems can be used in conjunction with batch cropping. For example, if lettuce is intercropped with tomatoes and cucumbers, one crop of lettuce can be harvested before tomato plant canopy development limits light availability (Resh 1995).

14.13   Pest and disease control

A number of plant pest and disease problems have been encountered in aquaponic systems. Pests observed on tomatoes include spider mite (Landesman 1977; Nair *et al.* 1985), russet mite (Rakocy, unpublished data), hornworm (Lewis *et al.* 1978; Sutton & Lewis 1982; Nair *et al.* 1985), western locust (Sutton & Lewis 1982), fall armyworm, pinworm, aphid, and leaf minor (Nair *et al.* 1985). Diseases observed on tomatoes include blight (Lewis *et al.* 1980) and bacteria wilt (McMurtry *et al.* 1990). In UVI's systems, lettuce is affected by fall armyworm, corn earworm, and two species of pathogenic root fungus (*Pythium dissotocum* and *P. myriotylum*). The root diseases that plague conventional hydroponics may be a threat to aquaponics. Four viral, two bacterial and twenty fungal pathogens have been associated with root diseases in hydroponically grown vegetables (Stanghellini & Rasmussen 1994). Most of the destructive root diseases in hydroponics have been attributed to the fungal genera *Pythium, Phytophthora, Plasmopara, Olpidium,* and *Fusarium.*

Pesticides should not be used to control insects on aquaponic plant crops. Even pesticides that are registered would pose a threat to the fish and would not be permitted in a fish culture system. Similarly, most therapeutants for treating fish parasites and diseases should not be used either. Vegetables may absorb and concentrate them. Even the common practice of adding salt to treat fish diseases or reduce nitrite toxicity would be deadly to vegetables. Nonchemical methods of plant pest and disease control are required such as biological control (resistant cultivars, predators, antagonistic organisms, pathogens), physical barriers, traps, treatment of the nutrient solution (filtration, UV sterilization), manipulation of the physical environment and other specialized cultural practices. Opportunities for biological control methods are greater in enclosed greenhouse environments than exterior installations. McMurtry (1989) used *Encarsia formosa* and *Chrysopa carnea* to control greenhouse white fly (*Trialeurodes vaporariorum*) and *Hippodamia convergens* to control potato aphid (*Macrosiphum euphorbiac*). In UVI's systems, caterpillars are effectively controlled by twice-a-week spraying with *Bacillus thuringiensis,* a bacterial pathogen that is specific to caterpillars. The fungal root pathogens that are encountered in summer dissipate in winter in response to lower water temperature and manipulation of suspended solids levels. An outbreak of *Pythium* coincided with a period during which the efficiency of suspended solids removal was dramatically increased.

Prohibition on the use of pesticides makes crop production in aquaponic systems more difficult. However, this restriction assures that crops from aquaponic systems will be raised in an environmentally sound manner, free of pesticide residues. A major advantage of aquaponic systems is that crops are less susceptible to attack from soil-borne diseases. It also appears that aquaponic systems may be more resistant to diseases that affect standard hydroponics. This resistance may be due to the presence of some organic matter in the culture water that creates a stable, ecologically balanced, growing environment with a wide diversity of microorganisms, some of which may be antagonistic to plant root pathogens.

14.14 Approaches to system design

Several approaches can be used to design an aquaponic system. The simplest approach is to duplicate a standard system, or scale a standard system down or up, keeping the components proportional. Changing aspects of a standard system is not recommended because changes often lead to unintended consequences. However, the design process often starts with a production goal for either fish or plants. In those cases, there are some guidelines that can be followed.

14.14.1 Use a standard system that is already designed

The easiest approach is to use a standard system design that has been tested and is in common use with a good track record. It is early in the development of aquaponics, but standard designs will emerge. The UVI system has been well documented and is studied and used commercially in several locations, but there are other systems with potential. Standard designs will include specifications for layout, tank sizes, pipe sizes, pipe placement, pumping rates, aeration rates, infrastructure needs, and so on. There will be operation manuals and projected production levels and budgets for various crops. Using a standard design will reduce risk.

14.14.2 Design for available space

If a limited amount of space is available, such as in an existing greenhouse, then that space will define the size of the aquaponic system. The easiest approach is to take a standard design and scale it down. If a scaled-down tank or pipe size falls between commercially available sizes, it is best to select the larger size. However, the water flow rate should equal the scaled-down rate for best results. The desired flow rate can be obtained by buying a higher capacity pump and installing a bypass line and valve, which circulates a portion of the flow back to the sump and allows the desired flow rate to go from the pump to the next stage of the system. If more space is available than the standard design requires, then the system could be scaled up within limitations or more than one scaled-down system could be installed.

14.14.3   Design for fish production

If the primary objective is to produce a certain amount of fish annually, the first step in the design process will be to determine the number of systems required, the number of rearing tanks required per system, and the optimum rearing tank size. The number of harvests will have to be calculated based on the length of the culture period. Assume that the final density is 60 kg/m^3 (0.5 lb/gallon) for an aerated system. Take the annual production per system and multiply it by the estimated feed conversion ratio (the kilograms of feed required to produce one kilogram of fish). Convert the pounds of annual feed consumption to grams (454 g/lb) and divide by 365 days to obtain the average daily feeding rate. Divide the average daily feeding rate by the desired feeding rate ratio, which ranges from 60 to 100 g/m^2/day for raft culture, to determine the required plant production area. For other systems such as NFT, the feeding rate ratio should be decreased in proportion to the water volume reduction of the system as discussed in the component ratio section. Use a ratio near the low end of the range for small plants such as Bibb lettuce and a ratio near the high end of the range for larger plants such as Chinese cabbage or romaine lettuce. The solids-removal component, water pump, and blowers should be sized accordingly.

14.14.4   Sample problem

This example illustrates only the main calculations, which are simplified (e.g., mortality is not considered) for the sake of clarity. Assume that you have a market for 227 kg (500 lb) of live tilapia per week in your city and that you want to raise lettuce with the tilapia because there is a good market for green leaf lettuce in your area. The key questions are: How many UVI aquaponic systems do you need to harvest 227 kg (500 lb) of tilapia weekly? How large should the rearing tanks be? What is the appropriate number and size of the hydroponic tanks? What would the weekly lettuce harvest be?

(1) Each UVI system contains four fish-rearing tanks (fig. 14.3). Fish production is staggered so that one fish tank is harvested every six weeks. The total growing period per tank is twenty-four weeks. If 227 kg (500 lb) of fish are required weekly, six production systems (twenty-four fish-rearing tanks) are needed.

(2) Aquaponic systems are designed to achieve a final density of 60 kg/m^3 (0.5 lb/gallon). Therefore the water volume of the rearing tanks should be 3.79 m^3 (1,000 gallons).

(3) In 52 weeks, there will be 8.7 harvests (52/6 = 8.7) per system. Annual production for the system therefore is 2.0 tonnes (4,350 lb); i.e., 227 kg per harvest × 8.7 harvests.

(4) The usual feed conversion ratio is 1.7. Therefore, annual feed input to the system is 3,360 kg (7,395 lb); i.e., 2.0 tonnes × 1.7 = 3,360 kg.

(5) The average daily feed input is 9.22 kg (20.3 lb); i.e., 3,360 kg per year/365 days = 9.22 kg.

(6) The average daily feed input converted to grams is 9,216 g (20.32 lb); i.e., 9.22 kg × 1000 g/kg = 9,216 g.

(7) The optimum feeding rate ratio for raft aquaponics ranges from 60 to 100 g/m^2/day. Select 80 g/m^2/day as the design ratio. Therefore, the required lettuce growing area is 115.2 m^2 (9,216 g/day divided by 80 g/m^2/day = 115.2 m^2).

(8) The growing area in square feet is 1,240 (115.2 m^2 × 10.76 ft^2/m^2 = 1,240 ft^2).

(9) Select a hydroponic tank width of 1.22 m (4 ft). Therefore, the total length of the hydroponic tanks is 94.5 m (310 ft); i.e., 115.2 m^2 /1.22 m = 94.5 m.

(10) Select four hydroponic tanks. They are 23.6 m (77.5 ft) long (94.5/4 = 23.6 m). They are rounded up to 24.4 m (80 ft) in length, which is a practical length for a standard greenhouse and allows the use of ten 2.36-m (8-ft) sheets of polystyrene per hydroponic tank.

(11) Green leaf lettuce produces good results with plant spacing of 60 plants per sheet (20/m^2)(1.85/ft^2). The plants require a four-week growth period. With staggered production, one hydroponic tank is harvested weekly. Each hydroponic tank with ten polystyrene sheets produces 600 plants. With six aquaponic production systems 3,600 plants are harvested weekly.

In summary, weekly production of 227 kg (500 lb) of tilapia results in the production of 3,600 green leaf lettuce plants (150 cases). Six aquaponic systems, each with four 3.78-m^3 (1,000-gallon) rearing tanks (water volume), are required. Each system will have four raft hydroponic tanks that are 24.4-m long and 1.22-m wide (80-ft long by 4-ft wide).

14.14.5   Design for plant production

If the primary objective is to produce a certain quantity of plant crops annually, the first step in the design process will be to determine the area required for plant production. The area needed will be based on plant spacing, length of the production cycle, number of crops per year or growing season, and the estimated yield per unit area and per crop cycle. Select the desired feeding rate ratio and multiple by the total area to obtain the average daily feeding rate that is required. Multiply the average daily feeding rate by 365 days to determine annual feed consumption. Estimate the feed conversion ratio (FCR) for the fish species that will be cultured. Convert FCR to feed conversion efficiency. For example, if FCR is 1.7:1, then the feed conversion efficiency is 1 divided by 1.7

or 0.59. Multiply the annual feed consumption by the feed conversion efficiency to determine net annual fish yield. Estimate the average fish weight at harvest and subtract the anticipated average fingerling weight at stocking. Divide this number into the net annual yield to determine the total number of fish produced annually. Multiply the total number of fish produced annually by the estimated harvest weight to determine total annual fish production. Divide total annual fish production by the number of production cycles per year. Take this number and divide by 60 kg/m^3 (0.5 lb/gallon) to determine the total volume that must be devoted to fish production. The required water volume can be partitioned among multiple systems and multiple tanks per system with the goal of creating a practical system size and tank array. Divide the desired individual fish weight at harvest by 60 kg/m^3 (0.5 lb/gallon) to determine the volume of water required per fish. Divide the volume of water required per fish into the water volume of the rearing tank to determine the fish-stocking rate. Increase this number by 5 to 10% to allow for expected mortality during the production cycle.

The solids removal component, water pump, and blowers should be sized accordingly.

14.14.6   Sample problem

Assume that there is a market for 1,000 Bibb lettuce plants weekly in your city. These plants will be sold individually in clear plastic clamshell containers. A portion of the root mass will be left intact to extend shelf life. Bibb lettuce transplants are cultured in a UVI raft system for three weeks at a density of 29.3 plants/m^2 (2.72/ft^2) . Assume that tilapia will be grown in this system. The key questions are: How large should the plant growing area be? What will be the annual production of tilapia? How large should the fish-rearing tanks be?

(1) Bibb lettuce production will be staggered so that 1,000 plants can be harvested weekly. Therefore, with a three-week growing period, the system must accommodate the culture of 3,000 plants.
(2) At a density of 29.3 plants/m^2, the total plant growing area will be 102.3 m^2 (3,000 plants divided 29.3/m^2 = 102.3 m^2). This area is equal to 1,100 square feet; i.e., 102.3 m^2 × 10.76 ft^2/m^2 = 1,100 ft^2.
(3) Select a hydroponic tank width of 2.44 m (8 ft). Therefore, the total hydroponic tank length will be 41.9 m (137.5 ft); i.e., 102.3 m^2 /2.44 m = 41.9 m.
(4) Two raft hydroponic tanks are required for the UVI system. Therefore, the minimum length of each hydroponic tank will be 20.95 m (68.75 ft); i.e., 41.9 m /2 = 20.95 m. Polystyrene sheets come in 2.44 (8 ft) lengths and 1.22 m (4 ft) widths so the total number of sheets lengthwise per hydroponic tank will be 8.59 (20.95 m divided by 2.44 m/sheet = 8.59). To avoid wasting material, round up to 9 sheets lengthwise. Therefore, the hydroponic tanks will be 21.96 m (72 ft) long; i.e., 9 sheets × 2.44 m per sheet = 21.96 m).

(5) The total plant growing area will then be 107 m² (1,152 ft²); i.e., 21.96 m × 2.44 m per tank × 2 tanks = 107 m².

(6) At planting density of 29.3 plants/m², a total of 3,135 plants will be cultured in the system. The extra plants will provide a safety margin against mortality and plants that do not meet marketing standards.

(7) Assume that a feeding rate ration of 60 g/m²/day provides sufficient nutrients for good plant growth. Therefore, daily feed input to the system will be 6,420 g (14.1 lb); i.e., 60 g/m²/day × 107 m² = 6,420 g.

(8) Annual feed input to the system will be 2,340 kg (5,146 lb); i.e., 6.42 kg/day × 365 days = 2,340 kg.

(9) Assume the feeding conversion ratio is 1.7. Therefore, the feed conversion efficiency is 0.59; i.e., 1 kg of gain divided by 1.7 kg of feed = 0.59.

(10) The total annual fish production gain will be 1,380 kg (3,036 lb); i.e., 2,340 kg × 0.59 feed conversion efficiency = 1,380 kg.

(11) Assume that the desired harvest weight of the fish will be 500 g (1.1 lb) and that 50 g (0.11 lb) fingerlings will be stocked. Therefore, individual fish will gain 450 g (500 g harvest weight minus 50 g stocking weight = 450 g). The weight gain per fish will be approximately 454 g (1 lb).

(12) The total number of fish harvested will be 3,036; i.e., 1,380 kg of total gain divided by 0.454 kg of gain per fish = 3,036 fish.

(13) Total annual production will be 1,518 kg (3,340 lb) when the initial stocking weight is considered (3,036 fish × 0.50 kg/fish = 1,518 kg).

(14) If there are four fish-rearing tanks and one tank is harvested every 6 weeks, there will be 8.7 harvests per year (52 weeks divided by 6 weeks = 8.7).

(15) Each harvest will be 175 kg (384 lb); i.e., 1,518 kg per year divided by 8.7 harvests per year = 175 kg/harvest).

(16) Final harvest density should not exceed 60 kg/m³ (0.5 lb/gallon). Therefore, the water volume of each rearing tank should be 2.92 m³ (768 gallons). The tank should be larger to provide a 2.4-cm (6-in) freeboard (space between the top edge of the tank and the water levels). A standard tank size of 3.8 m³ (1,000 gallons) is recommended.

(17) Assuming a mortality of 10% during the growth cycle, the tanks should be stocked with 385 juveniles each time; i.e., (175 kg divided by 0.5 kg/fish) × 1.1 (1/0.9 = 1.1 mortality factor).

In summary, two hydroponic tanks each 21.94-m (72-ft) long by 2.44-m (8-ft) wide will be required to produce 1,000 Bibb lettuce plants per week. Four fish-rearing tanks with a water volume of 3.8 m³ (1,000 gallons) per tank will be required. Approximately 175 kg (384 lb) of tilapia will be harvested every six weeks, and annual tilapia production will be 1,518 kg (3,340 lb).

14.15   Economics

The economics of aquaponic systems depend on specific site conditions and markets. It would be inaccurate to make sweeping generalizations because material costs, construction costs, operating costs, and market prices vary by location. For example, an outdoor tropical system would be less expensive to construct and operate than a controlled-environment greenhouse system in a cold temperate climate. Nevertheless, the economic potential of aquaponic systems looks promising based on studies examining the UVI system in the Virgin Islands and in Alberta, Canada.

The UVI system is capable of producing approximately 5,000 kg (11,000 lb) of tilapia and 1,400 cases of lettuce, or 5,000 kg (11,000 lb) basil, annually, based on studies in the Virgin Islands. Enterprise budgets for tilapia production combined with either lettuce or basil have been developed. The US Virgin Islands represent a small niche market with very high prices for fresh tilapia, lettuce, and basil, as more than 95% of vegetables supplies and nearly 80% of fish supplies are imported. The budgets were prepared to show revenues, costs, and profits from six production units. A commercial enterprise consisting of six production units is recommended because one fish-rearing tank (out of twenty-four) could be harvested weekly, thereby providing a continuous supply of fish for market development.

The enterprise budget for tilapia and lettuce show that the annual return to risk and management (profit) for six production units is US$185,248. The sale prices for fish, US$1.14/kg (US$ 2.50/lb), and lettuce, US$20.00/case, have been established through many years of market research at UVI. Most of the lettuce consumed in the Virgin Islands is imported from California. It is transported by truck across the United States to East Coast ports and then shipped by ocean freighters to the Caribbean islands. Local production capitalizes on the high price that transportation adds to imported lettuce. Local production surpasses the quality of imported lettuce due to its freshness. Although this enterprise budget is unique to the US Virgin Islands, it indicates that aquaponic systems can be profitable in certain niche markets.

The enterprise budget for tilapia and basil shows that the annual returns to risk and management for six production units are US$693,726. Aquaponic systems are very productive in producing culinary herbs such as basil. A conservative sales price for fresh basil with stems in the US Virgin Islands is US$4.55/kg (US$10.00/lb). However, this enterprise budget is not realistic in terms of market demand. The population (108,000 people) of the US Virgin Islands cannot absorb 30,000 kg (66,000 lb) of fresh basil annually, although there are opportunities for provisioning ships and export to neighboring islands. A more realistic approach for a six-unit operation is to devote a portion of the growing area to basil to meet local demand while growing other crops in the remainder of the system.

The breakeven price for the aquaponic production of tilapia in the Virgin Islands is US$0.67/kg (US$1.47/lb) compared to a sales price of US$1.14/kg (US$2.50/lb). The breakeven prices are US$6.15/case for lettuce (sales price US$20.00/case) and US$0.34/kg (US$0.75/lb) for basil (sales price US$4.55/kg). The breakeven prices for tilapia and lettuce do not compare favorably to commodity prices. However, the cost of construction materials, electricity, water, labor, and land are very high in the US Virgin Islands. Breakeven prices for

Table 14.8 Preliminary production and economic data from the UVI aquaponic system at the Crop Diversification Center South, Alberta, Canada.[1] (Data courtesy of Dr. Nick Savidov.)

Crop	Annual Production		Wholesale Price		Total Value	
	kg/m^2	mt/250 m^2	Unit	US$	US$/m^2	$/2,690 ft^2
Tomatoes	29.2	7.3	6.8 kg	17.28	74.2	18,542
Cucumbers	60.8	15.2	1 kg	1.58	95.8	23,946
Egg Plant	11.1	2.8	5 kg	25.78	57.4	14,362
Genovese Basil	30.2	7.5	84 g	5.59	2,008.2	502,044
Lemon Basil	13.0	3.3	84 g	6.31	976.9	244,222
Osmin Basil	6.8	1.7	84 g	7.03	572.8	143,208
Cilantro	18.5	4.6	84 g	7.74	1,703.8	425,959
Parsley	22.8	5.7	84 g	8.46	2,300.6	575,162
Portulaca	17.2	4.3	84 g	9.17	1,874.4	468,618

[1] Ecomonic data based on Calgary wholesale market prices for the week ending July 4, 2003.

tilapia and lettuce could be considerably lower in other locations. The breakeven price for basil compares favorably to commodity prices because fresh basil has a short shelf life and cannot be shipped great distances.

A UVI aquaponics system in an environmentally controlled greenhouse at the Crops Diversification Center South in Alberta, Canada, was evaluated for the production of tilapia and a number of plant crops. The crops were cultured for one production cycle and their yields were extrapolated to annual production levels. Based on prices at the Calgary wholesale market, annual gross revenue was determined for each crop per unit area and per system with a plant growing area of 250 m^2 (2,690 ft^2; table 14.8).

Annual production levels based on extrapolated data from short production cycles are subject to variation. Similarly, wholesale prices will fluctuate during the year based on supply and demand. Nevertheless, data indicate that culinary herbs in general can obtain a gross income more than twenty times greater than that of fruiting crops such as tomatoes and cucumbers. It appears that just one production unit could provide a livelihood for a small producer. However, these data do not show capital, operating, and marketing costs, which will be considerable. Furthermore, the quantity of herbs produced could flood the market and depress prices. Competition from current market suppliers will also lead to price reductions.

14.16   Prospects for the future

Aquaponics has become very popular in recent years, but it is still in its infancy and is being practiced mainly at the hobby and backyard levels. It is estimated that there are 1,500 aquaponic systems in the United States and many times this level in Australia. However, the number of commercial operations is still relatively small in the United States. Hydroponic growers generally do not consider aquaculture as a nutrient source for their operations. Aquaculturists, on the other hand, frequently mention the possibility of incorporating hydroponics into their closed recirculating systems to mitigate waste discharge and earn extra income. Data from successful, large-scale trials are needed to attract investor capital and spur commercial development.

Although the design principles of aquaponic systems and the choice of hydroponic components and fish and plant combinations may seem challenging, aquaponic systems are quite simple to operate when fish are stocked at a rate that provides a good feeding rate ratio for plant production. Aquaponic systems are easier to operate than hydroponic systems or recirculating fish production systems because less monitoring is required and there is generally a wider safety margin for ensuring good water quality. Small aquaponic systems can provide an excellent hobby. Systems can be as small as an aquarium with a tray of plants covering the aquarium top. Large commercial operations composed of many production units and occupying several acres are certainly possible if markets can absorb the output. The educational potential of aquaponic systems is already being realized in hundreds of schools where students learn a wide range of subjects that are demonstrated through the construction and operation of aquaponic systems. Regardless of scale or purpose, the culture of fish and plants through aquaponics is a gratifying endeavor that provides food.

14.17 References

Baum, C.M. (1981) Gardening in fertile waters. *New Alchemy Quarterly* 5:3–8.

Burgoon, P.S. & Baum, C. (1984) Year round fish and vegetable production in a passive solar greenhouse. In *Sixth International Congress on Soilless Culture, Proceedings of a Conference*, 29 April 29–May 5, 1984, pp. 151–71. Lunteren, The Netherlands.

Chen, Y. & Solovitch, T. (1988) Effects of humic substances on plant growth. In *ISHS Acta Horitcultureae 221: Symposium on Horticulture Substrates and their Analysis*, p. 412.

Chun, C. & Takakura, T. (1993) Control of root environment for hydroponic lettuce production: rate of root respiration under various dissolved oxygen concentrations. In *1993 International Summer Meeting of the American Society of Agricultural Engineers*. Spokane, Washington.

Collier, G.F. & Tibbitts, T.W. (1982) Tip burn of lettuce. *Horticultural Reviews* 4:49–65.

De Wit, M.C.J. (1978) Morphology and function of roots and shoot growth of crop plants under oxygen deficiency. In *Plant Life and Anaerobic Environments* (Ed. by D.D. Hook and R.M.M.), pp. 330–50. Ann Arbor Science. Ann Arbor, Michigan.

Douglas, J.S. (1985) *Advanced Guide to Hydroponics*. Pelham Books, London.

Ferguson, O. (1982) Aqua-ecology: The relationship between water, animals, plants, people, and their environment. Rodale's NETWORK, Summer, Rodale Press, Emmaus, Pennsylvania.

Gerber, N.N. (1985) Plant growth and nutrient formulas. In *Hydroponics Worldwide: State of the Art in Soilless Crop Production* (Ed. by A.J. Savage), pp. 58–69. International Center for Special Studies, Honolulu, Hawaii.

Gloger, K.G., Rakocy, J.E., Cotner, J.B., Bailey, D.S., Cole W.M. & Shultz, K.A. (1995) Waste treatment capacity of raft hydroponics in a closed recirculating fish culture system. *Book of Abstracts*, p. 126 (Abstract #216). World Aquaculture Society, Baton Rouge.

Head, W. (1984) *An assessment of a closed greenhouse aquaculture and hydroponic system*. Doctoral Dissertation. Oregon State University, Corvallis.

Jensen, M.H. & Collins, W.L. (1985) Hydroponic vegetable production. *Horticultural Reviews* 7:483–558.

Kimball, B.A. (1982) Carbon dioxide and agricultural yield: An analysis of 360 prior observations. Agriculture Resource Service, Water Conservation Laboratory Report, USDA.

Landesman, L. (1977) Over wintering tilapia in a recirculating system. In *Essays on Food and Energy*, pp. 121–7. Howard Community College and Foundation for Self-Sufficiency, Inc.

Lennard, W.A. (2006) Aquaponic integration of Murray Cod (*Maccullochella peelii peelii*) aquaculture and lettuce (*Lactuca sativa*) hydroponics. Doctoral Thesis, School of Applied Sciences, Department of Biotechnology and Environmental Biology, Royal Melbourne Institute of Technology. Melbourne, Victoria Australia.

Lewis, W.M., Yopp, J.H., Schram, Jr., H.L. & Brandenburg, A.M. (1978) Use of hydroponics to maintain quality of recirculated water in a fish culture system. *Transactions of the American Fisheries Society* 197:92–9.

Lewis, W.M., Yopp, J.H., Brandenburg, A.M. & Schnoor, K.D. (1980) On the maintenance of water quality for closed fish production systems by means of hydroponically grown vegetable crops. In *Symposium on New Developments in the Utilization of Heated Effluents and Recirculation Systems for Intensive Aquaculture* (Ed. by K. Tiews), pp. 121–9. European Inland Fisheries Advisory Commisssion (EIFAC), Stravanger, Norway.

Losordo, T.M. (1997) Tilapia culture in intensive recirculating systems. In *Tilapia Aquaculture in the Americas* (Ed. by B.A. Costa-Pierce and J.E. Rakocy), pp.185–211. World Aquaculture Society, Baton Rouge, Louisianna.

Luther, T. (1990) Bioponics—organic hydroponic gardening. *The Growing EDGE* 1(3):40–3.

Luther, T. (1991a) Bioponics part 2. *The Growing EDGE* 2(2):37–43, 65.

Luther, T. (1991b) Bioponics part 3: Temperature and pH as limiting factors. *The Growing EDGE* 2(3):40–3, 61.

Luther, T. (1991c) Bioponics part 4. *The Growing EDGE* 2(4):35–40, 59.

Luther, T. (1993) Bioponics part 5: Enzymes for hereditary potential. *The Growing EDGE* 4(2):36–41.

MacKay, K.T. & Van Toever, W. (1981) An ecological approach to a water recirculating system for salmonids: Preliminary experience. In *Proceedings of the Bio-Engineering Symposium for Fish Culture* (Ed. by L.J. Allen and E.C. Kinney), pp. 249–58. Fish Culture Section of the American Fisheries Society, Bethesda, Maryland.

Markin, J. (1982) Tomato production in gravel beds as a means of maintaining water quality and recovering waste nutrients in recirculating catfish production systems. M.S. Thesis. Auburn University, Auburn, Alabama.

McMurtry, M.R. (1989) *Performance of an integrated aqua-olericulture system as influenced by component ratio*. Doctoral Dissertation, North Carolina State University, Raleigh, North Carolina.

McMurtry, M.R., Nelson, P.V., Sanders, D.C. & Hughes, L. (1990) Sand culture of vegetables using recirculating aquaculture effluents. *Journal of Applied Agricultural Research* 5(4):280–4.

Naegel, L.C.A. (1977) Combined production of fish and plants in recirculating water. *Aquaculture* **10**:17–24.

Nair, A., Rakocy, J.E. & Hargreaves, J.A. (1985) Water quality characteristics of a closed recirculating system for tilapia culture and tomato hydroponics. In *Proceedings of the Second International Conference on Warm Water Aquaculture—Finfish* (Ed. by R. Day and T.L. Richards), pp. 223–54. Laie, Hawaii.

Parker, D., Anouti, A. & Dickerson, G. (1990) Experimental results: Integrated fish/plant production systems. Report N. ERL 90-34. Environmental Research Laboratory, University of Arizona, Tucson, Arizona.

Pierce, B.A. (1980). Water reuse aquaculture systems in two greenhouses in northern Vermont. Proceedings of the World Mariculture Society **11**:118–27.

Rakocy, J.E. (1984). A recirculating system for tilapia culture and vegetable hydroponics. In *Proceedings of the Auburn Symposium on Fisheries and Aquaculture* (Ed. by R.C. Smitherman and D. Tave), pp. 103–14. Auburn University, Auburn, Alabama.

Rakocy, J.E. (1989a) Hydroponic lettuce production in a recirculating fish culture system. *Virgin Islands Agricultural Experiment Station, Island Perspectives* **3**:4–10.

Rakocy, J.E. (1989b) Vegetable hydroponics and fish culture: a productive interphase. *World Aquaculture* **20**(3):42–7.

Rakocy, J.E. (1995) Aquaponics: the integration of fish and vegetable culture in recirculating systems. In *Proceedings of the 30th Annual Meeting of the Caribbean Food Crops Society* (Ed. by M.C. Palada and C.C. Clarke), pp. 101–8. St. Thomas, USVI.

Rakocy, J.E. & Hargreaves, J.A. (1993) Integration of vegetable hydroponics with fish culture: A review. In *Techniques for Modern Aquaculture* (Ed. by J.K. Wang), pp. 112–36. American Society of Agricultural Engineers, St. Joseph, Michigan.

Rakocy J.E. & Nair, A. (1987) Integrating fish culture and vegetable hydroponics: Problems & prospects. *Virgin Islands Perspective-Agriculture Research Notes* **2**: 16–8.

Rakocy, J.E., Hargreaves, J.A. & Bailey, D.S. (1991) Comparative water quality dynamics in a recirculating system with solids removal and fixed-film or algal biofiltration. *Journal of the World Aquaculture Society* **22**(3):49A.

Rakocy, J.E., Hargreaves, J.A. & Bailey, D.S. (1993) Nutrient accumulation in a recirculating aquaculture system integrated with vegetable hydroponic production. In *Techniques for Modern Aquaculture*, (Ed. by J.K. Wang), pp 148–158. American Society of Agricultural Engineers, St. Joseph, Michigan.

Rakocy, J.E. (1994) Waste management in integrated recirculating systems. Proceedings of the 21st United States-Japan Meeting on Aquaculture. *Bulletin of National Research Institute of Aquaculture, Supplement* **1**:75–80.

Rakocy, J.E., Bailey, D.S., Shultz, K.A. & Cole, W.M. (1997) Evaluation of a commercial-scale aquaponic unit for the production of tilapia and lettuce. In *Tilapia Aquaculture: Proceedings of the Fourth International Symposium on Tilapia in Aquaculture* (Ed by K. Fitzsimmons), pp. 357–72. Orlando, Florida.

Resh, H.M. (1995) Hydroponic food production: A definitive guidebook of soilless food-growing methods. Newconcepts Press, Inc., Mawhah, New Jersey.

Seawright, D.E. (1995) *Integrated aquaculture-hydroponic systems: nutrient dynamics and designer diet development*. Doctoral Dissertation, University of Washington, Seattle, Washington.

Stanghellini, M.E. & Rasmussen, S.L., 1994. Hydroponics: A solution for zoosporic pathogens. *Plant Diseases* **78**:1129–38.

Sutton, R.J. & Lewis, W.M. (1982) Further observations on a fish production system that incorporates hydroponically grown plants. *Progressive Fish-Culturist* **44**:55–9.

Twarowska, J.G., Westerman, P.W. & Losordo, T.M. (1997) Water treatment and waste characterization evaluation of an intensive recirculating fish production system. *Aquacultural Engineering* **16**:133–47.

Van Gorder, S. (1991) Optimizing production by continuous loading of recirculating systems. In *Design of High Density Recirculating Systems*, A Workshop Proceedings. pp. 10–15. Lousiana State University, Baton Rouge, Louisiana.

Verwer, F.L. & Wellman, J.J.C. (1980) The possibilities of Grodan rockwool in horticulture. *International Congress on Soilless Culture* **5**:263–78.

Watten, B.J. & Busch, R.L. (1984) Tropical production of tilapia (*Sarotherodon aurea*) and tomatoes (*Lycopersicon esculentum*) in a small-scale recirculating water system. *Aquaculture* **41**:271–83.

Wren, S.W. (1984) *Comparison of hydroponic crop production techniques in a recirculating fish culture system.* Master's Thesis, Texas A&M University, College Station, Texas.

Zweig, R.D. (1986) An integrated fish culture hydroponic vegetable production system. *Aquaculture Magazine* **12**(3):34–40.

Chapter 15

In-pond Raceways

Michael P. Masser

The concept of the in-pond raceway (IPR) was simple. It would be a system, like cages, that could be adapted to almost any body of water but with the advantage of controlled water movement to improve the water quality and allow for increased stocking density, thereby increasing total production per unit area. Many researchers and aquaculture entrepreneurs have looked at the vast amount of water impounded or potentially available in rivers, bays, and estuaries and have seen opportunities to produce fish. Cages, especially in private impounded waters, seemed to be problematic because of disease outbreaks, poor localized water quality, and slow growth. These problems often appeared to be related to poor water circulation through the cage (Masser & Woods 2008).

Flow-through systems, such as raceways (see chapter 9) tend to have fewer water quality issues than most other culture systems. The constant exchange of water removes metabolites from the culture area and allows increased stocking rates and production rates per unit area compared to cages and open ponds. Traditional raceways have to be located on springs or creeks with sufficient gradient, and these constant flowing water sources are exceedingly site limited. The largest drawback of traditional raceways, aside from limited suitable locations, is the constant discharge of wastes into the receiving surface waters of the state. The discharge of these wastes has caused public concern and led to the enactment of strict environmental regulations. Additionally, the high water flow rates common in raceway systems reduce the concentration of these waste products making them difficult to capture or mediate before leaving the site.

Aquaculture Production Systems, First Edition. Edited by James Tidwell.
© 2012 John Wiley & Sons, Inc. Published 2012 by John Wiley & Sons, Inc.

The idea of placing a fish in some type of box-like enclosure suspended in a body of water and moving water through it is not new, and several designs have been developed and patented (Collamer 1923; Fremont 1972; Fast 1977; Long 1990; Caillouet 1995). These designs all utilized some type of pumping system to move water through a box or raceway.

15.1   Development of the in-pond raceway

In the early 1990s, an IPR design was developed at Auburn University. The design used airlift pumps to move water through a box-like culture area. The raceway was rectangular and suspended from a floating pier. Airlifts were placed at one end of multiple raceways and water was pumped through the airlifts into the raceways at the surface. Water was discharged from the raceways along the bottom on the opposite end from the airlifts. The water discharged through a solids-settling chamber and then flowed back into the impoundment (Masser 1997; fig. 15.1).

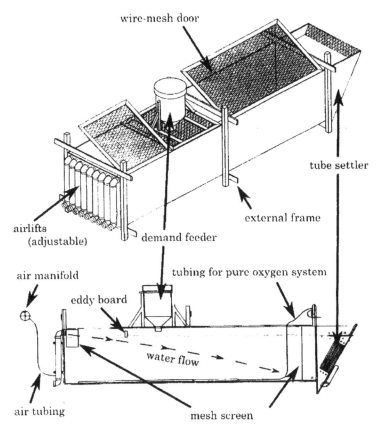

Figure 15.1 Diagram of the in-pond raceway (IPR) and its components.

The construction of an IPR was relatively simple as well. The raceway box itself was constructed from 1.3-cm thick (0.5-in thick) marine plywood or 0.7 cm (1/4 inch) plastic sheeting. These were attached to a treated wooden frame and suspended from a floating dock. Cage mesh-type materials (plastic-coated, welded wire mesh 2.54 cm × 1.27 cm) were used across the airlift openings and the discharge areas to keep fish inside the raceway. Cage mesh materials were also used to partition the raceway into sections for polyculture or segregation of different size classes of fish (Masser 1997).

The airlifts were constructed of 7.6-cm (3-in) diameter PVC pipe. They were attached to plywood or plastic sheets and aligned so that they filled the space across the front of the raceways. The bank or array of airlifts was set in a vertical track so they could be raised or lowered to regulate water flow. When properly designed and positioned, a single 7.6 cm (3 in) airlift could pump approximately 230 L (60 gallons) of water per minute, and nine airlifts were used per raceway. Air was supplied to the airlifts from a regenerative air blower. A one-horsepower blower could supply sufficient volume for approximately twenty-seven individual airlifts. Water exchange rates could be adjusted from as low as three exchanges per hour to as many as nine exchanges per hour, depending upon the height of the water discharge tubes above the water surface. Capacity to reduce the flow rate is important when fingerlings are small and during periods of cooler water temperatures (e.g., overwinter). An eddy board was placed across the width of the raceway approximately 1.8 m (5 ft) from the airlift discharge. This board deflected the water flow downward and created an eddy behind it, which functioned as the feeding area. Feed placed behind the eddy board would stay trapped by the eddy currents and not be washed out of the raceway (fig. 15.2).

The advantage of the airlift system was not only high volumes of water moved but also some degree of aeration. Although airlifts are not very efficient oxygen transfer systems in relative terms, they become more efficient when dissolved

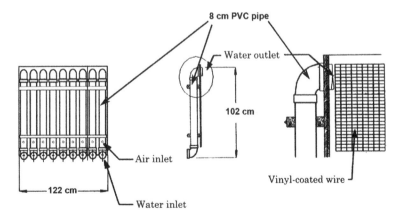

Figure 15.2 Diagram of IPR airlift array.

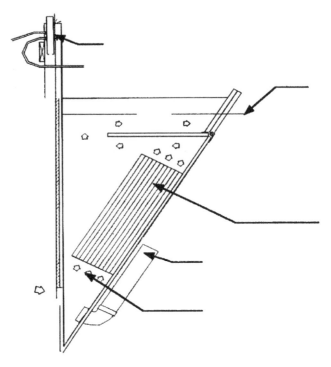

Figure 15.3 Diagram of IPR tube settler.

oxygen concentrations are low (Boyd 1990). In the Auburn study, the airlifts maintained dissolved oxygen concentrations in the high-density raceways above 3 mg/L, even if the impoundment they floated in had oxygen concentrations below 2 mg/L (Masser 2004).

Solid waste collection was attempted through either quiescent-cone or tube-type settlers attached to the back of the raceway (fig. 15.3). These passive methods captured a portion of the solid wastes excreted by the fish, which were then removed from the system. This lessened the negative impact on the water quality of the impoundment. Analysis of the dried solids confirmed that high percentages of nitrogen and phosphorous were collected (Hawcroft 1994; Yoo *et al.* 1995; Bernardez 1995; Martin 1997).

15.2   Stocking and feeding

Species that have been cultured in an IPR include channel catfish (*Ictalurus punctatus*), blue catfish (*Ictalurus furcatus*), and the channel × blue hybrid, as well as Nile tilapia (*Oreochromis niloticus*), hybrid striped bass (*Morone saxatilis* × *M. chrysops*), bluegill sunfish (*Lepomis macrochirus*), yellow perch (*Perca flavescens*), and rainbow trout (*Oncorhynchus mykiss*). Channel catfish and Nile tilapia were successfully polycultured in experiments at Auburn University (Bernardez 1995). Stocking rates in the IPR varied from 326 to 543 fish per cubic meter (9 to 15 fish per cubic foot) of culture area. Throughout the experiments at Auburn University with channel catfish, no differences in growth rates or food conversion rates were observed at these stocking rates (Masser 1997).

Feeding a nutritionally complete diet is essential in an IPR, because the fish get no natural food items from the environment. Floating feed was given to allow observation of feeding activity and to keep feed trapped behind the eddy board. When hand-fed, the daily feeding rates were based on commonly used feeding tables for the species cultured. However, demand feeders proved to be just as efficient (i.e., feed conversion) as hand-feeding channel catfish and reduced labor requirements (Bernardez 1995).

15.3 Backup systems and disease treatments

The IPR is an energy-dependent system. If a blower fails and water flow stops, the raceways will become oxygen limited in a matter of minutes. Backup or emergency systems are essential if the system is to be successful. Power and equipment failure backup, as well as notification systems similar to those used in recirculating aquaculture systems (RAS), have been utilized in IPR systems. A simple oxygen supply system was incorporated into the Auburn IPR design using cylinders of pure oxygen connected to an electric solenoid switch. The solenoid remained closed, stopping oxygen release from the bottles unless the electricity went off. When the electricity supply was interrupted, the solenoid opened to allow pure oxygen to be released into the raceways through micropore tubing (i.e., the same as used in hauling tanks; Hawcroft 1994; Yoo *et al.* 1995). This system can successfully maintain fish in the IPR for several hours or until electric service is restored.

Another benefit of oxygen backup systems in IPR is during disease treatments. With the blowers off and water flow stopped, the emergency oxygen system can be used to maintain the fish. The raceway then becomes a static tank that can be dosed with precise bath-type treatments (e.g., potassium permanganate or formalin). These treatments are inexpensive and effective methods for combating certain bacterial and protozoal diseases.

15.4 Comparison to other culture systems

The general advantages and disadvantages of the IPR compared to other systems are summarized in table 15.1. In the following sections, *traditional raceways* refers to land-based flow-through raceways, while RAS refers to recirculating aquaculture systems.

In terms of production, the Auburn IPR averaged a yield of approximately 136 kg/m^3 (8.5 lb/ft^3) of catfish and tilapia (Wilcox 1998; Masser 1999). These densities have been achieved by some cage and RAS operations but are not

Table 15.1 Comparison of IPR with other culture systems.[1]

Culture system production, environmental, and economic comparison data	IPR	Cage	IPR	Pond	IPR	Traditional Raceways	IPR	RAS
Production Features								
Higher stocking density and carrying capacity	+		+		=	=	+	
Feed conversion ratio (FCR)	+		+		=	=	=	=
Production yields per unit area	+		+		=	=	+	
Overwintering of certain species (e.g., catfish)	=	=		+		+		+
Economic Features								
Reduced labor requirements in regard to accessibility, feeding, grading, harvesting, and treatment of diseases	+		+		=	=	+	
More economically efficient & produces fish at reduced costs	+		+			+	+	
Environmental Features								
Better water quality, particularly dissolved oxygen (DO) and ammonia concentrations	+		+			+		+
Environmentally superior for solid waste collection and removal	+		+		+			+

Notes:
[1] Better system indicated by +; Similar systems indicated by = .

common. The feed conversion ratio (FCR) averaged 1.5:1 in research trials at Auburn (Wilcox 1998; Masser 1999).

Economic feasibility of IPRs is based on Auburn University trial data (table 15.2), which showed IPRs to be more economically efficient than ponds and cages. The IPR design and greater accessibility simplified feeding, grading, harvest, and disease treatments compared to other systems. This reduced labor requirements compared to ponds or RAS. The IPR also had lower production costs than are typical with RAS. The modified IPR in west Alabama even appears to have lower production costs than standard commercial catfish ponds (Brown *et al.* 2010).

The IPR, with its system of passive settling devices, is environmentally superior in terms of solid waste removal when compared to cages, ponds, and traditional raceways, which do little to capture wastes; however, IPRs are still not exceptionally efficient. As in ponds, cages, and traditional raceways, IPRs rely on natural processes within the receiving impoundment to process liquid wastes and solid wastes that are not removed by the passive settlers. Recirculating systems, with active solids removal and biological filters, are much more efficient at removal of waste products (Losordo *et al.* 1999).

Table 15.2 Economic comparisons between IPR, cages, and open-pond catfish culture (0.4 ha pond; Bernardez 1995).[1]

	Open-pond[2]	Cage	IPR
Assumptions			
Yield (kg)	1,730	1,286	2,433
Death loss (%)	6	10	10
Feed conversion	1.8	1.6	1.45
% Protein feed	32	36	36
Economic parameters (US$)			
Variable costs	3,135.63	2,391.27	4,160.25
Fixed costs	787.72	850.16	1,111.26
Total costs	3,923.35	3,241.43	5,271.51
Breakeven price (US$ per kg)			
To cover variable costs	1.81	1.86	1.71
To cover total costs	2.27	2.52	2.17

Notes:
[1] Pond construction and management costs have not been included in the budgets.
[2] Open-pond production yields are based on actual average production values observed in the catfish industry in Alabama.

15.5 Sustainability issues

The IPR system appears to be more environmentally sustainable than cages, raceways, and intensive open-pond production ponds. The IPR system is slowly being adapted to larger-scale operations and appears to be highly efficient at producing fish (Brown *et al.* 2010). Using hybrid catfish, a modified IPR in west Alabama has reached harvest densities of over 200 kg/m^3 and estimated production costs below the current average pond production costs. If this adoption continues, the IPR could replace traditional ponds for culture of many finfish species.

15.6 Future trends

Research still needs to be conducted on the IPR. Particularly, cost-effective solid and liquid waste reduction methods need to be evaluated and further developed. The current adaptations needed for scaling up the IPR system to more commercial sizes need to be further studied and economically evaluated.

15.7 References

Bernardez, R.G. (1995) *Evaluation of an in-pond raceway system and its economic feasibility for fish production.* Master's Thesis, Auburn University.
Boyd, C.E. (1990) *Water Quality in Ponds for Aquaculture.* Alabama Agricultural Experiment Station, Auburn University.

Brown, T.W., Chappell, J.A. & Hanson, T.R. (2010) In-pond raceway system demonstrates economic benefits for catfish production. *Global Aquaculture Advocate* **13**(4):18–21.

Caillouet, E.W. (1995) Floating fish cultivating system and related method (*US Patent Serial No. 5,450,818*).

Collamer, C.B. (1923) Aquarium (*US Patent Serial No. 1,444,367*).

Fast, A.W. (1977) Floating fish rearing system (*US Patent Serial No. 4,044,720*).

Fremont, H.J. (1972) Floating fish growing tank (*US Patent Serial No. 3,653,358*).

Hawcroft, B.A. (1994) *Development and Evaluation of an In-Pond Raceway and Water Removal System*. Master's Thesis, Auburn University.

Long, C.E. (1990) Raceway culturing of fish (*US Patent Serial No. 4,915,059*).

Losordo, T.M., Masser, M.P. & Rakocy, J.E. (1999) Recirculating aquaculture tank production systems: A review of component options. *Southern Regional Aquaculture Center Publication No. 453.*

Martin, J.M. (1997) Effluent control of an in-pond raceway system. Master's Thesis, Auburn University.

Masser, M.P. (1997) In-pond raceways. *Southern Regional Aquaculture Center Publication No. 170.*

Masser, M.P. (1999) Cages and in-pond raceways as sustainable aquaculture in watershed ponds. In *Proceedings of the Ninth National Extension Wildlife, Fisheries, and Aquaculture Conference* (Ed. by R.M. Timm & S.L. Dann), pp. 171–80.

Masser, M.P. (2004) Cages and in-pond raceways. In *Biology and Culture of Channel Catfish* (Ed. by C.S. Tucker & J.A. Hargreaves), pp. 530–44. Elsevier Scientific, Amsterdam.

Masser, M.P. & Woods, P. (2008) Cage culture problems. *Southern Regional Aquaculture Center Publication No. 165.*

Wilcox Jr., M.D. (1998) *Effects of an in-pond raceway on fish production performance and water quality*. Master's Thesis, Auburn University.

Yoo, K.H., Masser, M.P. & Hawcroft, B.A. (1995) An in-pond raceway system incorporating removal of fish wastes. *Aquacultural Engineering* **14**(2):175–87.

Chapter 16

On the Drawing Board

James H. Tidwell

Aquaculture has its work cut out for it. It is expected to increase production levels over 75% in the next twenty-five years (from 48 million tonnes in 2005 to 85 million tonnes in 2030) with a minimum impact on the environment and a maximum benefit to society (Subasinghe 2007). This is no small task. As we look into aquaculture's crystal ball, certain trends can be seen to be developing. These include: (a) a continuing trend toward intensification, (b) a continuing trend toward new species development and diversification, (c) continued development of new production systems and diversification, (d) increased influence of market forces on production, and (e) increased government regulation with (we hope) improved regulatory procedures (Subasinghe 2007).

16.1 Future trends

16.1.1 Fish meal and fish oil supplies

These trends seem fairly inevitable. However, there are other concerns that will impact aquaculture's future. One of the issues that have received much attention, discussion, and press in recent years is the use of fish meal and fish oil in aquaculture diets. Indeed, aquaculture has continued to absorb an increasing proportion of the supply of these important, if not essential, feed ingredients. In fact, when the trends of consumption are projected into the future (a statistically

Aquaculture Production Systems, First Edition. Edited by James Tidwell.
© 2012 John Wiley & Sons, Inc. Published 2012 by John Wiley & Sons, Inc.

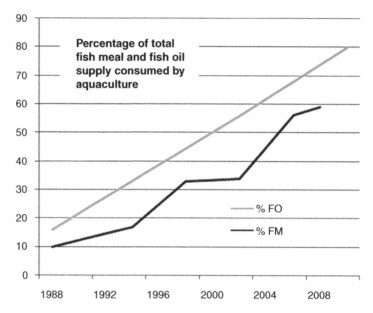

Figure 16.1 The percentage of fish oil and fish meal supply consumed by the aquaculture industry from 1988 to 2008.

dangerous practice) we can get results that indicate that aquaculture will consume *all* fish meal production (fig. 16.1) in a very few years. This projection has been termed the "fish meal trap" (New 1999). To be more accurate, it might be better described as the fish oil trap, as fish oil supplies are projected to become limiting before fish meal does (fig. 16.1).

In 2006, aquaculture consumed 3 million tonnes or 56% of world fish meal production (Tacon & Metian 2009). That same year, aquaculture utilized 87% of the world's fish oil production (Tacon & Metian 2009). Also of concern is the additional 5 to 6 million tonnes of low value/trash fish used as direct feeds for many aquacultured species in Asia (Tacon *et al.* 2006). These numbers and projections are indeed troubling. As we have seen, this "trap" could represent a major impediment to aquaculture growth and expansion, which are needed to provide the increasing demands for foodfish. Many environmental groups have used these figures to make claims that aquaculture is causing the collapse of these fish-meal fisheries and actually produces less fish than it utilizes. However, recent reexamination of these models and calculations shows that feed-based aquaculture produces *twice* as much fish as it uses (Tacon & Metian 2009). If the large numbers of aquacultured species that do not depend on manufactured diets are included, aquaculture as an industry actually produces three to five times as much fish as it consumes (fig. 16.2).

If we look at trends for these "industrial" fisheries targeted toward fish meal production (fig. 16.3), we see that they are some of the best managed fisheries in the world (Tidwell & Allan 2001). With continued management they can continue to produce approximately 30 million tonnes per year, sustainably, for

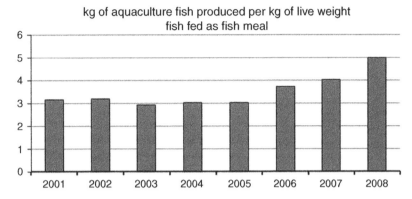

Return on Fish Meal Investment

kg of aquaculture fish produced per kg of live weight
fish fed as fish meal

Figure 16.2 The amount of total fish generated by aquaculture divided by the whole fish equivalent of fish meal (i.e., the return on the fish meal investment) from 2001 to 2008.

years to come (barring external perturbations such as global warming). Even if aquaculture continues to grow, management controls will not allow harvest pressures to be increased or allow these fisheries to be depleted.

Also, much has been made of "fishing down the food web" (Pauly *et al.* 1998). By some estimates over 90% of the oceans' large predators have been removed by human fishing activities. Populations of predatory marine mammals have also decreased. With these factors considered, proper cropping of these short-lived,

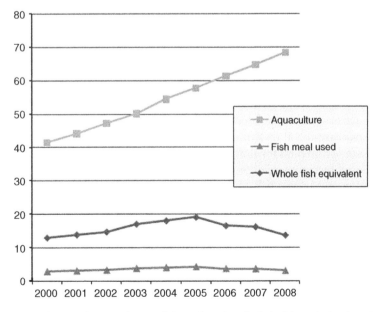

Figure 16.3 Total aquaculture production, fish meal used, and whole fish equivalent based on fish meal used by aquaculture from 2000 to 2008.

highly fecund forage species might actually be needed to prevent over population in the absence of predatory pressures

Another criticism by environmental groups has been that these harvested "fish meal" fish could be better utilized as direct food for humans rather than feeding them to other fishes. As with many issues, the issues and answers are complex. There have been examples where the increased demand for pelagic fish by the animal feed industry has decreased the availability of fresh fish for poor communities (Hasan 2007). However, other studies have shown that reduction (fish meal) fisheries can also benefit locals by contributing to animal production enterprises, which generate jobs and improve living standards and food security (Hecht & Jones 2007). Actual impact differs largely on the region being considered. In Africa and Asia, the species used for reduction to fish meal, or direct feeding to aquacultured animals, do have potential for direct human consumption, while those species used in Europe to produce fish meal are not really suitable for any other uses (Huntington 2007).

Will this "fish meal/fish oil trap" stifle aquaculture development? Or will a better understanding of the specific nutritional requirements of aquaculture species, coupled with more information on alternative ingredients, reduce aquaculture's dependency on fish meal production? I tend to believe the latter. Research indicates that once we understand a species' nutritional requirements and tolerances, the fish meal and fish oil content of aquafeeds can be reduced substantially. For salmon it is estimated that at least 50% of the fish meal and 50 to 80% of fish oil can be replaced with vegetable substitutes. For marine fish, 30 to 80% of fish meal and 60% of fish oil used could come from alternative sources (Royal Commission on Environmental Pollution 2005). This can include use of terrestrial protein crops such as soybean meal, but examples also include increased utilization of by-catch from commercial fisheries, as well as wastes and offal from fish processing (Hardy *et al.* 2005). When we do not yet know the specific nutritional requirements of a culture species, nutritionists tend to "over formulate." That means we include very high protein and fish meal levels to ensure they meet the animals' requirements (which are not yet known). However, once the research is conducted to delineate the animals' nutritional requirements, species-specific diets can be formulated, which greatly improve nutrient retention efficiency and increase the ability to utilize alternative ingredients. An example is research at our lab on diets for the largemouth bass (*Micropterus salmoides*). The diets being used by commercial producers contained 40% fish meal, while our research has determined that fish meal inclusion could be reduced to $\leq 8\%$ without decreasing growth or feed conversion efficiency (Cochran *et al.* 2009).

16.1.2   Climate change

What impacts will climate change have on aquaculture production? While still debated in the forums of politics and television news, the overwhelming majority of scientists agree that the trends are real. What they do not yet agree on is the pace and subsequent effects. Because of the complicated interrelationships within the planet's ecology, there will likely be a patchwork of impacts. However, in the big picture, it is obvious that conditions *will* in fact change and that these changes will influence how aquaculture develops in the near and distant future. According to Soto and Brugere (2008), climate change "is characterized by its unpredictability and the large uncertainty that has to be factored in all models and the reaction of ecosystems." Other related factors that will undoubtedly affect aquaculture's future include recent rapid global increases in energy prices and grain prices.

According to Nomuva (FAO-Fisheries) and Steiner (United Nations Environment Programme, UNEP) in 2009, during a climate change discussion in Bonn, Germany, "the challenges and threat" of climate change must be a top priority for world leaders. Part of that priority is ensuring healthy aquatic ecosystems. Likely impacts include changes in air and sea surface temperatures, changes in rainfall amounts and patterns, acidification of ocean waters, rises in sea levels, and resulting changes in estuaries and shoreline communities, and increased frequency and intensities of climate patterns (such as El Niño) and storms (such as cyclones and hurricanes; UNFCCC-COP-15 2009).

Based on these changes, the zones considered optimal for the production and reproduction of different species will likely move (Soto & Brugere 2008). It is predicted by some that aquaculture will in fact be disproportionately affected by climate change, based on the fact that it raises poikilothermic (cold blooded) animals, and the fact that the vast majority of production is in open and semi-open systems, which operate at ambient temperatures and have no capacity for temperature control (UNFCCC-COP-15 2009). However, even within aquaculture there can be winners and losers. In a warmer world, tropical ecosystem productivity may likely be decreased, while higher latitudes could actually benefit (Soto & Brugere 2008). However, there are still many unknowns such as species invasions (in fact, who is resident and who is the invader may be redefined as temperature zones shift) and the occurrence and virulence of diseases change (Soto & Brugere 2008).

So what can be done? The United Nations Environment Programme (UNEP) gives a number of recommendations (see the end of this chapter for a list of recommended actions) for both capture fisheries and aquaculture (UNEP 2009). The first step is to develop a better understanding of the interworkings of both the production system and its environment by implementing an Ecosystem Approach to Aquaculture (EAA). By understanding the interaction of the farm, the water body (and its watershed), and the global market (and requisite transportation), it is felt that resilience to change can be increased (Soto & Brugere 2008). Next, we can reduce the dependence on raising carnivorous species. We can also decrease the impact of those we do produce by genetically selecting for strains that are able to utilize lower protein and fish meal levels and higher levels of by-products. We can also improve the diets themselves through improved manufacturing processes (Soto & Brugere 2008). We should increase the production of "extractive" species that can actively remove waste products such as nitrogenous wastes, carbon, and phosphorous, thereby improving the environment (bivalves and microalgaes are examples; fig. 16.4).

Figure 16.4 Mussels, which can act as extractive species to remove algae and excess nutrients near cage or net pen production.

The next step is integrating these extractive species into the production systems for food species to lessen the environmental impact and improve the economic viability, or both. Another example of combining systems is that while many types of pond-based aquaculture actually suffer from excessive production of algal biomass (Brune *et al.* 2003; fig. 16.5), there are efforts elsewhere to mass-produce algae in ponds for biofuels. A continuous partial harvest of algae from aquaculture systems could potentially improve the pond environment for aquaculture production while simultaneously producing a secondary algae crop for biofuels production.

Another major initiative related to climate change will likely be the continued development of new culture species. As conditions change in a particular locale, switching to another species (better adapted to the new conditions), rather than moving the production system, may be more practical. Research into the genetic improvement of both new and traditional species will also be especially important. This approach has the potential to shift the animals' temperature tolerance, oxygen tolerance, disease resistance, and dietary requirements—all of which are important in improving aquaculture's adaptation to climate change.

A third area needed to enhance the adaptability of aquaculture-based crops to climate change is in the area of public policy. How have we dealt with

Figure 16.5 A pond with and excessively dense algae bloom.

uncertainty in other crops? Improved forecasting, crop insurance, and risk management are all concepts that have been used in terrestrial crops for years and could play an increasing role in aquatic crops. Many of the future changes in aquaculture production will be driven by sustainability. This has to include economic sustainability as well as environmental sustainability. A balance of both is important.

16.1.3 Aquaculture to reduce CO_2

Aquaculture may even serve as a cure, or at least a treatment, for the ills of other industries. While there are major concerns over the future pricing and availability of petroleum, there are still substantial coal reserves in many parts of the world. However, coal has the drawback of releasing carbon, which has been locked away for millions of years, back into the atmosphere as CO_2 (fig. 16.6). One approach being investigated is to use algae as a biological "scrubber" to strip the CO_2 from the flue gas of coal-fired power plants, before it is released to the atmosphere. These algae, whose growth is enhanced by the increased CO_2 concentration, can then be harvested to produce biofuels. It is reported that a hectare devoted to algae production can yield 50,000 liters of biofuel, compared to 350 liters from a terrestrial crop such as sunflower seeds (Penwarden 2006). While this approach would actually represent carbon recycling, another approach is to use marine algae (*Coceoli thophorids*) to absorb the CO_2,

Figure 16.6 A coal-fired power plant.

convert it to calcium carbonate (limestone), and permanently sequester the carbon into a form, which could potentially be used as a building material (Romanosky 2000).

16.1.4   Improved genetics

Another future development will likely be "designer strains" of animals specifically suited for the production systems they are being raised in. At present, only 1 to 2% of the animals used in aquaculture are genetically improved (Gjedevem 1997), with most aquacultured species being only slightly removed from the wild. As animals become selected for improved production in aquaculture, the process will be further refined until strains specific to different production systems are developed. Just as we have different strains of chickens depending on whether they will be raised outside (i.e., pastured poultry) or indoors in high-density chicken houses, we will have specific genetic strains of shrimp used in ponds that are different from shrimp that are used in super-intensive indoor systems.

16.1.5   Production intensification

As stated previously, there is a general trend toward production intensification. In ponds, production intensification has been an incremental process over the past fifty years. Production was originally limited to approximately 500 kg/ha, primarily by the availability of natural foods. As supplemental feeds were added, production was increased threefold to 1,500 kg/ha when it then became constrained by the pond's ability to provide oxygen reliably (when feeding rates exceeded 30 kg/ha/day). With the advent of mechanical aeration, pond production rates again tripled with an average carrying capacity exceeding 4,500 kg/ha. The limiting factor in pond intensification is now primarily the accumulation of nitrogenous waste products. Further intensification of pond production will require methods to more rapidly and/or efficiently remove or convert nitrogenous waste products within the pond system.

Different approaches to accomplish this are under development. Currently, most pond systems rely on nitrifying bacteria to convert the waste product ammonia to less toxic nitrate through nitrification. The nitrate is then assimilated by the pond's algal population. Algae can also directly assimilate ammonia in its ammonium form. However, these processes become limited by the availability of light (to drive photosynthesis), oxygen, or other factors such as micronutrients, to rates that limit the further intensification of pond systems (Hargreaves & Tucker 2003). A new pond-management approach has gained traction in recent years. In this approach, the pond ecology is shifted away from autotrophic algae toward heterotrophic bacteria whose populations are not light limited (see chapter 12). This shift is encouraged by manipulation of the carbon/nitrogen ratios in the system. Most of these systems are carbon limited, so when sufficient carbon is supplied (often in the form of a carbohydrate, such as sugar) the system can rapidly and efficiently remove nitrogen from the water by converting it directly into bacterial biomass (Avnimelech 1999). The bacterial "floc" that is formed can often serve as supplemental food for some filter feeding, scavenger, or detritivore species. Results of this system indicate that production may again be tripled with rates of near 15,000 kg/ha being reported (Boyd & Clay 2002).

Other approaches to this same problem include maximizing the efficiency of algae-based systems. In a traditional pond system, heavy phytoplankton populations limit light penetration to just a few centimeters so that photosynthesis is restricted to just this photic zone. To address this restriction, a new modified pond system has been developed that maximizes the potential of algal assimilation of nitrogen by directing the flow of the water across shallow zones to maximize light contact and photosynthetic rates (Brune *et al.* 2003). This system, known as the partitioned aquaculture system (PAS), is described in much more detail in chapter 13. Other approaches that might be considered in the future include hybridizing components from other systems into the pond production systems. Examples could be adding biofilters to the pond systems, adding a hydroponic component to flow the water through, or adding a floating island of plants within the pond itself.

16.1.6   Increased use of marine systems

Although the majority of aquaculture production for food currently occurs in freshwater ponds, much of the future of the industry looks back to the sea. A large part of the financial support for aquaculture research in the United States is now being directed to marine systems and species. So what do we see, in terms of new marine aquaculture production technologies?

As with other types of production, much of the development will focus on sustainability and the reduction of environmental impact. Also, it will likely be a matter of evolution, not revolution. We will see existing systems modified, improved, or combined in new and innovative ways to address problems (either real or perceived).

16.1.7   Evolution of sea cages

Cage culture has evolved over millennia and it continues to evolve. It began with capture fisheries. If a fisherman caught more than he could use or sell, he needed to "warehouse his inventory." If this was a time or place where refrigeration was not an option, what could be done? For historic ocean-capture fisheries like cod, the best option was to process fish and pack them in salt. For freshwater fishermen in the rivers and lakes of Asia, it was to hold the live animals in baskets (fig. 16.7). While holding them, you might as well try to put some extra weight on them by feeding them. Then why not diversify to capturing juveniles during the spawning season and growing them out on foods you provide?

As these concepts moved to protected coastal waters, they evolved from fenced-in enclosures of mud bottoms to deeper areas where currents and depth

Figure 16.7 Bamboo baskets that were used to catch, store, and later raise fish as precursors to fish cages.

Figure 16.8 An ocean net pen in Corsica.

could flush away or dilute waste products (fig. 16.8). However, directly under these stationary cages waste products can accumulate, so there has been a push to move the cages out into what are known as "high energy" environments to better supply the cages with high-quality water and disperse any waste products. This, too, is difficult. As we move out of the protected fjords and into the unprotected coastal areas, we move into a more punishing environment.

So how do we "evolve" the system to deal with these issues? One approach has been to stay in the protected waters, but try to deal with the wastes. Mariculture Systems in the United States markets a system where the permeable net pens are replaced by impermeable enclosures. Water is pumped into these floating reservoirs. The intakes for the incoming water can be positioned to take advantage of differences in temperatures and water quality parameters at different depths. The water is pumped into the cone-shaped enclosures in a way as to set up a constant swirl whose speed can be manipulated based on variables such as fish size. The swirl will also, through the Venturi effect, move solids such as fish feces to the bottom of the culture vessel allowing the solids to be collected and removed or treated. This system has taken the traditional net pen system and hybridized it with many of the components of the recirculating aquaculture systems discussed in chapter 11.

Another approach to dealing with the limitations of net pen culture has been to fully "seal" the fish inside the net pen enclosure. Traditional designs extended to or above the water's surface where the fish were open to the air to allow feeding and so on. By adding another net pen on the top, the fish were now fully sealed inside the enclosure. With the fish now fully sealed inside the cage, it

Figure 16.9 A fully enclosed ocean net pen. Illustration courtesy of Richard Langan.

can be lowered underwater to escape the harsh wave action of the surface, or it can be lowered even deeper, should environmental factors like oil spills or toxic algae blooms make this action advantageous. This fully enclosed design is seen in such models as the Aquapod (Ocean Farm Technologies), OceanGlobe (KSAS), and SeaStation (OceanSpar; fig. 16.9). With these designs come hardware and software for monitoring, feeding, and even self-generating power (fig. 16.10).

So what does the future hold? Poal Lader, a SINITEF Fisheries and Aquaculture research scientist sees a "farm" that runs itself. Instead of free-ranged fish, he sees free-ranged farms; large autonomous fish farms that can move about. There would be a base station that would be moored to the bottom. It would monitor the conditions in the region in terms of oxygen levels, temperature, weather, nutrients, and other environmental variables. This information would be shared with enclosed culture modules that could move around to take advantage of the best conditions. It could even be used to transport the fish closer to their intended markets as they approach harvest size.

This vision of the future may not be as far away as it sounds. Cliff Goady, Director of the Massachusetts Institute of Technologies, Sea Grant Offshore Aquaculture Engineering project has developed large, low rpm electric propellers, which allow enclosed net pens to be moved and maneuvered underwater. They have been tested under commercial conditions off the coast of Puerto Rico.

16.1.8   Trophic integration

Related to sustainability and hybrid systems is the concept of trophic integration. The basic idea is to combine the culture of extractive species, such as shellfish

Figure 16.10 A enclosed net pen with monitoring and energy generation. Photo courtesy of the Atlantic Marine Aquaculture Center, University of New Hampshire.

or plants, with fed species. The theory is that the extractive species will act as a "nutrient scrubber" to clean up the wastes generated by the fed species. The environment will be protected and the growth and production of the extractive species will be increased by the increased supply of nutrients. Much of the work in this area has been in marine systems and has been directed toward ameliorating the impact of sea cages, such as those used for salmon production. Research has been going on for about twenty years in North America and other regions such as Israel and South Africa. Most have involved small-scale study systems, but commercial-scale systems are now operating. However, in China, the concept is hundreds of years old and operates on the scale of entire bays, with fish being co-cultured with scallops, abalone, and kelp. In Canada, mussels and kelp are raised around salmon cages. In South Africa, abalone are cultured in onshore tanks and their wastewater is used to culture seaweeds. Such an approach reduces the release of pollutants and nutrients into the environment. It also yields additional crops for additional income. This increases diversification, which can be important for improved cash flow and risk management.

16.1.9   Integration of aquatic and terrestrial systems

Another concept of integrated aquaculture is to combine aquaculture with terrestrial agriculture. Integrated agriculture-aquaculture (IAA) systems have been used in Asia for centuries. There are many examples of variations on this concept with livestock-fish integrations of chicken-, duck-, and pig-based systems, usually combined with carp species. There are also rice-fish, rice-shrimp, and rice-prawn systems.

All of these integrated systems show how complimentary species can be used to benefit both the producer and the environment. However, they can also be more complicated to operate than fed monoculture systems. Regulatory and/or financial incentives may be required in some situations to encourage their adoption.

16.1.10   Artificial floating islands

In the future, we will continue to see a blurring of the lines among systems and the hybridization of components from different systems, and even among major categories of systems. An example could be the use of artificial floating islands (AFI). These are vegetated floating platforms that have been used in ponds, lakes, and reservoirs (fig. 16.11). These AFI units can provide several ecological

Figure 16.11 A floating island.

services, including improvement of water quality, enhancement of habitat, and control of shoreline erosion. Their use in Japan dates back to the 1920s, and interest in them has increased in recent years. However, little of the research on AFIs has been published in English, limiting their adoption in many areas.

There are two major categories of AFIs: dry AFIs (which have the plant roots contained within an enclosure) and wet AFIs (which have the plant roots extending down into the water). Vegetation in these systems includes cattails (*Typha latifola*), yellow iris (*Iris psuedacorus*), and common reed (*Phragmites australis*). Oshima *et al.* (2001) showed that AFI could reduce both total-nitrogen and chlorophyll-A in a pond system. To my knowledge, they have not yet been tested in an aquaculture system.

16.1.11   Species diversification

Another major aspect of the future for aquaculture will be the continuing development of new species. Even though a relatively staggering number of species are already cultivated (>400), the number will undoubtedly continue to increase. This will be driven not only by the continuing development of new production technologies, but also by the expansion of aquaculture into new geographic regions. Since the use of nonindigenous species is discouraged and even prohibited in some places, there will be increased pressure to develop native species for commercial production. An example has been the multidisciplinary and multi-institutional program in Brazil to develop the indigenous freshwater prawn *Macrobrachium amazonicum* as an alternative to the nonnative *Macrobrachium rosenbergii* (Moraes-Valenti & Valenti 2010).

Other forces will also drive an increase in the number of species being raised. Not all aquaculture animals are grown as food. The ornamental fish trade exceeds US$15 billion/year in total economic impact (FAO 2005). The nature of the business is that it is highly desirable to continue to develop new "products" to be sold. As stated by Hill and Yanong (2002), "the aquarium hobby thrives on novelty." This can be addressed not only by developing new genetic strains of traditional species, but also by the use of new culture species. It is estimated that there are already over 8,000 marine species used in the ornamental trade (FAO 2005).

We will also see new species being used in our more traditional culture systems. For example, in Canada the development of new aquaculture species is predicted to provide an economic impact of US$880 million by 2020. These species include Atlantic cod, Atlantic halibut, Arctic char, and sablefish. A major contributor to their rapid development is the use or adaptation of hatchery, nursery, and growout technologies previously developed for traditional species.

16.1.12   Aquaculture of nonfood products

Another driving force for species diversification will be new uses for aquatic animals and plants beyond food and ornamental applications. Production of

Figure 16.12 Photobioreactors for raising algae. Photo courtesy of Sam Morton, Center for Applied Energy Research, University of Kentucky.

aquatic species for biofuels has already redirected a number of aquaculture production systems from their previous use in food fish production. Examples include leases offered to owners of catfish ponds in Mississippi to be used for algae production and the shift of tank-based Kent SeaTech to Kent BioEnergy. The partitioned aquaculture system (PAS) discussed in chapter 13 is now also being used to produce algae for biofuels.

Other examples of diversification include production of algae for long-chain omega-3 fatty acids. As demand for these compounds has increased, the need to identify and develop alternative sources has also increased. Even when harvested as fish oil, the ultimate producer of these fatty acids in the food chain was actually algae. Direct production of algae for these fatty acids could be potentially more efficient. New production systems have been developed to produce pure cultures of certain algae strains on a commercial scale. One of the most popular is the photobioreactor (fig. 16.12). There are a number of different styles including vertical tubes, horizontal tubes, flat panels, spiral tubes, raceway systems, and spheres.

Other areas of expansion and diversification of aquaculture systems include the production of therapeutic compounds (i.e., "farmaceuticals"). It is estimated that more than 50% of the drugs in use today originated from natural sources (i.e., plants, animals, or microorganisms). Many were originally isolated from soil microbes (such as penicillin). In fact, more than 120 pharmaceuticals derived from soil organisms are still prescribed today (Fenical 2006). However, we are now realizing that there may be a much larger "pool" of natural compounds to evaluate in marine ecosystems. Of the thirty-six Phyla of life, seventeen are found in terrestrial environments, but twice as many (thirty-four) live in marine ecosystems. Early work on development of drugs from marine sources began in the 1950s (Bergmann & Feeney 1951). The pace of evaluation rapidly increased in the 1990s, when the National Cancer Institute initiated a dedicated drug discovery program (Cragg *et al.* 2005). By 2006, there were over thirty marine-derived molecules in preclinical development or being used clinical trials against cancer (Fenical 2006). In addition, there are also marine-sourced drugs in clinical trials for acute pain, asthma, Alzheimer's disease, malaria, infections, and inflammation.

Discovering a compound that shows bioactivity is not itself sufficient to justify its entrance into the drug development process. There must also be a cost-effective source of the candidate compounds. Sipkema *et al.* (2005) conducted an economic and technical analysis of potential production methods for producing pharmaceutically active compounds from sponges. They found that traditional mariculture methods (net bags on longlines) and tank culture were both superior to cell culture for the large-scale production.

16.1.13   Conclusion

As has been said, the only thing that is constant is change. The growth of aquaculture is essential to provide a steady supply of high-quality protein and healthy fatty acids for a rapidly increasing human population. However, the production methods used will need to be environmentally and economically sustainable, under conditions that will likely change substantially in the next fifty years. Major perturbations will likely include climate change and volatility in the costs of inputs such as feed and energy. It is very likely that the aquaculture production systems of the twenty-second century will include new and unique combinations of today's technologies as well as new technologies we have not yet imagined. It should be an interesting ride.

What can we do now?

- Implement comprehensive and integrated ecosystem approaches to managing coasts, oceans, fisheries, and aquaculture; to adapting to climate change; and to reducing risk from natural disasters.
- Move to environmentally friendly and fuel-efficient fishing and aquaculture practices.
- Eliminate subsidies that promote overfishing and excess fishing capacity.
- Provide climate change education in schools and create greater awareness among all stakeholders.
- Undertake assessments of local vulnerability and risk to achieve climate proofing.
- Integrate aquaculture with other sectors.
- Build local ocean-climate models.

- Strengthen our knowledge of aquatic ecosystem dynamics and biogeochemical cycles such as ocean carbon and nitrogen cycles.
- Encourage sustainable, environmentally friendly, biofuel production from algae and seaweed.
- Encourage funding mechanisms and innovations that benefit from synergies between adaptation and mitigation in fisheries and aquaculture.
- Conduct scientific and other studies (e.g., economic) to identify options for carbon sequestration by aquatic ecosystems that do not harm these and other ecosystems.
- Consider appropriate regulatory measures to safeguard the aquatic environment and its resources against adverse impacts of mitigation strategies and measures.

16.2   References

Bergmann, W. & Feeney, R. (1951) Contributions to the study of marine products. XXXII The nucleotides of sponges. I. *Journal of Organic Chemistry* **16**(6):981–7.

Boyd, C.E. & Clay, J.W. (2002) *Evaluation of Belize Aquaculture, Ltd: A Superintensive Shrimp Aquaculture System*. Report prepared under the World Bank, Network of Aquaculture Centers in Asia-Pacific, World Wildlife Fund and Food and Agriculture Organization of the United Nations Consortium on Shrimp Farming and the Environment.

Brune, D.E., Schwartz, G., Eversole, A.G., Collier, J.A. & Schwedler, T.E. (2003) Intensification of pond aquaculture and high rate photosynthetic systems. *Aquacultural Engineering* **28**:65–86.

Cochran, N.J., Coyle, S.D. & Tidwell, J.H. (2009) Evaluation of reduced fish meal diets for second year growout of the largemouth bass, *Micropterus salmoides*. *Journal of the World Aquaculture Society* **40**(6):735–43.

Cragg, G.M., Kingston, D.G.I. & Newman, D.J. (2005) *Anticancer Agents from Natural Products*. CRC Press, Boca Raton.

FAO (2005) *Ornamental Fish*. Topics Fact Sheet, Fisheries and Aquaculture Topics. FAO, Rome. http://www.fao.org/fishery/topic/13611/en.

Fenical, W. (2006) Marine pharmaceuticals: Past, present, and future. *Oceanography* **19**(2):110–19.

Gjedrem, T. (1997) Selective breeding to improve aquaculture production. *World Aquaculture* **28**(1):33–45.

Hardy, R.W., Sealy, W.M. & Gatlin, D.M. (2005) Fisheries by-catch and by-product meal as protein sources for rainbow trout *Oncorhynchus mykiss*. *Journal of the World Aquaculture Society* **36**(3):393–400.

Hasan, M.R. (2007) Use of wild fish and/or other aquatic species to feed cultured fish and its implications to food security. *FAO Aquaculture Newsletter* **37**:30–2.

Hecht, T. & Jones, C.L.W. (2007) *The use of wild fish as feed in aquaculture and its implications for food security and poverty alleviation in Africa and the Near East*. Review prepared for FAO, Rome.

Hill, J.E. & Yanong, R.P.E. (2002) Freshwater ornamental fish commonly cultured in Florida. University of Florida Institute of Food and Agricultural Sciences, *Circular* **54**:1–6.

Huntington, T. (2007) *Regional Review for Europe: Use of Wild Fish and/or Other Aquatic Species to Feed Cultured Fish and Its Implications to Food Security and Poverty Alleviation.* Review prepared for FAO, Rome.

Moraes-Valenti, P. & Valenti, W.C. (2010) Culture of the Amazon River Prawn, *Macrobrachium amazonicum.* In *Freshwater Prawns: Biology and Farming* (Ed. by M.B. New, W.C. Valenti, J.H. Tidwell, L.R. D'Abramo & M.N. Kutty), pp. 485–501. Wiley-Blackwell, Oxford.

New, M.B. (1999) Global aquaculture: Current trends and challenges for the 21st Century. *World Aquaculture* 30(1):8–13, 63–79.

Oshima, H., Karasawa, K & Nakamure, K. (2001) Water purification experiment by artificial floating island. *Proceedings of the Japan Society on Water Environment* 35(146). (In Japanese).

Pauly, D., Christensen, V., Dalsgaard, J., Foese, R. & Torres Jr., F. (1998) Fishing down marine food webs. *Science* 279(5352):860–3.

Penwarden, M. (2006) Carbon capture: The algae alternative. Website Matter Network. http://www.matternetwork.com. Accessed June 10, 2008.

Romanosky, R. 2000. Cal State to explore the use of marine algae to soak up carbon dioxide. DOE's sequestration program continues to expand. Fossil Energy Techline. US Department of Energy. December 11, 2000.

Royal Commission on Environmental Pollution (2005) *Turning the Tide.* The 25th Report Addressing the Impact of Fisheries on the Marine Environment. The Stationery Office, Norwich.

Sipkema, D., Osinga, R., Schatton, W., Mendola, D., Tramper, J. & Wijffels, R.H. (2005) Large-scale production of pharmaceuticals by marine sponges: Sea, cell or synthesis? *Biotechnology and Bioengineering* 90(2):201–22.

Soto, D. & Brugere, C. (2008) The challenges of climate change for aquaculture. *FAO Aquaculture Newsletter* 40:30–2.

Subasinghe, R.P. (2007) Aquaculture status and prospects. *FAO Aquaculture Newsletter* 38:4–7.

Tacon, A.G. J. & Metian, M. (2009) Fishing for aquaculture: Non-food use of small pelagic forage fish-a global perspective. *Reviews in Fishery Science* 17(3):305–17.

Tacon, A.G.J., Hasan, M.R. & Sugasinghe, R.P. (2006) Use of fishery resources as feed inputs for aquaculture development: Trends and policy implications. *FAO Fisheries Circular No. 1018.* FAO, Rome.

Tidwell, J.H. & Allan, G. (2001) Fish a food: Aquaculture's contribution. *EMBO Reports* 2:958–63.

United Nations Environment Programme (UNEP) (2009) *Fisheries and Aquaculture in Our Changing Climate.* Multiagency Policy Brief.

United Nations Framework Convention on Climate Change (UNFCCC-COP) (2009) Conference of Parties. 15th Session 2009, Proceedings. Geneva, Switzerland.

Index

Aquaculture Production Systems, First Edition. Edited by James Tidwell.
© 2012 John Wiley & Sons, Inc. Published 2012 by John Wiley & Sons, Inc.

Printed and bound by CPI Group (UK) Ltd, Croydon, CR0 4YY

16/04/2025

14658590-0004